Gaseous Loss of Nitrogen from Plant-Soil Systems

edited by
J.R. FRENEY *and* J.R. SIMPSON
CSIRO, Division of Plant Industry, Canberra, Australia

1983 **MARTINUS NIJHOFF / Dr W. JUNK PUBLISHERS**
a member of the KLUWER ACADEMIC PUBLISHERS GROUP
THE HAGUE / BOSTON / LANCASTER

Distributors

for the United States and Canada: Kluwer Boston, Inc., 190 Old Derby Street, Hingham, MA 02043, USA
for all other countries: Kluwer Academic Publishers Group, Distribution Center, P.O.Box 322, 3300 AH Dordrecht, The Netherlands

Library of Congress Cataloging in Publication Data

Library of Congress Cataloging in Publication Data
Main entry under title:

Gaseous loss of nitrogen from plant-soil systems.

   (Developments in plant and soil sciences ; v. 9)
   Includes index.
   1. Nitrogen--Environmental aspects. 2. Nitrogen
cycle. 3. Plant-soil relationships. I. Freney,
J. R. (John Raymond) II. Simpson, J. R. III. Series.
QH545.N5G37 1983   631.4'16   83-2326
ISBN 90-247-2820-7

ISBN 90-247-2820-7 (this volume)

83 007035

Contents

GASEOUS LOSS OF NITROGEN FROM PLANT-SOIL SYSTEMS

Developments in Plant and Soil Sciences

Volume 9

Preface

A growing interest has been shown recently in the dymanics of nitrogen in agricultural and natural ecosystems. This has been caused by increasing demands for food and fibre by a rapidly expanding world population, and by a growing concern that increased land clearing, cultivation and use of both fertilizer and biologically fixed nitrogen can have detrimental effects on the environment. These include effects on water quality, eutrophication of surface waters and changes in atmospheric composition all caused by increased cycling of nitrogenous compounds.

The input and availability of nitrogen frequently affects the productivity of farming systems more than any other single management factor, but often the nitrogen is used inefficiently. Much of the fertilizer nitrogen applied to the soil is not utilised by the crop: it is lost either in solution form, by leaching of nitrate, or in gaseous forms as ammonia, nitrous oxide, nitric oxide or dinitrogen. The leached nitrate can contaminate rivers and groundwaters, while the emitted ammonia can contaminate surface waters or combine with atmospheric sulfur dioxide to form aerosols which affect visibility, health and climate.

There is also concern that increased evolution of nitrous oxide will deplete the protective ozone layer of the stratosphere. The possibility of a link between the intensity of agricultural use of nitrogen, nitrous oxide emissions and amounts of stratospheric ozone has focussed attention on these interactions. This, in turn, has highlighted the gaps in our knowledge of the biological processes leading to nitrous oxide production in soils and its destruction in the atmosphere. As a consequence of the research stimulated by such concern and interest, a number of important discoveries have been made in recent years.

New approaches to the direct measurement of gaseous fluxes in the field have facilitated the accurate estimation of exchanges of ammonia, nitric oxide and nitrous oxide and produced a better understanding of the chemical and physical factors involved. This research should assist in developing better management practices which conserve nitrogen in agricultural systems and maximise economic returns by the efficient conversion of fertilizer or soil nitrogen into plant and animal proteins.

This monograph attempts to review all relevant studies on gaseous emissions from plant-soil systems, and is compiled by scientists who have contributed to the recent scientific advances in this subject. Topics covered include the underlying microbial, chemical and physical processes, methodology for determining rates of gaseous emission, data on these rates in major agricultural systems, the ecological significance of such exchanges, the fate of the emitted gases, and management practices which could reduce the adverse effects and losses. It is the aim of our contributors that this book will promote a better understanding of the importance of gaseous emissions from plant-soil eco-systems and encourage the more efficient use of nitrogen in agriculture, with less contamination of our environment.

March 1983

J.R. FRENEY
J.R. SIMPSON

Contributors

Beauchamp, E.G., Department of Land Resource Science, University of Guelph, Ontario, Canada.

Catchpoole, V.R., CSIRO, Division of Tropical Crops and Pastures, Cunningham Laboratory, St. Lucia, Queensland, Australia.

Chalk, P.M., School of Agriculture and Forestry, University of Melbourne, Parkville, Victoria, Australia.

Craswell, E.T., International Fertilizer Development Center, Muscle Shoals, Alabama, USA.

Denmead, O.T., CSIRO, Division of Environmental Mechanics, Canberra, A.C.T., Australia.

Farquhar, G.D., Department of Environmental Biology, Research School of Biological Sciences, Australian National University, Canberra, A.C.T., Australia.

Fillery, I.R.P., International Fertilizer Development Center, Muscle Shoals, Alabama, USA, Stationed at International Rice Research Institute, Los Baños, Philippines.

Freney, J.R., CSIRO, Division of Plant Industry, Canberra, A.C.T., Australia.

Galbally, I.E., CSIRO, Division of Atmospheric Physics, Aspendale, Victoria, Australia.

Harper, L.A., United States Department of Agriculture, Agricultural Research Service, Southern Piedmont Research Center, Watkinsville, Georgia, USA.

Hauck, R.D., Tennessee Valley Authority, Division of Agricultural Development, Muscle Shoals, Alabama, USA.

Rolston, D.E., Department of Land, Air and Water Resources, University of California, Davis, California, USA.

Roy, C.R., Australian Radiation Laboratory, Department of Health, Yallambie, Victoria, Australia.

Ryden, J.C., The Grassland Research Institute, Hurley, Maidenhead, Berkshire, United Kingdom.

Simpson, J.R., CSIRO, Division of Plant Industry, Canberra, A.C.T., Australia.

Smith, C.J., Center for Wetland Resources, Louisiana State University, Baton Rouge, Louisiana, USA.

Steele, K.W., Ministry of Agriculture and Fisheries, Ruakura Soil and Plant Research Station, Hamilton, New Zealand.

Vallis, I., CSIRO, Division of Tropical Crops and Pastures, Cunningham Laboratory, St. Lucia, Queensland, Australia.

Vlek, P.L.G., International Fertilizer Development Center, Muscle Shoals, Alabama, USA.

Weir, B., Department of Environmental Biology, Research School of Biological Sciences, Australian National University Canberra, A.C.T., Australia.

Wetselaar, R., CSIRO, Division of Water and Land Research, Canberra, A.C.T., Australia.

1. Volatilization of ammonia

J.R. FRENEY, J.R. SIMPSON and O.T. DENMEAD

1.1. Introduction

Ammonia is ubiquitous in Nature, being formed from the biological degradation of proteins in soil organic matter, plant residues and animal wastes. Its presence is readily detectable near barns, stables and feedlots where plant and animal residues are concentrated but it is also formed in many other situations from less concentrated sources, e.g. in fields and forests (Lemon and Van Houtte 1980). It is constantly being formed in soils at rates which depend on the level of microbial activity and the susceptibility of organic N compounds to biological attack. It is also being added to soils in increasing amounts as fertilizer.

Ammonia has been recognized as a labile form of nitrogen in agricultural systems since early last century. Early theories pointed to NH_3 as the principal source of N for plants but Boussingault (1856) proved that other inorganic N sources were important.

Evolution of NH_3 after fertilizer application to soils did not attract much attention until the 1950's when a number of attempts were made to measure its importance. The field data obtained were often more qualitative than quantitative, demonstrating losses of NH_3 after surface applications of urea or urine. It was only in the 1970's that comprehensive and reliable field techniques were developed for measuring NH_3 volatilization directly without disturbing the microenvironment. Studies using these techniques have enabled us to assess the importance of both NH_3 loss and the factors controlling it in several different agricultural ecosystems. The work has indicated that, depending on circumstances, NH_3 volatilization can be only a minor loss or it can represent the dominant fate of the N applied as urea, urine or aqueous NH_3.

The reliable field studies of NH_3 volatilization rates that have been made have concentrated on fertilized agricultural systems. There have been few measurements on grazed grasslands and no direct measurements on undisturbed natural ecosystems (Dawson 1977, Vlek *et al.* 1981). Therefore we have had to draw heavily on the results of laboratory experiments in order to collate the factors which influence the procesess of NH_3 volatilization.

1.2. The volatilization process

1.2.1. Fundamental equilibria

Since NH_3 is a gas at normal atmospheric temperatures and pressures, it might be expected that any NH_3 present in soils, waters, fertilizers and manures would quickly volatilize to the atmosphere. However, it is a basic gas which reacts readily with protons, metals, and acidic compounds to form ions, compounds or complexes of varying stability and is thereby protected in solution or solid forms. Ammonia also has a very strong affinity for water and its reactions in water are fundamental determinants of the rate of volatilization. Following Simpson (1981) and Vlek *et al.* (1981), the various reactions which govern ammonia loss may be represented as

$$\text{Source}_1 \qquad \text{Source}_2 \qquad NH_3 \text{ (gas in atmosphere)}$$
$$\downarrow \qquad\qquad \downarrow \qquad\qquad \uparrow \downarrow \qquad\qquad (1)$$
$$\text{Absorbed } NH_4^+ \rightleftarrows NH_4^+ \text{ (in solution)} \rightleftarrows NH_3 \text{ (in solution)} \rightleftarrows NH_3 \text{ (gas in soil)}$$

The rate of NH_3 volatilization may be controlled by the rate of removal and dispersion of NH_3 into the atmosphere, by changing the concentration of ammonium or NH_3 in solution, or by displacing any of the equilibria in some other way.

The driving force for NH_3 volatilization from a moist soil or a solution is normally considered to be the difference in NH_3 partial pressure between that in equilibrium with the liquid phase and that in the ambient atmosphere (Koelliker and Miner 1973, Freney *et al.* 1981, Denmead *et al.* 1982). The equilibrium vapour pressure of NH_3 is controlled by the NH_3 concentration in the adjacent solution, which in the absence of other ionic species, is affected by the ammonium ion concentration and pH.

A. Ammonia–ammonium equilibria
The equilibrium between ammonium and NH_3 in solution can be represented by Equation (2),

$$NH_3 + H_2O \overset{K_b}{\rightleftharpoons} NH_4^+ + OH^- , \qquad (2)$$

or by a combination of Equations (3) and (4),

$$NH_3 + H^+ \underset{K_a}{\overset{\rightarrow}{\rightleftharpoons}} NH_4^+ , \qquad (3)$$

and

$$H_2O \overset{K_w}{\rightleftharpoons} H^+ + OH^- , \qquad (4)$$

where

$$K_b = \frac{[NH_4^+] \cdot [OH^-]}{[NH_3]}, \qquad (5)$$

$$K_a = \frac{[NH_3] \cdot [H^+]}{[NH_4^+]} \tag{6}$$

and $\quad K_w = [OH^-] \cdot [H^+].$ (7)

It is apparent that the two representations of the equilibrium are the same and that the dissocation parameters, K_b, K_a and K_w, are related, viz.

$$K_a = \frac{K_w}{K_b} \; (mol/l) \tag{8}$$

The effect of pH on the equilibrium between ammonium and NH_3 can be derived from Equation (6), since

$$\log_{10} [H^+] - \log_{10} K_a = \log_{10} \frac{[NH_4^+]}{[NH_3]} \tag{9}$$

$$\therefore \; pK_a - pH = \log_{10} \frac{[NH_4^+]}{[NH_3]} \tag{10}$$

From (8), $\log_{10} K_a = \log_{10} K_w - \log_{10} K_b.$ (11)

$$\therefore \; pK_a = pK_w - pK_b, \tag{12}$$

and $(pK_w - pK_b) - pH = \log_{10} \dfrac{[NH_4^+]}{[NH_3]}.$ (13)

The relative concentrations of ammonium and NH_3 at a number of pH values have been determined from expressions (10) or (13) by several workers (e.g. Bates and Pinching 1950, Warren 1962) and the percentages of NH_3 in solution at pH 6, 7, 8 and 9 are approximately 0.1, 1, 10 and 50 (see also Fig. 1.1 from Court et al. 1964). Thus there is a far greater potential for NH_3 volatilization at higher pH values.

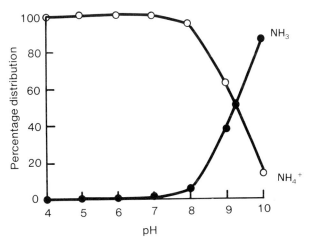

Fig. 1.1. Effect of pH on relative concentrations of ammonium and ammonia in solution (from Court et al. 1964).

The presence of carbon dioxide or other chemical species may affect the nature of this pH relationship (Vlek and Craswell 1981) but the potential for greater volatilization at high pH remains.

Accurate values for K_b have been obtained by Bates and Pinching (1950) from e.m.f. measurements. Values for K_b, pK_b, K_w (from Weast 1971) and pK_w are given in Table 1.1 for temperatures between 0 and 50°C.

It is apparent from this table that the dissociation constants for water (K_w) and aqueous ammonia (K_b) are affected differentially by temperature, and therefore $pK_w - pK_b$ (i.e. pK_a) varies markedly with temperature. Emerson *et al.* (1975) provide an expression for calculating the value of pK_a or K_a at a particular temperature, viz.

$$\log_{10} K_a = -0.09018 - \frac{2729.92}{T}, \tag{14}$$

where T is the absolute temperature (°K).

B. Gas – liquid equilibria

As mentioned above, the concentration of NH_3 gas in equilibrium with a given solution depends on the concentration of NH_3 in that solution, i.e. the partition of NH_3 between the liquid and gaseous phases follows the requirements of Henry's law, and therefore

$$\frac{\text{Concentration of } NH_3 \text{ in the liquid phase}}{\text{Concentration of } NH_3 \text{ in the gaseous phase}} = \text{constant} \tag{15}$$

Table 1.1. Effect of temperature on the ionization constant for water and the dissociation constant for aqueous ammonia.

Temperature	Water ionization constant (K_w)	pK_w	Ammonia dissociation constant (K_b)	pK_b	$pK_w - pK_b$
0	1.138×10^{-15}	14.944	1.374×10^{-5}	4.862	10.082
5	1.845×10^{-15}	14.734	1.479×10^{-5}	4.830	9.904
10	2.917×10^{-15}	14.535	1.570×10^{-5}	4.804	9.731
15	4.508×10^{-15}	14.346	1.652×10^{-5}	4.782	9.564
20	6.808×10^{-15}	14.617	1.710×10^{-5}	4.767	9.400
25	1.007×10^{-14}	13.997	1.774×10^{-5}	4.751	9.246
30	1.469×10^{-14}	13.833	1.820×10^{-5}	4.740	9.093
35	2.089×10^{-14}	13.680	1.849×10^{-5}	4.733	8.947
40	2.917×10^{-14}	13.535	1.862×10^{-5}	4.730	8.805
45	4.018×10^{-14}	13.396	1.879×10^{-5}	4.726	8.670
50	5.470×10^{-14}	13.262	1.892×10^{-5}	4.723	8.539

From Bates and Pinching (1950) and Weast (1971)

at a particular temperature. The partition of NH_3 between the two phases varies markedly with temperature. Hales and Drewes (1979) and the Subcommittee on Ammonia (1979) provided an expression for the variation of the 'constant' with temperature, viz.

$$\log_{10} H = 1477.7/T - 1.6937 \tag{16}$$

where H (M/M), the Henry *constant*, is the ratio between the dissolved molar (M) concentration of NH_3 in pure water and the molar gaseous concentration, and T is the absolute temperature (° K).

Many workers find it more convenient to consider the partial pressure of NH_3 in equilibrium with the concentration of NH_3 in solution. In that case

$$[NH_3] = K_H \cdot p(NH_3), \tag{17}$$

where K_H is the Henry *coefficient* (M/bar), $p(NH_3)$ is the partial pressure of NH_3 (bar) and $[NH_3]$ is the dissolved molar concentration of NH_3. The Henry *coefficient* and *constant* are related by the expression (18):

$$K_H = H/RT, \tag{18}$$

where R is the ideal gas constant (0.0831; litre bar/K).
From equations (6) and (17) the relationship between the partial pressure of NH_3 ammonium ion in solution at equilibrium is given by

$$p(NH_3) = \frac{K_a}{K_H} \frac{[NH_4^+]_{\text{solution}}}{[H^+]}. \tag{19}$$

Combining Equations (14), (16), (18) and (19), we obtain

$$p(NH_3) = RT\,(10^{1.60352 - 4207.62/T}) \cdot \frac{[NH_4^+]_{\text{solution}}}{[H^+]}. \tag{20}$$

Thus temperature, ammonium concentration and pH markedly affect the partial pressure of NH_3 in equilibrium with a solution. Fig 1.2, from Farquhar *et al.* (1980) gives two examples of the effect of temperature and ammonium concentration on the partial pressure of NH_3.

Denmead *et al.* (1982) used a reparameterization of the relationships given by the Subcommittee on Ammonia (1979) to calculate the equilibrium partial pressure, $p(NH_3)$, from the ammoniacal N concentration (i.e. the sum of ammonium–N and NH_3–N in solution), pH and temperature. Their relationships are:

$$[NH_3]_{\text{solution}} = \frac{[NH_3 + NH_4^+]_{\text{solution}}}{1 + 10^{(0.09018 + 2729.92/T - \text{pH})}} \tag{21}$$

$$\text{and } p(NH_3) = \frac{0.00488\,[NH_3]_{\text{solution}} \cdot T}{10^{(1477.8/T - 1.6937)}}, \tag{22}$$

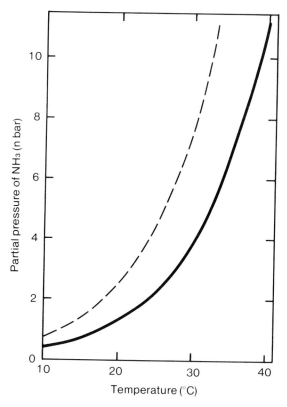

Fig. 1.2. Effect of temperature on partial pressure of ammonia gas in equilibrium with an ammonium solution at a fixed pH (6.8). ——, $[NH_4^+]_{solution} = 46\,\mu M$; ––– $[NH_4^+]_{solution} = 92\,\mu M$ (from Farquhar *et al.* 1980).

where the concentration of NH_3 is in g/m^3, pressures are in mb and temperatures in $^\circ K$. Equation (22) shows that the equilibrium partial pressure increases linearly with the concentration of NH_3 in solution, and exponentially with temperature as above.

It is apparent that the presence of carbon dioxide, other forms of alkalinity or acidity, and ion pair formation will affect the equilibrium between NH_3 and ammonium, and thus influence NH_3 loss (Marion and Dutt 1974, Hales and Drewes 1979, Vlek and Stumpe 1978).

1.2.2. Retention in soil

Within the soil mass, NH_3 is continually sorbed and/or desorbed from soil minerals, organic matter and water (Gardner 1965, Parr and Papendick 1966a, Rolston *et al.* 1971, 1972). The relative importance of the mineral and organic matter in retaining NH_3 depends on the soil but usually the organic fraction is more reactive than the mineral matter (Mortland 1958, Sohn and Peech 1958, Burge and Broadbent 1961, Young 1964, Broadbent and Stevenson 1966). Under acid conditions, NH_3 is retained mainly by reaction with clay minerals to form adsorbed ammonium whereas under strongly alkaline conditions, reactions with organic components are more important (Mortland 1955, Parr and Papendick 1966a).

Ammonia reacts with exchangeable hydrogen ions on clay minerals and organic matter and with hydroxyl groups associated with silicon on the edges of clay minerals. It can coordinate with exchangeable metal ions and can also be held by hydrogen bonding (Mortland 1966). The mechanisms of adsorption of NH_3 on minerals range from chemical reactions, whereby the NH_3 is bound very strongly, to physical adsorption where it is bound very weakly. Physical adsorption on mineral surfaces takes place only when there is a positive pressure of NH_3 in the soil; when the pressure in the gaseous phase is lowered the adsorbed NH_3 is desorbed (Mortland 1966).

Adsorption is related to the surface area of sorbing material. Thus the sizes and kinds of minerals in soils affect the amounts of NH_3 adsorbed. In general, clay soils sorb more NH_3 than sandy soils because clay soils have a greater surface area per unit weight (Coffee and Bartholomew 1964b, Mortland 1966). The type of clay mineral is also important: montmorillonite- or vermiculite-type clay minerals have much greater surface areas ($\sim 800 \text{ m}^2/\text{g}$) than kaolinite type minerals (~ 5–$10 \text{ m}^2/\text{g}$) (Mortland 1966). Sorption on clay minerals is also influenced by the moisture content of the clays and the type of cation on the exchange complex (Brown and Bartholomew 1963). This effect may be due partly to the quantity and nature of the water of hydration.

Many workers have shown that soil organic matter reacts with NH_3 (see Broadbent and Stevenson 1966). Ammonia reacts with carboxyl and possibly other acidic groups of organic matter to form salts and with other organic components to form non-exchangeable reaction products. Phenolic hydroxyl groups appear to be important in the formation of these products but little is known about the mechanism of the fixation process even though a number of plausible mechanisms have been proposed (Broadbent and Stevenson 1966).

1.2.3. Transport

Ammonia must be transported to the soil surface before it can be lost to the atmosphere and a number of workers have shown that the rate of loss is related to the depth of the NH_3 below the soil surface (e.g. Baker *et al.* 1959, Movsumov 1965, Jusuf and Soepardi 1968, Shankaracharya and Mehta 1969, Fenn and Kissel 1976). Transport can be accomplished by movement in the liquid or gaseous phases, and the relative importance of these two phases will depend on the water content of the soil (Gardner 1965). Distribution of NH_3

between the two phases will be related by Henry's Law and affected by the temperature of the solution (section 1.2.1). Movement in the gaseous phase will usually be by molecular diffusion, but in the liquid phase it may be by diffusion of the NH_3 molecules through the water or by convection if the water is moving with respect to the soil.

Diffusion in the liquid phase is affected by the tortuous nature of the diffusion path in soils, the soil water content, the higher viscosity of the water near the surfaces of clay minerals, the concentration of NH_3 in solution, the temperature, and the fact that diffusion can only occur in that fraction of the pore space that is filled with water. Diffusion in the gas phase is determined by the porosity and tortuosity of the soil and the concentration of NH_3 in the pore space (Gardner 1965). The relative importance of convection and diffusion in the transport of NH_3 has been assessed by Gardner (1965) who concluded that considerably more NH_3 could be lost by convection than diffusion ($\sim 9:1$). However, when the NH_3 is located near the soil surface diffusion alone could account for appreciable losses.

Ammonium ions can also be transported to the soil surface by convection and diffusion and be converted there to NH_3 which can be lost from the soil. This transport is likely to be more important in flooded than in non-flooded soils. Patrick and Reddy (1976) and Reddy et al. (1976, 1980) have shown that ammonium diffuses from the anaerobic soil layer of flooded soils to the overlying aerobic soil layer. Experiments with ^{15}N have shown that more N is lost from a flooded soil than is actually present as ammonium in the aerobic layer at any one time (Tusneem and Patrick 1971, Broadbent and Tusneem 1971). Reddy et al. (1980) determined the diffusion coefficient for ammonium in different flooded soils and found that it ranged from 0.059 to 0.216 cm²/day. The diffusion coefficient increased with increasing soil moisture content and ammonium concentration.

1.2.4. Volatilization

As discussed in Koelliker and Miner (1973), and Denmead et al. (1982), volatilization of NH_3 from solution occurs in response to a difference in vapour pressure between the solution and the ambient air. Increasing wind speed increases the rate of volatilization from flooded systems, promoting more rapid transport of NH_3 away from the water surface, and there is some evidence from our recent unpublished work that the same can happen in dryland systems. Freney et al. 1981 and Denmead et al. (1982) incorporate these separate influences on volatilization rate into a bulk aerodynamic formula of the type

$$F = k \, (p_0 - p_z) \tag{23}$$

where F is the flux density of NH_3 away from the surface, k is an exchange coefficient v ˙   is the

partial ˙   our

pre˙

˙ı   ˳

equat.

betweei.

layers abc

liquid phasc

chemical react

with sparingly sc   ˳,

this resistance appc   ı the

approximate square oı   ˳.ater (1974)

suggest that for a gas suc.   ˳, with high solubility and rapid aqueous phase chemistry the liquid ph˳se resistance should be small and the gas phase resistance should control the exchange. They expect k to be linearly related to wind speed as is the case for water vapour (Webb 1965, Liss 1973, Liss and Slater 1974).

Freney *et al.* (1981) found that Equation (23), with k linearly related to wind speed, did give a reasonably good fit to their limited data for NH_3 volatilization from a flooded rice field ($r = 0.79$). An analysis of our recent work in a flooded field has given a similar result (R. Leuning, personal communication). In that work, simultaneous measurements of evaporation allowed calculation of the corresponding exchange coefficient for water vapour. Usually it was higher than that for NH_3, indicating some resistance to NH_3 transport in the liquid phase. The effect appeared to be due to the development of a temperature inversion in the water which suppressed NH_3 transport to the surface.

Equation (23) also provided a general description of the NH_3 volatilization process in irrigated maize (Denmead *et al.* 1982) although in that work the exchange coefficient was related to $u^{2.3}$ ($r = 0.90$). The authors suggested two possible reasons for enhanced volatilization in high winds: (i) better mechanical mixing of the NH_3 solution, which would reduce the liquid phase resistance; or (ii) enhanced vertical transport in the canopy air space which would reduce the gas phase resistance. This could have been due to deeper penetration of displacing gusts into the canopy in high winds as Denmead *et al.* (1982) suggest, or a decrease in the stability of the air near the ground.

1.3. Factors influencing ammonia volatilization

1.3.1. Inherent soil properties

A. Cation exchange capacity (CEC)
As ammonium is a positively charged ion it reacts readily with the cation

exchange complex in soils. This reaction reduces the amount of ammonium, and therefore of NH_3, in solution at a given pH (section 1.2.1; Ryan and Keeney 1975). If CEC is a dominant factor controlling volatilization, as believed by many workers, an inverse relationship should exist between the two. The results of early research (Martin and Chapman 1951, Volk 1959, Ernst and Massey 1960, Gasser 1964a) led some authors to conclude that CEC was the most important factor affecting NH_3 loss from soils (Anderson 1962, Gasser 1964a, b). Gasser (1964a), for instance, plotted his results along with those of Martin and Chapman (1951) and Volk (1959) and found that ammonia loss decreased as CEC increased. More recent reports have also shown a negative correlation between NH_3 loss and CEC (e.g. Loftis and Scarsbrook 1969, Chai and Hou 1975, Lippold *et al.* 1975, Ryan and Keeney 1975, Fenn and Kissel 1976, Daftarder and Shinde 1980, Lyster *et al.* 1980, Ryan *et al.* 1981). Estimates of a critical CEC value, above which losses are small, vary greatly with experimental conditions, but it appears that a minimum CEC of about 25 meq is required to reduce NH_3 loss substantially (Gasser 1964a, Fenn and Kissel 1976, Faurie and Bardin 1979a, Lyster *et al.* 1980).

Since CEC is a function of the amounts and types of clay minerals and organic matter present in soils, an inverse relationship between NH_3 evolution and these parameters might also be expected. Faurie *et al.* (1975) and Ryan and Keeney (1975) did in fact find a linear inverse relationship between ammonia loss and clay content, and less definite relationships have been observed by other workers (Baligar and Patil 1968, Loftis and Scarsbrook 1969).

The results of studies on the relationship between soil factors and NH_3 loss can easily be mis-interpreted. For example, the water-holding capacity and buffering capacity of soils increase in parallel with CEC and these parameters also influence NH_3 loss (Fenn and Kissel 1976). Temperature interacts in the relationship between CEC and NH_3 loss through its effect on ammonium–NH_3 equilibria and transport (Fenn and Kissel 1976).

It is assumed that an equilibrium is reached between ammonium in the soil solution and that adsorbed on the external and interlamellar surfaces of clay minerals (Nömmik 1957). The equilibria have been studied recently by Kowalenko and Cameron (1976) who showed that the ion exchange process between soluble ammonium and the external exchangeable ammonium could best be described by a non-linear Freundlich equilibrium relationship, of the form,

$$(NH_4^+ - N)_{exch} = a\,(NH_4^+ - N)_{sol}^{\,b}\,. \tag{24}$$

These authors assumed that the relationship between the external and interlamellar exchangeable ammonium followed a Langmuir kinetic model (Ellis and Knezek 1972).

B. pH

Both laboratory and field data indicate that the most severe losses of NH_3 occur when ammonium based fertilizers are applied to the surface of alkaline soils or when materials which produce microenvironments of high pH are added to more acidic soils.

Many workers have shown that NH_3 losses increase with increasing soil pH (e.g. Wahhab *et al.* 1957, Wagner and Smith 1958, Ernst and Massey 1960, Du Plessis and Kroontje 1964, Shankaracharya and Mehta 1969, Faurie and Bardin 1971, Ida and Mori 1971, Watkins *et al.* 1972, Chai *et al.* 1974, Mills *et al.* 1974, Filimonov and Strel'nikova 1974, Chai and Hou 1975, Lippold *et al.* 1975, More and Varade 1977, Faurie and Bardin 1979a, Ryan *et al.* 1981). This effect was well demonstrated by Jewitt (1942) who found NH_3 losses of 87%, 13% and zero when ammonium sulfate was applied to soils of pH 10.5, 8.6 and 7.0, respectively. The pH effect is to be expected from the fundamental equilibria discussed in section 1.2.1; higher pHs increase the concentration of NH_3 present in the soil solution and soil air, so increasing the potential for NH_3 loss from soil.

Ammonia volatilization can also be expected to occur from soils which have surface accretions of plant ash after burning of vegetation. The major effect may be due to the increased pH at the soil surface resulting from the ash, but related processes such as saturation of soil exchange sites may also be involved (Raison and McGarity 1978).

C. Buffer capacity

The dissociation of ammonium ions releases hydrogen ions in addition to NH_3 (Equation 3). Consequently, as NH_3 loss proceeds the medium becomes acidified, at a rate depending on the conversion of ammonium to NH_3 and H^+, and the buffer capacity (Avnimelech and Laher 1977). As the reaction proceeds the fraction of ammoniacal N in NH_3 form is reduced and an equilibrium is reached. For this reason NH_3 volatilization is usually more important in soils of high base status where the acidity produced by the dissociation can be neutralized by carbonate or other forms of alkalinity. Many workers have found large losses of NH_3 from calcareous soils when ammonium based fertilizers are applied (e.g. Lehr and Van Wesemael 1961, Larsen and Gunary 1962, Harding *et al.* 1963, Ivanov 1963, Harmsen and Kolenbrander 1965, Balba and Naseem 1968, Fenn and Kissel 1973, 1974, 1975, 1976, Fenn 1975, Fenn and Escarzaga 1976, 1977, Feagley and Hossner 1978, Fenn *et al.* 1981a, b, c, Ryan *et al.* 1981).

Avnimelech and Laher (1977) made a study of the relative importance of initial soil pH, buffer capacity and ammonium concentration on NH_3 volatilization from ammonium salts applied to alkaline soil. They concluded that initial pH is of prime importance only when the buffer capacity is high or when the concentration of ammonium in the soil solution is low, e.g. when the cation exchange capacity of the soil is high. At a given pH, NH_3 loss increases with increasing buffer capacity. When the buffer capacity of the soil is low, proportionate losses of NH_3 will decrease as the concentration of added ammonium is increased.

D. Calcium carbonate

As already mentioned there are many reports on the stimulation of NH_3 volatilization by calcium carbonate; a strong correlation has been found between NH_3 loss and calcium carbonate content of soils. Later work showed that ammonia loss was more closely related to clay-sized calcium carbonate than to total calcium carbonate (Ryan et al. 1981).

The effect of calcium carbonate was first considered to be due to its alkalinity and buffering capacity but then it became apparent that other processes were involved (Van Schreven 1956, Volk 1961, Terman and Hunt 1964, Harmsen and Kolenbrander 1965, Fenn and Kissel 1973, 1974, 1975, Fenn 1975). Van Schreven (1956) found that NH_3 loss varied with the anion present in the fertilizer applied, and Terman and Hunt (1964) concluded that volatilization losses increased as the solubility of the non-nitrogenous reaction products decreased.

Terman and Hunt (1964), Fenn and Kissel (1973), Terman (1979) and Fenn et al. (1981c) explained the results as follows: when ammonium fluoride, ammonium sulfate and diammonium phosphate react with calcium carbonate, calcium salts of low solubility and ammonium carbonate are formed. The precipitation of the calcium fluoride, sulfate and phosphate provides a driving force and the reaction proceeds with the formation of much ammonium carbonate, e.g.

$$(NH_4)_2SO_4 + CaCO_3 \rightarrow CaSO_4 \downarrow + (NH_4)_2CO_3. \tag{25}$$

The ammonium carbonate is thought to decompose as follows,

$$(NH_4)_2CO_3 + H_2O \rightarrow 2NH_3 + CO_2 + 2H_2O, \tag{26}$$

although Feagley and Hossner's (1978) results indicate that the reaction proceeds with the formation of ammonium bicarbonate rather than ammonium carbonate. Because soluble calcium salts are formed when ammonium nitrate, chloride or iodide reacts with calcium carbonate, an equilibrium rather than a driving force exists and a much lower concentration of ammonium carbonate is formed, e.g.

$$2NH_4NO_3 + CaCO_3 \rightleftarrows Ca(NO_3)_2 + (NH_4)_2CO_3. \tag{27}$$

Thus, when these compounds are used, NH_3 loss should depend on the initial pH and buffer capacity of the soil, which control the stability of the ammonium carbonate.

Other workers have studied the reactions of ammonium hydroxide, ammonium carbonate and urea with calcareous soils or base saturated clays, and the effects of adding different compounds on NH_3 loss from these materials. For example, Du Plessis and Kroontje (1966) found that increasing carbon dioxide concentrations reduced NH_3 loss from ammonium hydroxide applied to base saturated clays. They suggested that precipitation of exchangeable calcium as the carbonate eliminated adsorbed calcium and allowed ammonium to occupy cation exchange sites.

This study was extended to soils by Fenn and Miyamoto (1981) who found that the amounts of extractable calcium and magnesium were considerably reduced by the addition of urea, ammonium carbonate or ammonium hydroxide. This decrease was most probably caused by precipitation of calcium and magnesium carbonates. Since a large fraction of the potentially exchangeable cations were precipitated, exchange sites were freed for reaction with ammonium. When urea was injected into soil at depths of 2.5 cm or more these reactions reduced upward movement of ammonium and subsequent NH_3 losses.

1.3.2. Interacting processes

The rate of volatilization can be increased by the addition of potential NH_3 sources (e.g. plant litter, animal excreta, fertilizer) which increase the concentration of ammonium or NH_3 in solution, or decreased by the removal of NH_3 through immobilization, nitrification, plant uptake or leaching (see Fig. 1.3).

Ammonia loss is also affected by the nature of the source although there may be an interaction with soil type, and ammonia loss may not always be greater, for example, from urea than from ammonium sulfate (e.g. Matocha 1976, Balba and Naseem 1968, Lyster et al. 1980, Faurie and Bardin 1979a).

A. Urease activity

Ammonium salts and NH_3 fertilizer sources affect the concentration of ammonium or NH_3 in soil immediately on application but organic sources e.g. urea, have to be mineralized to ammonium before they can affect the loss process.

Urea is hydrolysed in most soils and the reaction is catalyzed by the enzyme urease (urea amidohydrolase, EC 3.5.1.5) which is widely distributed in plants, microorganisms and soils. The hydrolysis of urea is reported to proceed via carbamic acid (Gorin 1959, Blakely et al. 1969, Bremner and Mulvaney 1978).

While some urease apparently exists in all soils, the activity varies from soil to soil. Bremner and Mulvaney (1978) tabulated urease activity values for surface soils from different parts of the world and showed that the amount of urea hydrolysed per hour per g soil ranged from 0 to 2964 μg. Urease activity

14

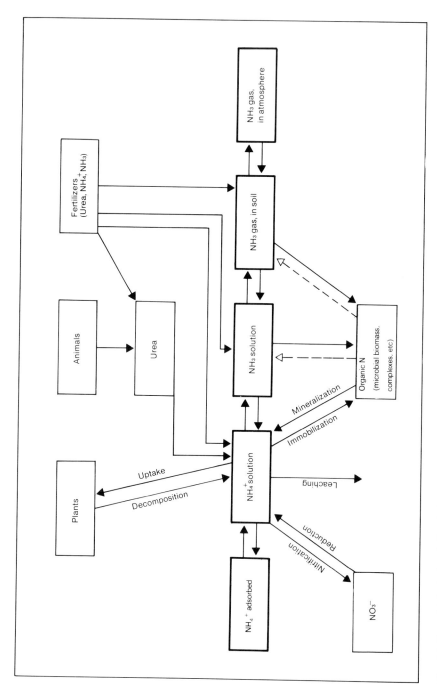

Fig. 1.3. Some biological and chemical reactions affecting ammonia volatilization.

tends to increase with organic matter content, and sandy or calcareous soils tend to have lower activities than heavy-textured or noncalcareous soils. Urease activity decreases markedly with soil depth, and the decrease is closely related to organic matter content (Simpson 1968, Bremner and Mulvaney 1978). Plants and plant litter on the soil surface contribute to the urease activity in soil (Meyer *et al.* 1961, McGarity and Hoult 1971). Urease activity increases markedly with temperature between 10 and 70° C, with urea concentration until the enzyme is saturated with substrate, and with soil cation exchange capacity (Bremner and Mulvaney 1978). Soil moisture (e.g. Sankhayan and Skukla 1976) and soil pH (e.g. Delaune and Patrick 1970) have also been shown to affect urease activity.

B. *Plant uptake*
Assimilation of N by plants reduces the potential for NH_3 loss (Mills *et al.* 1974, Klubek *et al.* 1978, Reddy and Patrick 1980, Buresh *et al.* 1981). This effect will of course be governed by the time of fertilizer application relative to stage of growth; actively growing plants with well developed root systems should be more effective in reducing the ammonium concentration of the soil solution than recently established plants. Even so, corn seedlings (*Zea mays*) have been shown to reduce NH_3 losses (Mills *et al.* 1974).

C. *Leaching*
Because the ammonium ion is strongly sorbed on clay minerals and organic matter, it is not readily leached from soils except under certain conditions. Leaching could occur in sandy soils with low cation exchange capacity or in heavier soils when the cation exchange capacity is exceeded; this can occur during rain after urea application, or when large amounts of ammonium in sewage water are added to light-textured soils (Khanna 1981).

1.3.3. Environmental factors

Ammonia emission from natural surfaces in the field follows a marked diurnal cycle resembling that of solar radiation but the cycle probably results from associated phenomena, such as changes in temperature, or wind speed, or the effects of water evaporation on solution chemistry (e.g. Denmead *et al.* 1974, 1978, Beauchamp *et al.* 1978, Freney *et al.* 1981).

A. *Water*
Soil water content affects volatilization in a number of different ways depending on the NH_3 source, its time and method of application and its depth of placement (e.g. Kresge and Satchell 1960, Chao and Kroontje 1964, Oganov and Ibragimov 1966, Balba and Naseem 1968, Jusuf and Soepardi 1968,

Shankaracharya and Mehta 1969, Fernando and Bhavanandan 1971, Ida and Mori 1971, Rajaratnam and Purushothaman 1973, Lippold *et al.* 1975, Zardalishvili *et al.* 1976, Fenn and Kissel 1976, Prasad 1976a, b, Fenn and Escarzaga 1976, 1977, More and Varade 1977, 1978, Denmead *et al.* 1978, Connell *et al.* 1979, Faurie and Bardin 1979b, Lyster *et al.* 1980). When anhydrous NH_3 is the source, the water content of the absorbing material, be it soil, clay, or plant litter, markedly affects the retention of NH_3 (e.g. Jackson and Chang 1947, Blue and Eno 1954, Stanley and Smith 1956, Mortland 1958, Brown and Bartholomew 1963, Coffee and Bartholomew 1964a, Parr and Papendick 1966a,b, Shimpi and Savant 1975). For instance, Shimpi and Savant (1975) found that the amount of NH_3 retained by a noncalcareous medium black soil decreased from ~ 320 to ~ 160 mg $N(100\,g$ soil$)^{-1}$ as the moisture content of the soil increased from oven dry to 30%, but then increased again as the water content was increased further to 100%. It has been suggested that there is competition between NH_3 and water molecules for adsorption sites in soils and clays and that these two species mutually displace each other from the sorption sites (e.g. Brown and Bartholomew 1963, James and Harward 1964, Young and McNeal 1964, Parr and Papendick 1966a). Shimpi and Savant (1975) suggest that the initial decrease in NH_3 retention with increasing moisture is due to the failure of the NH_3 molecules to displace water molecules already held (see also Ashworth 1978), and the subsequent increase is due to the dissolution of NH_3 in the soil water.

Soil moisture also affects the retention of anhydrous NH_3 during application through its effect on the mechanical properties of the soil and the closure of the injection channel (Parr and Papendick 1966b).

The initial soil water content can also influence NH_3 loss after applications of ammonium sulfate and ammonium nitrate to calcareous soils. Sufficient water must be present to dissolve the added compounds before reaction and volatilization can occur (Fenn and Kissel 1973, Fenn and Escarzaga 1976). In their experiments, Fenn and Escarzaga (1976) found that soils with no free water retained essentially all applied ammonium N. Fenn and Escarzaga (1977) found that when ammonium salts were applied to initially wet soils and initially dry soils, followed in each case by additions of water, NH_3 loss was greater from the initially wet soil. They suggested that this might have occurred because during infiltration of the initially dry soil, the ammonium solution would penetrate both the large and the small pores, whereas in the initially wet soil, most of the NH_4^+ would remain in the large pores. When evaporation occurred subsequently, capillary movement to the surface would occur preferentially through the large pores so that in the initially wet soil, the ammonium would be replaced and moved to the surface more effectively than in the dry soil.

A further effect of water on NH_3 volatilization is apparent when urea is the fertilizer. The loss of NH_3 from this system is affected by the rate of hydrolysis of urea to ammonium carbonate. The *chemical* hydrolysis is very slow under all conditions likely to be encountered in the field (Wei-tsung Chin and Kroontje 1963) and the *biological* hydrolysis catalyzed by urease is negligible in air-dry soil (Fenn and Escarzaga 1976). However, losses can occur when the soil is dry if sufficient moisture is absorbed from the air or dew to partially dissolve the urea (e.g. Harper *et al.*, this book). Dissolution of the fertilizer and subsequent NH_3 loss can be controlled by coating of the fertilizer, e.g. in the case of urea, by elemental sulfur (Matocha 1976, Prasad 1976a).

As urea is retained by soil particles only by polar orientation or hydrogen bonding, it moves rapidly when water is applied, though slightly behind the wetting front (Fenn and Miyamoto 1981). Ammonia losses from surface-applied urea will thus decrease with increasing amounts of applied water (Fenn and Miyamoto 1981).

Other biological processes, such as ammonification which contributes to the build-up of NH_3 (Faurie and Bardin 1979b), or nitrification which reduces the concentration of NH_3 (Cornforth and Chesney 1971, Lyster *et al.* 1980, Fleisher and Hagin 1981), are affected by the initial soil water content and by subsequent wetting and drying (Faurie and Bardin 1979b).

As the upward movement of water helps transport ammonia to the soil surface, a relationship might be expected between NH_3 loss and water loss from soils. Indeed, a number of researchers have shown that NH_3 loss and water loss are directly related (e.g. Jewitt 1942, Martin and Chapman 1951, Wahhab *et al.* 1957, Denmead *et al.* 1976, Fenn and Escarzaga 1977). Kresge and Satchell (1960) found that NH_3 loss could be reduced by limiting evaporation, and Jewitt (1942) and Wahhab *et al.* (1957) found that NH_3 loss ceased when water loss ceased. Interrupting the capillary flow of water affected the rate of NH_3 loss (Fenn and Escarzaga 1977).

The amount of rain which falls immediately after application of urea is another factor which influences ammonia volatilization (Volk 1970, Carrier and Bernier 1971, Morrison and Foster 1977). It is recommended that fertilizer be incorporated into the soil by cultivation if rain does not fall immediately after surface application (Meyer *et al.* 1961).

B. Temperature
Increasing temperature increases the relative proportion of NH_3 to ammonium present at a given pH, decreases the solubility of NH_3 in water, and increases diffusion of NH_3 through the soil. Therefore, the higher the temperature the greater the potential for NH_3 loss. Temperature also affects the solubility of the fertilizers added to soil, the urease activity and the rate of microbial transformations of NH_3. These various effects are described by Martin and Chapman (1951), Fenn and Kissel (1974), Wahhab *et al.* (1957), Faurie and Bardin (1979b), Prasad (1976a), Farquhar *et al.* (1980), Filimonov and

Strel'nikova (1974) and Bremner and Mulvaney (1978).

C. Wind speed

As described in section 1.2.4, increasing wind speed should increase the volatilization rate by promoting more rapid transport of NH_3 away from the air-water or air-soil interface. Few data are available to confirm this but relationships with wind speed have been observed for ammonia loss from rice paddies (Freney et al. 1981) and irrigation water (Denmead et al. 1982).

Bouwmeester and Vlek (1981) incorporated the effects of wind speed into a model for predicting NH_3 volatilization from flooded soils from a knowledge of floodwater chemistry and meterological conditions. Their model predicts that increasing wind speed will increase volatilization rate, but the effect diminishes at higher wind speeds. The analysis also indicates that over the ranges of variation likely to be encountered in the field, wind speed, water temperature and pH will all have roughly similar effects on volatilization rate.

The effects of wind speed are confounded by other factors. Bouwmeester and Vlek (1981), for example, suggest that at high pHs, volatilization rates become insensitive to further increases in wind speed due to depletion of NH_3 in the surface layer of the floodwater. The replenishment of NH_3 in the liquid phase then controls the transfer process. Their model assumes that this would be by molecular diffusion only. In the work of Denmead et al. (1982) the mixing of the solution due to the turbulent flow down the irrigation furrows appears to have been sufficient to avoid the development of NH_3 concentration gradients in the solution and a depleted zone near the surface. In section 1.2.4, we reported recent work which indicated that temperature inversions developed in the water can also suppress NH_3 transport to the surface.

D. Atmospheric ammonia concentration

As discussed in section 1.2.4, the rate of volatilization of NH_3 depends on the difference in NH_3 partial pressure between the soil solution or flood water and the atmosphere. Atmospheric NH_3 concentrations are usually very low and there is no evidence to suggest that they seriously limit volatilization rates in the field. There is some evidence, however, that plants play an important role in this context. The maintenance of low ambient concentrations depends in part on the existence of an NH_3 compensation point, i.e. an atmospheric NH_3 concentration above which plants will absorb NH_3 from the air, and below which they will release it (Farquhar et al., 1980, this book). Denmead et al. (1976) found that even though there was a substantial release of NH_3 from the ground in a grass-clover pasture, almost none of it escaped to the atmosphere above the canopy. The effect of plant absorption was to reduce the NH_3 concentration of the air from > 16 μg m^{-3} near the soil surface to 1 μg m^{-3} at the top of the canopy. Lemon and van Houtte (1980) report a similar phenomenon in a quackgrass field, but in that case the plants reduced the NH_3 concentration from 40 $\mu g \ m^{-3}$ above the plant canopy to 3 $\mu g \ m^{-3}$ within it.

1.3.4. Agronomic factors

Ammonia volatilization can be influenced by farming operations. The effects can be categorized into those which influence (i) source strength (e.g. fertilizer addition), (ii) equilibria (e.g. addition of urea which produces microenvironments of high pH) or (iii) transport processes. Volatilization is affected not only by the rate of application and chemical nature of the source (Section 1.3.2) but also by the physical form of the fertilizer and its method of placement. Carrier and Bernier (1971), Matocha (1976) and Prasad (1976a), for instance, demonstrated substantially less NH_3 volatilization from sulfur coated ureas than from conventional ammoniacal fertilizers regardless of method of application. Others have shown that NH_3 losses can be reduced by placing the ammonium based fertilizer below the soil surface (e.g. Movsumov 1965, Jusuf and Soepardi 1968, Shankaracharya and Mehta 1969, Fenn and Kissel 1976) or by cultivation to incorporate the fertilizer within the soil (e.g. Baligar and Patil 1968, Matocha 1976).

The work of Fenn and Miyamoto (1981) suggested that NH_3 loss from surface applications of urea to calcareous soils could be reduced by the addition of calcium salts. Fenn et al. (1981c) subsequently studied the reaction and proposed a theory to explain the results. They found that NH_3 volatilization from soils after surface application of urea was considerably reduced by the addition of calcium or magnesium nitrates or chlorides. They suggest that this suppression may have resulted from: (i) precipitation of carbonate by calcium thus reducing ammonium carbonate accumulation (Equation 29), (ii) reduction in soil pH by depression of the dissociation of the calcium carbonate: calcium hydroxide buffer system, and (iii) formation of ammonium nitrate or ammonium chloride (Equation 29) which are acidic and therefore neutralize some of the ammonium carbonate formed, i.e.

$$CO(NH_2)_2 + 3H_2O \xrightarrow{urease} (NH_4)_2CO_3 \cdot H_2O \qquad (28)$$

$$(NH_4)_2CO_3 \cdot H_2O + CaX_2 \rightarrow CaCO_3 \downarrow + 2NH_4X \qquad (29)$$

where X = chloride or nitrate.

In acid soils the initial pH in the presence of calcium or magnesium may be too low for calcium carbonate precipitation. Precipitation will only occur if sufficient ammonium carbonate is produced to raise the pH above 7, and NH_3 loss should not be appreciable until this pH value is exceeded. These mechanisms were confirmed by the later work of Fenn et al. (1981b) who also showed that added calcium has an additional effect viz. it reduces the rate of urea hydrolysis. This may be due to the formation of a calcium-urea complex.

Fenn *et al.* (1981a, b) studied the effects of Ca/N ratios on NH_3 volatilization with the expectation that the inclusion of soluble calcium or magnesium salts with N fertilizers could be economical if Ca/N ratios of < 1 inhibited NH_3 loss. Ammonia loss progressively decreased with increasing Ca/urea – N ratios; additions of 25% and 50% Ca reduced NH_3 losses in the greenhouse from 69% of the applied N to 11 and 6%, respectively.

Irrigated crops growing on N-deficient soils often require late applications of N to realize their full yield potential. This is particularly so for crops with a high N demand, such as maize (*Zea mays*) and potatoes (*Solanum tuberosum*). Heavier applications of nitrogen at planting frequently do not solve the problem, due probably to large losses of N through denitrification when the soil is saturated after the early irrigations.

Late N can be supplied by dissolving anhydrous NH_3 in water which is then applied to the crop in sprinkler systems or by furrow or flood irrigation. These methods of applying NH_3 have advantages of simplicity, convenience and low cost. The NH_3 and water percolate together into the root zone, thus making the N available to the plants even in the absence of rainfall (Warnock 1966). However, there are disadvantages caused by adsorption and retention of NH_3 at the soil surface, and by the volatility of NH_3. These result in uneven distribution along the rows or with distance from the sprinkler, retention of NH_3 above the root zone, loss of NH_3 to the atmosphere, and reduced efficiency. Declining concentrations of NH_3 as irrigation water passes along the furrows are frequently observed (e.g. Chapman 1956, Humbert and Ayres 1956, Denmead *et al.* 1982).

Ammonia application through sprinkler systems allows greater exposure to the air and there is thus a greater chance for evaporation of water and NH_3 loss (Warnock 1966). Henderson *et al.* (1955) and Jackson *et al.* (1959) found that losses of NH_3 could be greater than 60% when applied through a sprinkler system. Both Jackson *et al.* (1959) and Miyamoto *et al.* (1975) found that NH_3 loss increased with increasing concentrations of applied NH_3, and Jackson *et al.* (1959) found that the losses decreased with increasing distance from the sprinkler. It was suggested that this pattern may be due to increased volatilization from small droplets and the preponderance of small droplets close to the sprinkler.

Loss of NH_3 during furrow irrigation can also be enormous. Denmead *et al.* (1982) report that when water-run NH_3 was applied to a maize crop, as much as 30% of the amount applied could be volatilized per hour.

The quality of the irrigation water also influences NH_3 loss in both water-run and sprinkler applications as it changes the relative amounts of N present as ammonium and NH_3 (Miyamoto *et al.* 1975).

Since solution vapour pressure and wind speed are important determinants of volatilization rate in these systems, practical measures to reduce volatilization-tion losses should aim at reducing their effects. Miyamoto *et al.* (1975) and Miyamoto and Ryan (1976) found that reducing the solution pH by addition of sulfuric acid resulted in substantial reductions in the rates of volatilization, as would be expected from the dependence of vapour pressure on pH implicit in Equations (21) and (22). Addition of acids to the irrigation water would seem to be a useful practice where cheap by-product acids are available.

Another practical measure is to delay the application of NH_3 to as late as possible in the crop's development since foliar density influences the wind speed close to the ground. In experiments with water-run NH_3 applied to a maize crop, Denmead *et al.* (1982) found that an increase in leaf area index from 1 to 1.9 reduced volatilization by a factor of 7.

Indications are that shortening the length of furrows so that irrigation can be completed during night-time could also lead to more efficient N application. At night, water temperatures are reduced, wind speeds are generally low and the night-time inversion suppresses turbulent transfer. In the experiments reported by Denmead *et al.* (1982), volatilization rates during night-time applications of water-run NH_3 were only about half those measured for similar aqueous NH_3 concentrations by day. Further suggestions for reducing loss are to irrigate on cool, humid, calm days, with a low concentration of NH_3, and to keep the exposed surface of flowing water to a minimum by reducing turbulence of the water and by using short, narrow, deep furrows (Sokoloff 1951, Adams and Anderson 1961).

1.4. Importance of ammonia volatilization from ecosystems

On the global scale NH_3 volatilization has been estimated to supply between 113×10^6 and 244×10^6 t N/year to the atmosphere (Burns and Hardy 1975, Söderlund and Svensson 1976). The primary sources of this NH_3 appear to be microbial decomposition of organic material at the earth's surface, animal excreta, fertilizer manufacture and use, and the burning of coal. Healy *et al.* (1970), Galbally *et al.* (1980) and Lenhard and Gravenhorst (1980) have attempted to evaluate the relative importance of these sources in the United Kingdom, Australia and the Federal Republic of Germany, respectively. Each group concluded that the principal source in their region was animal excreta.

1.4.1. Agricultural soils

Changes in N fertilizer practices have probably accentuated NH_3 volatilization losses from agricultural soils. There has been a marked increase in rates of N fertilizer application and a trend away from the use of ammonium salts. Currently, anhydrous or aqueous NH_3 and urea, which are the cheapest forms of N to produce by industrial synthesis, are becoming the dominant forms of N fertilizer, thus increasing the potential for NH_3 loss.

Even though the methods used for estimating NH_3 volatilization prior to the 1970s left much to be desired, the evidence of many experiments is that when urea is surface applied to grassland or bare soil, NH_3 losses of 25% or more of the applied N can occur. Large losses of NH_3 can also occur if ammonium salts are applied to the surface of calcareous soils.

More recent measurements, using techniques which do not disturb the physical environment, suggest that with suitable fertilizer practices losses by NH_3 volatilization can be much reduced. Two examples are given to illustrate this point:

(i) Early published estimates of NH_3 loss during injection of anhydrous NH_3 suggested that losses could amount to as much as 45 kg/ha (Blue and Eno 1954). However, this seems to be an extreme case. For example, when NH_3 was injected into a heavy clay soil of pH 8.2 the loss of NH_3 was limited to ~ 1 kg N/ha from an application of 107 kg N/ha (Denmead et al. 1977). It seems that, with suitable design of applicator and depth of injection, losses can be reduced to acceptable levels.

(ii) The use of water-run NH_3 for supplying late N for crops with high N demands has been discussed in section 1.3.4. There, attention was drawn to the work of Denmead et al. (1982) who found that volatilization losses could approach 30% per hour of the amount of N applied. A later experiment (by the same group) showed that losses of NH_3 could be limited to $<2\%$ of the amount applied by substituting urea for NH_3 in the irrigation water. As the irrigating solution had virtually no urease activity, no NH_3 was lost during the irrigation. Also, as the urea is not adsorbed by the soil to any appreciable extent (Fenn and Miyamoto 1981) it has the potential to move deeper into the soil than applied NH_3. The actual movement depends on the urease activity, the cation exchange capacity of the soil, and infiltration rate.

1.4.2. Grasslands

In grasslands, NH_3 can be lost directly from plants, from decomposing litter (plants, small animals and microorganisms) and from the excreta of grazing animals (Woodmansee et al. 1978, 1981, Chapters 8 and 9). Depending on the atmospheric NH_3 concentration and the internal compensation point, which varies with species and age of tissue plants can lose NH_3 (see Farquhar et al. 1979, 1980, Chapter 6). The importance of this phenomenon in grasslands has yet to be determined.

Floate and Torrance (1970) concluded from their studies that only a small proportion of the total N in decomposing litter is emitted as NH_3. However loss of NH_3 by volatilization can be considerable when pastures are grazed or when certain N fertilizers are added to grasslands. Woodmansee et al. (1981) record that more than 80% of the N in urine may be lost by volatilization. Denmead et al. (1974, 1976) measured average daily NH_3 emission rates of 13 and 26 mg N/m^2 from grazed pastures at stocking rates of 22 and 50 sheep/ha, respectively. From figures given by Healy et al. (1970) for urea excretion by sheep, these data indicate that about 26% of the N excreted in urine was volatilized as NH_3. That result agrees well with those of Vallis et al. (1982) for NH_3 losses from simulated urine patches on a *Setaria* pasture in southern Queensland. The greatest loss found by them was 28% of the applied N. When urea fertilizer was applied to a grazed pasture in a warm environment losses of NH_3 up to 42% of the applied N were found (Catchpoole et al. 1982).

Further discussion on the significance of NH_3 loss from grasslands is given in Chapters 8 and 9 of this book.

1.4.3. Forests

There seems to be little chance of NH_3 being volatilized from undisturbed forest floors because the pH there is usually low. However, if the ecosystem is disturbed by fertilization, clear cutting or other management practices then NH_3 may be lost.

There have been few studies on NH_3 loss from urea applications to forest soils and the results have varied widely. Losses have varied from as little as 3.5% to as much as 48% of the nitrogen applied (e.g. Acquaye and Cunningham 1965, Overrein 1968, 1969, Bernier et al. 1969, Volk 1970, Crane 1972, Carrier and Bernier 1971, Watkins et al. 1972, Nömmik 1973a, Mahendrappa and Ogden 1973, Morrison and Foster 1977). Crane (1972) and Mahendrappa (1975) observed that the large variation in results may be due to the different technique used for measuring absorption of NH_3. All of the field studies reported for NH_3 volatilization were done at or near the soil surface. Ammonia lost from the surface may be absorbed by the forest canopy and thus not lost from the forest system (Mahendrappa 1975).

Crane (1972) suggested that the size of the urea pellet may influence the rate of volatilization of NH_3 from forest soils. The volatilization can be retarded by increasing the urea pellet size without affecting the total loss (Nömmik 1973a, b, Watkins et al. 1972). This is probably caused by the slower dissolution rate of the larger pellet. Addition of phosphoric or boric acid to large-pellet urea greatly reduced the loss of NH_3 to the atmosphere (Nömmik 1973b).

Other forestry management practices which may increase NH_3 loss are (i) clearfelling, which eliminates the absorption of NH_3 by the forest canopy (Vitousek 1981), (ii) slash burning, which increases soil pH (Vitousek 1981), (iii)

24

animal grazing (Vitousek 1981), (iv) sludge disposal (Sidle and Kardos 1979) and (v) waste water disposal (Sopper 1975). However no data are available to confirm the importance of such effects.

1.4.4. Flooded soils

A number of workers have shown that NH_3 may be volatilized when ammonium-containing or ammonium-producing fertilizers are applied to flooded soils (e.g. Macrae and Ancajas 1970, Ventura and Yoshida 1977, Wetselaar *et al.* 1977, Mikkelsen *et al.* 1978, Bouldin and Alimagno 1976, Sahrawat 1980, Freney *et al.* 1981). Ammonia volatilization has been measured from acid as well as alkaline soils and the loss has varied from 0.5 to 60% of the N applied (Sahrawat 1980). As with the data from forests much of the variation is probably due to the different methods used for measuring NH_3 volatilization.

1.5. Conclusions

It is apparent from this chapter that NH_3 volatilization is an important mechanism for N loss from fertilizers and animal excreta in a number of agricultural systems. The process is a highly complex one, affected by a combination of biological, chemical and physical factors. Most of these factors appear to have been defined but there is an urgent need for further data on the extent of NH_3 volatilization from a wide variety of natural and Man-made ecosystems using techniques which do not disturb the micro-environments involved in each system.

1.6. References

Acquaye, D.K. and Cunningham, R.K. 1965 Losses of nitrogen by ammonia volatilization from surface-fertilized tropical forest soils. Trop. Agric. 42, 281–292.

Adams, J.R. and Anderson, M.S. 1961 Liquid nitrogen fertilizers for direct application. U.S. Dep. Agric. Agric. Handbook No. 198, 44 pp.

Anderson, J.R. 1962 Urease activity, ammonia volatilization and related microbiological aspects in some South African soils. Proc. 36th Congr. S. Afr. Sugar Tech. Assn. 97–105.

Ashworth, J. 1978 Reactions of ammonia with soil. II. Sorption of NH_3 on English soils and on Wyoming bentonite. J. Soil Sci. 29, 195–206.

Avnimelech, Y. and Laher, M. 1977 Ammonia volatilization from soils: equilibrium considerations. Soil Sci. Soc. Am. J. 41, 1080–1084.

Baker, J.H., Peech, M. and Musgrave, R.B. 1959 Determination of application losses of anhydrous ammonia. Agron. J. 51, 361–362.

Balba, A.M. and Naseem, M.G. 1968 The loss of ammonia by volatilization from nitrogenous fertilizers added to the soil. Beitr. Trop. Landwirtsch Veterinaermed. 3, 231–238.

Baligar, V.C. and Patil, S.V. 1968 Volatile losses of ammonia as influenced by rate and methods of urea application to soils. Indian J. Agron. 13, 230–233.

Bates, R.G. and Pinching, G.D. 1950 Dissociation constant of aqueous ammonia at 0 to 50° from E.m.f. studies of the ammonium salt of a weak acid. J. Am. Chem. Soc. 72, 1393–1396.

Beauchamp, E.G., Kidd, G.E. and Thurtell, G. 1978 Ammonia volatilization from sewage sludge applied in the field. J. Environ. Qual. 7, 141–146.

Bernier, B. Carrier, D. and Smirnoff, W.A. 1969 Preliminary observations on nitrogen losses through ammonia volatilization following urea fertilization of soil in a jack pine forest. Nat. Can. 96, 251–255.

Blakely, R.L., Webb, E.C. and Zerner, B. 1969 Jack bean urease (EC 3.5.1.5). A new purification and reliable rate assay. Biochemistry 8, 1894–1900.

Blue, W.G. and Eno, C.F. 1954 Distribution and retention of anhydrous ammonia in sandy soils. Soil Sci. Soc. Am. Proc. 18, 420–424.

Bouldin, D.R. and Alimagno, B.V. 1976 NH_3 volatilization losses from IRRI paddies following broadcast applications of fertilizer nitrogen Terminal Report, IRRI.

Bouwmeester, R.J.B. and Vlek, P.L.G. 1981 Rate control of ammonia volatilization from rice paddies. Atmos. Environ. 15, 131–140.

Boussingault, J.B. 1856 Recherches sur la végétation. Troisième mémoire. De l'action du salpêtre sur le développement des plantes. Ann. Chim. Phys. Ser. 3, 46, 5–41.

Bremner, J.M. and Mulvaney, R.L. 1978 Urease activity in soils. In: Burns, R.J. (ed.), Soil Enzymes pp. 149–196. Academic Press, London.

Broadbent, F.E. and Stevenson, F.J. 1966 Organic matter interactions. In: McVickar, M.H., Martin, W.P., Miles, I.E. and Tucker, H.H. (eds.), Agricultural Anhydrous Ammonia, pp. 169–187. Agricultural Ammonia Institute, Memphis and American Society of Agronomy, Madison.

Broadbent, F.E. and Tusneem, M.E. 1971 Losses of nitrogen from some flooded soils in tracer experiments. Soil Sci. Soc. Am. Proc. 35, 922–926.

Brown, J.M. and Bartholomew, W.V. 1963 Sorption of gaseous ammonia by clay minerals as influenced by sorbed aqueous vapour and exchangeable cations. Soil Sci. Soc. Am. Proc. 27, 160–164.

Buresh, R.J., De Laune, R.D. and Patrick, W.H., Jr. 1981 Influence of *Spartina alterniflora* on nitrogen loss from marsh soil. Soil Sci. Soc. Am. J. 45, 660–661.

Burge, W.D. and Broadbent, F.E. 1961 Fixation of ammonia by organic soils. Soil Sci. Soc. Am. Proc. 25, 199–204.

Burns, R.C. and Hardy, R.W.F. 1975 Nitrogen Fixation in Bacteria and Higher Plants. Springer-Verlag, New York.

Carrier, D. and Bernier, B. 1971 Loss of nitrogen by volatilization of ammonia after fertilizing in jack pine forest. Can. J. For. Res. 1, 69–79.

Catchpoole, V.R., Harper, L.A. and Myers, R.J.K. 1982 Annual losses of ammonia from a grazed pasture fertilized with urea. Proc. 14th Intern. Grassl. Congr. Lexington (in press).

Chai, H.H. and Hou, T.T. 1975 Studies on the volatilization of gaseous nitrogen from ammonium sulfate, urea and sodium nitrate applied to soils. Soils Fert. Taiwan 12–31.

Chai, H.H., Hou, T.T. and Chiang, C.T. 1974 Studies on the volatilization of gaseous nitrogen from ammonium sulfate, urea and sodium nitrate in soils. Taiwan Fertilizer Company Tech. Bull. No. 50.

Chao, T.T. and Kroontje, W. 1964 Relationships between ammonia volatilization, ammonia concentration and water evaporation. Soil Sci. Soc. Am. Proc. 28, 393–395.

Chapman, H.D. 1956 The application of fertilizers in irrigation water. Better Crops with Plant Food 40, 6–10, 43–47.

Coffee, R.C. and Bartholomew, W.V. 1964a Ammonia sorption and retention by plant residue materials. Soil Sci. Soc. Am. Proc. 28, 482–485.

Coffee, R.C. and Bartholomew, W.V. 1964b Some aspects of ammonia sorption by soil surfaces. Soil Sci. Soc. Am. Proc. 28, 485–490.

Connell, J.H., Meyer, R.D., Meyer, J.L. and Carlson, R.M. 1979 Gaseous ammonia losses following nitrogen fertilization. Calif. Agric. 33, 11–12.

Cornforth, I.S. and Chesney, H.A.D. 1971 Nitrification inhibitors and ammonia volatilization. Plant Soil 34, 497–501.

Court, M.N., Stephen, R.C. and Waid, J.S. 1964 Toxicity as a cause of the inefficiency of urea as a fertilizer. J. Soil Sci. 15, 42–48.

Crane, W.J.B. 1972 Urea-nitrogen transformations, soil reactions, and elemental movement via leaching and volatilization, in a coniferous forest ecosystem following fertilization. Ph.D. Thesis, University of Washington, 300 pp.

Daftarder, S.Y. and Shinde, S.A. 1980 Kinetics of ammonia volatilization of anhydrous ammonia applied to a Vertisol as influenced by farm yard manure, sorbed cations and cation exchange capacity. Commun. Soil Sci. Plant Anal. 11, 135–145.

Dawson, G.A. 1977 Atmospheric ammonia from undisturbed land. J. Geophys. Res. 82, 3125–3133.

Delaune, R.D. and Patrick, W.H., Jr. 1970 Urea conversion to ammonia in waterlogged soils. Soil Sci. Soc. Am. Proc. 34, 603–607.

Denmead, O.T., Freney, J.R. and Simpson, J.R. 1976 A closed ammonia cycle within a plant canopy. Soil Biol. Biochem. 8, 161–164.

Denmead, O.T., Freney, J.R. and Simpson, J.R. 1982 Dynamics of ammonia volatilization during furrow irrigation of maize. Soil Sci. Soc. Am. J. 46, 149–155.

Denmead, O.T., Nulsen, R. and Thurtell, G.W. 1978 Ammonia exchange over a corn crop. Soil Sci. Soc. Am. J. 42, 840–842.

Denmead, O.T., Simpson, J.R. and Freney, J.R. 1974 Ammonia flux into the atmosphere from a grazed pasture. Science (Wash. D.C.) 185, 609–610.

Denmead, O.T., Simpson, J.R. and Freney, J.R. 1977 A direct field measurement of ammonia emission after injection of anhydrous ammonia. Soil Sci. Soc. Am. J. 41, 1001–1004.

Du Plessis, M.C.F. and Kroontje, W. 1964 The relationship between pH and ammonia equilibria in soil. Soil Sci. Soc. Am. Proc. 28, 751–754.

Du Plessis, M.C.F. and Kroontje, W. 1966 The effect of carbon dioxide on the chemisorption of ammonia by base-saturated clays. Soil Sci. Soc. Am. Proc. 30, 693–696.

Ellis, B.G. and Knezek, B.D. 1972 Adsorption reactions of micronutrients in soils. In: Mortvedt, J.J. Giordano, P.M. and Lindsay, W.L. (eds.), Micronutrients in Agriculture, pp. 59–78. Soil Science Society America Inc., Madison.

Emerson, K., Russo, R.C., Lund, R.E. and Thurston, R.V. 1975 Aqueous ammonia equilibrium calculations: Effect of pH and temperature. J. Fish. Res. Board Can. 32, 2379–2383.

Ernst, J.W. and Massey, H.F. 1960 The effects of several factors on volatilization of ammonia formed from urea in the soil. Soil Sci. Soc. Am. Proc. 24, 87–90.

Farquhar, G.D., Wetselaar, R. and Firth P.M. 1979 Ammonia volatilization from senescing leaves of maize. Science (Wash. D.C.) 203, 1257–1258.

Farquhar, G.D., Firth, P.M., Wetselaar, R. and Weir, B. 1980 On the gaseous exchange of ammonia between leaves and the environment: determination of the ammonia compensation point. Plant Physiol. 66, 710–714.

Faurie, G. and Bardin, R. 1971 Effect de l'apport d'azote organique et inorganique à des sols de pelouses xérophiles: influence du phosphore. Soil Biol. Biochem. 3, 57–67.

Faurie, G. and Bardin R. 1979a La volatilization de l'ammoniac. I. Influence de la nature du sol et des composés azotes. Ann. Agron. 30, 363–385.

Faurie, G. and Bardin, R. 1979b La volatilization de l'ammoniac. II. Influence des facteurs climatiques et du couvert végétal. Ann. Agron. 30, 401–414.

Faurie, G., Josserand, A. and Bardin, R. 1975 Influence des colloides argileux sur la rétention d'ammonium et la nitrification. Rev. Ecol. Biol. Sol. 12, 201–210.

Feagley, S.E. and Hossner, L.R. 1978 Ammonia volatilization reaction mechanism between ammonium sulfate and carbonate systems. Soil Sci. Soc. Am. J. 42, 364–367.

Fenn, L.B. 1975 Ammonia volatilization from surface applications of ammonium compounds on calcareous soils. III. Effects of mixing low and high loss ammonium compounds. Soil Sci. Soc. Am. Proc. 39, 366–368.

Fenn, L.B. and Escarzaga, R. 1976 Ammonia volatilization from surface applications of ammonium compounds on calcareous soils. V. Soil water content and method of nitrogen application. Soil Sci. Soc. Am. J. 40, 537–541.

Fenn, L.B. and Escarzaga, R. 1977 Ammonia volatilization from surface applications of ammonium compounds to calcareous soils. VI. Effects of initial soil water content and quantity of applied water. Soil Sci. Soc. Am. J. 41, 358–363.

Fenn, L.B. and Kissel, D.E. 1973 Ammonia volatilization from surface applications of ammonium compounds on calcareous soils. I. General theory. Soil Sci. Soc. Am. Proc. 37, 855–859.

Fenn, L.B. and Kissel, D.E. 1974 Ammonia volatilization from surface applications of ammonium compounds on calcareous soils. II. Effects of temperature and rate of ammonium nitrogen application. Soil Sci. Soc. Am. Proc. 38, 606–610.

Fenn, L.B. and Kissel, D.E. 1975 Ammonia volatilization from surface applications of ammonium compounds on calcareous soils. IV. Effect of calcium carbonate content. Soil Sci. Soc. Am. Proc. 39, 631–633.

Fenn, L.B. and Kissel, D.E. 1976 The influence of cation exchange capacity and depth of incorporation on ammonia volatilization from ammonium compounds applied to calcareous soils. Soil Sci. Soc. Am. J. 40, 394–398.

Fenn, L.B. and Miyamoto 1981 Ammonia loss and associated reactions of urea in calcareous soils. Soil Sci. Soc. Am. J. 45, 537–540.

Fenn, L.B., Matocha, J.E. and Wu, E. 1981a Ammonia losses from surface-applied urea and ammonium fertilizers as influenced by rates of soluble calcium. Soil Sci. Soc. Am. J. 45, 883–886.

Fenn, L.B., Matocha, J.E. and Wu, E. 1981b A comparison of calcium carbonate precipitation and pH depression on calcium-reduced ammonia loss from surface-applied urea. Soil Sci. Soc. Am. J. 45, 1128–1131.

Fenn, L.B., Taylor, R.M. and Matocha, J.E. 1981c Ammonia losses from surface-applied nitrogen fertilizer as controlled by soluble calcium and magnesium: general theory. Soil Sci. Soc. Am. J. 45, 777–781.

Fernando, V. and Bhavanandan, V.P. 1971 Volatilization losses of ammonia from urea applied to the soil. Tea Q. 42, 48–56.

Filimonov, D.A. and Strel'nikova, R.A. 1974 Gaseous losses of ammonia following surface application of urea. Soviet Soil Sci. 6, 426–432.

Fleisher, Z. and Hagin, J. 1981 Lowering ammonia volatilization losses from urea application by activation of nitrification process. Fert. Res. 2, 101–107.

Floate, M.J.S. and Torrance, C.J.W. 1970 Decomposition of the organic materials from hill soils and pastures. 1. Incubation method for studying the mineralization of carbon, nitrogen and phosphorus. J. Sci. Food Agric. 21, 116–120.

Freney, J.R., Denmead, O.T., Watanabe, I. and Craswell, E.T. 1981 Ammonia and nitrous oxide losses following applications of ammonium sulfate to flooded rice. Aust. J. Agric. Res. 32, 37–45.

Galbally, I.E., Freney, J.R., Denmead, O.T. and Roy C.R. 1980 Processes controlling the nitrogen cycle in the atmosphere over Australia. In: Trudinger, P.A., Walter, M.A. and Ralph B.J. (eds.), Biogeochemistry of Ancient and Modern Environments, pp. 319–325. Australian Academy of Science, Canberra.

28

Gardner, W.R. 1965 Movement of nitrogen in soil. In: Bartholomew, W.V. and Clark, F.E. (eds.), Soil Nitrogen, pp. 550–572. Agronomy 10. American Society of Agronomy, Madison.

Gasser, J.K.R. 1964a Some factors affecting losses of ammonia from urea and ammonium sulfate applied to soils. J. Soil Sci. 15, 258–272.

Gasser, J.K.R. 1964b Urea as a fertilizer. Soils Fert. 27, 175–180.

Gorin, G. 1959 On the mechanism of urease action. Biochim. Biophys. Acta 34, 268–270.

Hales, J.M. and Drewes, D.R. 1979 Solubility of ammonia in water at low concentrations. Atmos. Environ. 13, 1133–1147.

Harding, R.B., Embleton, T.W., Jones, W.W. and Ryan, T.M. 1963 Leaching and gaseous losses of nitrogen from some nontilled California soils. Agron. J. 55, 515–518.

Harmsen, G.W. and Kolenbrander, G.J. 1965 Soil inorganic nitrogen. In: Bartholomew, W.V. and Clark, F.E. (eds.), Soil Nitrogen, pp. 43–92. Agronomy 10. American Society of Agronomy, Madison.

Healy, T.V., McKay, H.A.C., Pilbeam, A. and Scargill, D. 1970 Ammonia and ammonium sulfate in the troposphere over the United Kingdom. J. Geophys. Res. 75, 2317–2321.

Henderson, D.W., Bianchi, W.C. and Doneen, L.D. 1955 Ammonia loss from sprinkler jets. Agric. Eng. 36, 398–399.

Humbert, R.P. and Ayres, A.S. 1956 The use of aqua ammonia in the Hawaiian sugar industry. Proc. 9th Congr. Int. Soc. Sugar Tech. 1, 524–538. Elsevier, New York.

Ida, A. and Mori, I. 1971 Behaviour of nitrogen in mineral arable soil. 3. The effect of fertilizers, soil pH and soil moisture on gaseous nitrogen losses. Bulletin, Tokai-Kinki National Agric. Expt. Sta. No. 21. pp. 135–150.

Ivanov, P. 1963 Possible loss of nitrogen as a result of evaporation of ammonia when applying ammoniacal fertilizers to some soils. Izv. Inst. Pshenitsata Slunchogleda Krai Gr. Tolbukhin Akad. Selskostop. Nauki Bulg. 4, 17–24.

Jackson, M.L. and Chang, S.C. 1947 Anhydrous ammonia retention by soils as influenced by depth of application, soil texture, moisture content, pH value and tilth. J. Am. Soc. Agron. 39, 623–633.

Jackson, T.L., Alban, L.A. and Wolfe, J.W. 1959 Ammonia nitrogen loss from sprinkler applications. Oreg. Agric. Expt. Stn. Circ. 593. 5 pp.

James, D.W. and Harward, M.E. 1964 Competition of NH_3 and H_2O for adsorption sites on clay minerals. Soil Sci. Soc. Am. Proc. 28, 636–640.

Jewitt, T.N. 1942 Loss of ammonia from ammonium sulfate applied to alkaline soils. Soil Sci. 54, 401–409.

Jusuf, I. and Soepardi, G. 1968 Effect of some factors on ammonia volatilization from urea added to soil. Commun. Agricultura, Indonesia 1, 1–9.

Khanna, P.K. 1981 Leaching of nitrogen from terrestrial ecosystems – patterns, mechanisms and ecosystem responses. In: Clark, F.E. and Rosswall, T. (eds.), Terrestrial Nitrogen Cycles. Ecol. Bull. 33, 343–352.

Koelliker, J.K. and Miner, J.R. 1973 Desorption of ammonia from anaerobic lagoons. Trans. Am. Soc. Agric. Eng. 16, 148–151.

Kowalenko, C.G. and Cameron, D.R. 1976 Nitrogen transformations in an incubated soil as affected by combinations of moisture content and temperature and adsorption-fixation of ammonium. Can. J. Soil Sci. 56, 63–70.

Kresge, C.B. and Satchell, D.P. 1960 Gaseous loss of ammonia from nitrogen fertilizers applied to soils. Agron. J. 52, 104–107.

Larsen, S. and Gunary, E. 1962 Ammonia loss from ammoniacal fertilizers applied to calcareous soils. J. Sci. Food Agric. 13, 566–572.

Lehr, I.J. and Van Wesemael, J.C. 1961 The volatilization of ammonia from lime rich soils. Landbouwkd. Tijdschr. 73, 1156–1168.

Lemon, E. and Van Houtte, R. 1980 Ammonia exchange at the land surface. Agron. J. 72, 876–883.

Lenhard, U. and Gravenhorst, G. 1980 Evaluation of ammonia fluxes into the free atmosphere over Western Germany. Tellus 32, 48–55.

Lippold, H., Heber, P. and Forster, I. 1975 Ammonia losses from urea fertilizer. I. Laboratory studies on ammonia volatilization as influenced by soil pH, exchange capacity, temperature and water content. Arch. Acker Pflanzenbau BodenkDe. 19, 619–630.

Liss, P.S. 1973 Processes of gas exchange across an air-water interface. Deep-Sea Res. 20, 221–238.

Liss, P.S. and Slater, P.G. 1974 Flux of gases across the air-sea interface. Nature (Lond.) 247, 181–184.

Loftis, J.R. and Scarsbrook, C.E. 1969 Ammonia volatilization from ratios of formamide and urea solutions in soils. Agron. J. 61, 725–727.

Lyster, S., Morgan, M.A. and O'Toole, P. 1980 Ammonia volatilization from soils fertilized with urea and ammonium nitrate. J. Life Sciences Royal Dublin Society 1, 167–176.

McGarity, J.W. and Hoult, E.H. 1971 The plant component as a factor in ammonia volatilization from pasture swards. J. Br. Grassl. Soc. 26, 31–34.

Macrae, I.C. and Ancajas, R. 1970 Volatilization of ammonia from submerged tropical soils. Plant Soil 33, 97–103.

Mahendrappa, M.K. 1975 Ammonia volatilization from some forest floor materials following urea fertilization. Can. J. For. Res. 5, 210–216.

Mahendrappa, M.K. and Ogden, E.D. 1973 Patterns of ammonia volatilization from a forest soil. Plant Soil 38, 257–265.

Marion, G.M. and Dutt, G.R. 1974 Ion association in the ammonia–carbon dioxide–water system. Soil Sci. Soc. Am. Proc. 38, 889–891.

Martin, J.P. and Chapman, H.D. 1951 Volatilization of ammonia from surface-fertilized soils. Soil Sci 71, 25–34.

Matocha, J.E. 1976 Ammonia volatilization and nitrogen utilization from sulfur-coated ureas and conventional nitrogen fertilizers. Soil Sci. Soc. Am. J. 40, 597–601.

Meyer, R.D., Olson, R.A. and Rhoades, H.F. 1961 Ammonia losses from fertilized Nebraska soils. Agron. J. 53, 241–244.

Mikkelsen, D.S., De Datta, S.K. and Obcemea, W.N. 1978 Ammonia volatilization losses from flooded rice soils. Soil Sci. Soc. Am. J. 42, 725–730.

Mills, H.A., Barker, A.V. and Maynard, D.N. 1974 Ammonia volatilization from soils. Agron. J. 66, 355–358.

Miyamoto, S. and Ryan, J. 1976 Sulfuric acid for the treatment of ammoniated irrigation water: II. Reducing calcium precipitation and sodium hazard. Soil Sci. Soc. Am. J. 40, 305–309.

Miyamoto, S., Ryan, J. and Strohlein, J.L. 1975 Sulfuric acid for the treatment of ammoniated irrigation water: I. Reducing ammonia volatilization. Soil Sci. Soc. Am. Proc. 39, 544–548.

More, S.D. and Varade, S.B. 1977 Gaseous loss of ammonia from urea applied to calcareous black soils of Marathwada. Research Bull. Marathwada Agric. University 1, 1–2.

More, S.D. and Varade, S.B. 1978 Volatilization losses of ammonia from different nitrogen carriers as affected by soil-moisture, organic matter and method of fertilizer application. J. Indian Soc. Soil Sci. 26, 112–115.

Morrison, I.K. and Foster, N.W. 1977 Fate of urea fertilizer added to a boreal forest *Pinus banksiana* Lamb stand. Soil Sci. Soc. Am. J. 41, 441–448.

Mortland, M.M. 1955 Adsorption of ammonia by clays and muck. Soil Sci. 80, 11–18.

Mortland, M.M. 1958 Reactions of ammonia in soils. Adv. Agron. 10, 325–348.

Mortland, M.M. 1966 Ammonia interactions with soil minerals. In: McVickar, M.H., Martin, W.P., Miles, I.E. and Tucker, H.H. (eds.), Agricultural Anhydrous Ammonia, pp. 188–197. Agricultural Ammonia Institute, Memphis, and American Society of Agronomy, Madison.

Movsumov, Z.R. 1965 Loss of N from soil in the form of gaseous ammonia. Dokl. Akad. Nauk Az. SSR 21, No. 3, 87–91.

Nömmik, H. 1957 Fixation and defixation of ammonium in soils. Acta Agric. Scand. 7, 395–436.

Nömmik, H. 1973a Assessment of volatilization loss of ammonia from surface-applied urea on forest soil by N^{15} recovery. Plant Soil 38, 589–603.

Nömmik, H. 1973b The effect of pellet size on the ammonia loss from urea applied to forest soil. Plant Soil 39, 309–318.

Oganov, G.M. and Ibragimov, C.Z. 1966 Loss of nitrogen by volatilization in soils of the Lenkoran subtropics. Agrokhimiya No. 3, 10–16.

Overrein, L.N. 1968 Lysimeter studies on tracer nitrogen in forest soil. 1. Nitrogen losses by leaching and volatilization after addition of urea – N^{15}. Soil Sci. 106, 280–290.

Overrein, L.N. 1969 Lysimeter studies on tracer nitrogen in forest soil. 2. Comparative losses of nitrogen through leaching and volatilization after addition of urea-, ammonium- and nitrate-N^{15}. Soil Sci. 107, 149–159.

Parr, J.F. and Papendick, R.I. 1966a Retention of anhydrous ammonia by soil. II. Effect of ammonia concentration and soil moisture. Soil Sci. 101, 109–119.

Parr, J.F. and Papendick, R.I. 1966b Retention of ammonia in soils. In: McVickar, M.H., Martin, W.P., Miles, I.E. and Tucker, H.H., Agricultural Anhydrous Ammonia, pp. 213–236. Agricultural Ammonia Institute, Memphis, and American Society of Agronomy, Madison.

Patrick, W.H. Jr. and Reddy, K.R. 1976 Nitrification-denitrification reactions in flooded soils and water bottoms: Dependence on oxygen supply and ammonium diffusion. J. Environ. Qual. 5, 469–472.

Prasad, M. 1976a Gaseous loss of ammonia from sulfur-coated urea, ammonium sulfate, and urea applied to calcareous soil (pH 7.3). Soil Sci. Soc. Am. J. 40, 130–134.

Prasad, M. 1976b The release of nitrogen from sulfur-coated urea as affected by soil moisture, coating weight, and method of placement. Soil Sci. Soc. Am. J. 40, 134–136.

Raison, R.J. and McGarity, J.W. 1978 Effect of plant ash on nitrogen fertilizer transformations and ammonia volatilization. Soil Sci. Soc. Am. J. 42, 140–143.

Rajaratnam, J.A. and Purushothaman, V. 1973 Studies on nitrogen losses from soils in Malaysia. 1. Influence of soil moisture, rates and types of nitrogenous compound on ammonia volatilization. Malaysian Agric. Res. 2, 59–64.

Reddy, K.R. and Patrick, W.H., Jr. 1980 Losses of applied ammonium ^{15}N, urea ^{15}N, and organic ^{15}N in flooded soils. Soil Sci. 130, 326–330.

Reddy, K.R., Patrick, W.H., Jr. and Phillips, R.E. 1976 Ammonium diffusion as a factor in nitrogen loss from flooded soils. Soil Sci. Soc. Am. J. 40, 528–533.

Reddy, K.R., Patrick, W.H., Jr. and Phillips, R.E. 1980 Evaluation of selected processes controlling nitrogen loss in a flooded soil. Soil Sci. Soc. Am. J. 44, 1241–1246.

Riley, D. and Barber, S.A. 1971 Effect of ammonium and nitrate fertilization on phosphorus uptake as related to root-induced pH changes at the soil-root interface. Soil Sci. Soc. Am. Proc. 35, 301–306.

Rolston, D.E., Nielsen, D.R. and Biggar, J.W. 1971 Miscible displacement of ammonia in soil: determining sorption isotherms. Soil Sci. Soc. Am. Proc. 35, 899–905.

Rolston, D.E., Nielsen, D.R. and Biggar, J.W. 1972 Desorption of ammonia from soil during displacement. Soil Sci. Soc. Am. Proc. 36, 905–911.

Ryan, J.A. and Keeney, D.R. 1975 Ammonia volatilization from surface-applied wastewater sludge. J. Water Pollut. Control Fed. 47, 386–393.

Ryan, J., Curtin, D. and Safi. I. 1981 Ammonia volatilization as influenced by calcium carbonate particle size and iron oxides. Soil Sci. Soc. Am. J. 45, 338–341.

Sahrawat, K.L. 1980 Ammonia volatilization losses in some tropical flooded rice soils under field conditions. Il Riso 29, 21–27.

Sankhayan, S.D. and Skukla, U.C. 1976 Rates of urea hydrolysis in five soils of India. Geoderma 16, 171–178.

Shankaracharya, N.B. and Mehta, B.V. 1969 Evaluation of loss of nitrogen by ammonia volatilization from soil fertilized with urea. J. Indian Soc. Soil Sci. 17, 423–430.

Shimpi, S.S. and Savant, N.K. 1975 Ammonia retention in tropical soils as influenced by moisture content and continuous submergence. Soil Sci. Soc. Am. Proc. 39, 153–154.

Sidle, R.C. and Kardos, L.T. 1979 Nitrate leaching in a sludge-treated forest soil. Soil Sci. Soc. Am. J. 43, 278–282.

Simpson, J.R. 1968 Losses of urea nitrogen from the surface of pasture soils. Trans 9th Intern. Congr. Soil Sci. Adelaide 2, 459–466.

Simpson, J.R. 1981 A modelling approach to nitrogen cycling in agro-ecosystems. In: Wetselaar, R., Simpson, J.R. and Rosswall, T. (eds.), Nitrogen Cycling in South-East Asian Wet Monsoonal Ecosystems, pp. 174–179. Australian Academy of Science, Canberra.

Söderlund, R. and Svensson, B.H. 1976 The global nitrogen cycle. In: Svensson, B.H. and Söderlund, R. (eds.), Nitrogen, Phosphorus and Sulphur-Global Cycles. Scope Report 7. Ecol. Bull. 7, 23–73.

Sohn, J.B. and Peech, M. 1958 Retention and fixation of ammonia by soils. Soil Sci. 85, 1–9.

Sokoloff, V.P. 1951 Losses of Ammonia from Irrigation Water. John Hopkins University, Baltimore. 42 pp.

Sopper, W.E. 1975 Wastewater recycling on forest lands. In: Bernier, B. and Winget, C.H. (eds.), Forest Soils and Forest Land Management, pp. 227–243. Les Presses de l'Université Laval, Quebec.

Stanley, F.A. and Smith, G.E. 1956 Effect of soil moisture and depth of application on retention of anhydrous ammonia. Soil Sci. Soc. Am. Proc. 20, 557–561.

Subcommittee on Ammonia, National Research Council 1979 Ammonia. University Park Press, Baltimore 384 pp.

Terman, G.L. 1979 Volatilization losses of nitrogen as ammonia from surface-applied fertilizers, organic amendments, and crop residues. Adv. Agron. 31, 189–223.

Terman, G.L. and Hunt, C.M. 1964 Volatilization losses of nitrogen from surface-applied fertilizers, as measured by crop response. Soil Sci. Soc. Am. Proc. 28, 667–672.

Tusneem, M.F. and Patrick, W.H., Jr. 1971 Nitrogen transformations in waterlogged soil. Agr. Expt. Sta. Lousiana State University Bull. 657, 75.

Vallis, I., Harper, L.A., Catchpoole, V.R. and Weier, K.L. 1982 Volatilization of ammonia from urine patches in a subtropical pasture. Aust. J. Agric. Res. 33, 97–107.

Van Schreven, D.A. 1956 The effect of some ammonium-containing fertilizers on the loss of nitrogen after application to a calcareous soil of The Northeastern Polder. Proc. 6th Int. Congr. Soil Sci. D. 65–73.

Ventura, W.B. and Yoshida, T. 1977 Ammonia volatilization from a flooded tropical soil. Plant Soil 46, 521–531.

Vitousek, P.M. 1981 Clear-cutting and the nitrogen cycle. In: Clark, F.E. and Rosswall, T. (eds.), Terrestrial Nitrogen Cycles. Ecol. Bull. 33, 631–642.

Vlek, P.L.G. and Craswell, E.T. 1981 Ammonia volatilization from flooded soils. Fert Res. 2, 227–245.

Vlek, P.L.G. and Stumpe, J.W. 1978 Effects of solution chemistry and environmental conditions on ammonia volatilization losses from aqueous systems. Soil Sci. Soc. Am. J. 42, 416–421.

Vlek, P.L.G., Fillery, I.R.P. and Burford, J.R. 1981 Accession, transformation, and loss of nitrogen in soils of the arid region. Plant Soil 58, 133–175.

Volk, G.M. 1959 Volatile loss of ammonia following surface application of urea to turf or bare soils. Agron. J. 51, 746–749.

Volk, G.M. 1961 Gaseous loss of ammonia from surface-applied nitrogenous fertilizers. J. Agric. Food Chem. 9, 280–283.

Volk, G.M. 1970 Gaseous loss of ammonia from prilled urea applied to slash pine. Soil Sci. Soc. Am. Proc. 34, 513–516.

Wagner, G.H. and Smith, G.E. 1958 Nitrogen losses from soils fertilized with different nitrogen carriers. Soil Sci. 85, 125–129.

Wahhab, A., Randhawa, M.S. and Alam, S.Q. 1957 Loss of ammonia from ammonium sulphate under different conditions when applied to soils. Soil Sci. 84, 249–255.

Warnock, R.E. 1966 Ammonia application in irrigation water. In: McVickar, M.H., Martin, W.P., Miles, I.E. and Tucker, H.H. (eds.), Agricultural Anhydrous Ammonia, pp. 115–124, Agricultural Ammonia Institute, Memphis, and American Society of Agronomy, Madison.

Warren, K.S. 1962 Ammonia toxicity and pH. Nature (Lond.) 195, 47–49.

Watkins, S.H., Strand, R.F., DeBell, D.S. and Esch, J., Jr. 1972 Factors influencing ammonia losses from urea applied to northwestern forest soils. Soil Sci. Soc. Am. Proc. 36, 354–357.

Weast, R.C. 1971 Handbook of Chemistry and Physics. Chemical Rubber Company, Cleveland.

Webb, E.K. 1965 Aerial microclimate. In: Waggoner, P.E. (ed.), Agricultural Meteorology. Meteorol. Monographs 6, 27–58. American Meteorological Society.

Wei-tsung Chin and Kroontje, W. 1963 Urea hydrolysis and subsequent loss of ammonia. Soil Sci. Soc. Am. Proc. 27, 316–318.

Wetselaar, R., Shaw, T., Firth, P., Oupathum, J. and Thitipoca, H. 1977 Ammonia volatilization losses from variously placed ammonium sulphate under lowland rice field conditions in central Thailand. Proc. Int. Seminar on Soil Environment and Fertility Management in Intensive Agriculture, pp. 282–288.

Woodmansee, R.G., Vallis, I. and Mott, J.J. 1981 Grassland nitrogen. In: Clark, F.E. and Rosswall, T. (eds.), Terrestrial Nitrogen Cycles. Ecol. Bull. 33, 443–462.

Woodmansee, R.G., Dodd, J.L., Bowman, R.A., Clark, F.E. and Dickinson, C.E. 1978 Nitrogen budget of a shortgrass prairie ecosystem. Oecologia (Berl.) 34, 363–376.

Young, J.L. 1964 Ammonia and ammonium reactions with some Pacific Northwest soils. Soil Sci. Soc. Am. Proc. 28, 339–345.

Young, J.L. and McNeal, B.L. 1964 Ammonia and ammonium reactions with some layer-silicate minerals. Soil Sci. Soc. Am. Proc. 28, 334–339.

Zardalishvili, O. Yu., Aladashvili, N.Z. and Tetruashvili, V.G. 1976 Gaseous nitrogen losses and plant uptake of fertilizer nitrogen. Agrokhimiya No. 1, 27–30.

2. Biological denitrification

I.R.P. FILLERY

2.1. Introduction

Biological denitrification or dissimilatory reduction of nitrate and/or nitrite to gaseous N oxides and N_2 gas has long been considered an important mechanism of N loss from soil. Since denitrification occurs only in the absence of oxygen, or at particularly low oxygen concentrations, it has been widely accepted as the cause of the poor efficiency of N use in flooded soils, where a well-developed anoxic layer is a characteristic feature (De Datta 1981). It is apparent from estimates of N loss from nonflooded plant-soil systems, that denitrification can also play an important role in N cycling in these soils. Allison (1955, 1966) and Hauck (1971) have reviewed the literature pertaining to N balance in plant-soil systems and conclude that between 10 and 30% of the applied N is commonly lost, most probably by gaseous loss mechanisms, since conditions were often not conducive to leaching. Denitrification is believed to account for most of this loss.

A major limitation of the N balance approach in estimating denitrification loss is that alternate pathways of gaseous loss exist, in particular NH_3 volatilization from soil (Terman 1979), floodwater (Vlek and Craswell 1981) and gaseous N loss via leaves (Wetselaar and Farquhar 1980). These pathways of N loss have not been traditionally accepted to be of major importance. However, recent research on NH_3 volatilization from flooded soils, and research on gaseous N loss via leaves shows these mechanisms to be more important than earlier considered. Failure to account for loss via these mechanisms will clearly result in inflated values for denitrification, where these are determined by difference and an overestimate of the importance of denitrification as a N loss mechanism in plant-soil systems.

This information highlights the need for research on denitrification in plant-soil systems. Significantly, very few attempts have been made to quantify, directly, the extent of denitrification in the field. This fact reflects the difficultly in measuring small rates of production of N_2 in an atmosphere already containing about 79% N_2. A detailed discussion on the techniques used for measuring denitrification in soil is presented in Chapter 4. An alternative approach is to evaluate the process of denitrification under controlled conditions, either in atmospheres lacking N_2, in which case denitrification is measured directly by evaluating production of gaseous N oxides and N_2, or under air by determining the loss of nitrate. An extensive literature has been built-up using these approaches, but the question remains, how effectively can this research contribute to our understanding of biological denitrification in plant-soil systems?

This review will detail our current knowledge of the process of biological denitrification including aspects related to the biochemistry of denitrification, in addition to information on soil and environmental factors as they affect biological denitrification.

2.2. Microorganisms linked to denitrification

A broad group of bacteria can use nitrate instead of oxygen as the terminal electron acceptor in respiration (Payne 1973) but do not further reduce the nitrite so produced to gaseous N compounds. These organisms are described as nitrate respirers.

Fewer genera of bacteria are capable of denitrification itself. Payne (1973) listed a total of 15 genera. Focht and Verstraete (1977) updated the list and included several genera recently identified by Kessel (1976) as denitrifiers, including bacteria from the genera, *Cytophaga*, *Flavobacterium*, *Propiono-bacterium* and *Vibrio*. Not included in Focht and Verstraete's list are the N_2 fixing bacteria from the genera, *Azospirillum* (Neyra et al. 1977, Neyra and van Berkum 1977) and *Rhizobium* (Rigaud et al. 1973, Zablotowicz et al. 1978, Daniel et al. 1980). The identification of the latter bacteria as denitrifiers and knowledge that nitrate respiration at least, may be linked to N_2 fixation under anaerobic conditions, raises questions about the cycling of N in ecosystems where these bacteria play an important role in the biological fixation of N_2. The list of genera known to denitrify is presented in Table 2.1.

Table 2.1. Reported genera of denitrifying bacteria

Genus		
Acinetobacter	Alcaligenes	Halobacterium
Gluconobacter	Bacillus	Hyphomicrobium
Micrococcus	Moraxella	Paracoccus
Pseudomonas	Rhodopseudomonas	Azospirillum
Spirillum	Thiobacillus	Xanthomonas
Cytophaga	Flavobacterium	Vibrio
Propionobacterium	Rhizobium	

The two major groups of denitrifying organisms isolated from soil by Gamble *et al.* (1977) were representative of the *Pseudomonas fluorescens* biotype II and the genus *Alcaligenes*. This report confirmed the earlier findings of Valera and Alexander (1961), and Vives and Parés (1975) cited by Focht and Verstraete (1977) that *Alcaligenes* and *Pseudomonas* genera were the most ubiquitous denitrifiers in soil. Significantly, none of the denitrifiers commonly used in biochemical studies, including *Pseudomonas denitrificans*, *Pseudomonas perfectomarinus* or *Paracoccus denitrificans* were found to be widespread in soil (Gamble *et al.* 1977). The importance of this finding remains to be determined. Recent research on the role of NO in denitrification (Garber and Hollocher 1981) implies that significant biochemical differences may exist in the different denitrifying organisms. Little attention has been given to the likelihood that 'laboratory denitrifiers' may differ biochemically and physiologically from soil denitrifiers. This problem requires examination.

Nearly all denitrifying bacteria are strict aerobes; being unable to grow by fermentative means. However, there are several interesting exceptions. Garcia (1977) has isolated denitrifying strains of the genus *Bacillus* that were able to ferment carbohydrates. These organisms also grew in the presence of high nitrite concentrations and used NO as an electron acceptor. One strain also proved to be thermophilic.

2.3. Biochemistry of denitrification

2.3.1. Intermediates in the reduction of nitrate to N_2 gas

The sequence of products most often observed in soil denitrification studies includes nitrite, N_2O and N_2 in that order (Kluyver and Verhoeven 1954, Wijler and Delwiche 1954, Nömmik 1956, Cady and Bartholomew 1960, 1961, Schwartzbeck *et al.* 1961, Cooper and Smith 1963, Fillery 1979).

Nitrite is the immediate product of nitrate reduction in both nitrate respiring and denitrifying bacteria. While the nitrate respirers do not reduce nitrite further to N oxides, some are able to reduce nitrite to NH_3. However, this reduction is not considered to be linked to energy conservation.

The quantity of N_2O released during denitrification is highly dependent on the physical and chemical composition of the media supporting denitrification. In most studies N_2O persists for only short periods and is rapidly reduced to N_2. There are, however, soil conditions where this gas accumulates markedly. These conditions will be outlined in detail in a later section.

Small quantities of NO have also been detected during denitrification (Nömmik 1956, Cady and Bartholomew 1961, 1963, Fillery 1979). However, the present concensus is that the NO arises from chemical reduction (chemodenitrification) processes not from biological denitrification. The rationale is that substantial NO production from anaerobic soil is generally observed at low pHs, or conditions which support chemodenitrification reactions involving nitrite, and that comparable quantities of NO are generally produced in sterilized systems amended with nitrite. Chemodenitrification cannot account for the production of NO at near neutral pHs, however, leaving open the question of the origin of NO in slightly acid to neutral soils (Fillery 1979, Galbally and Roy 1978).

Information obtained from biochemical studies using cell-free preparations strongly supports the concept of NO as an intermediate. Nitric oxide reductases, which catalyze the reduction of NO to N_2O and N_2, have been isolated in several denitrifiers. In other studies, NO has been found to be the product of several cellular-bound nitrite reductases, obtained from denitrifying bacteria. Many have, however, challenged the significance of these reports, claiming that such studies do not always accurately reflect the functional pathways in intact microorganisms.

This fact has led several groups to critically reappraise the role of NO in intact denitrifiers. St. John and Hollocher (1977), using ^{15}N labelled substrate, were unable to determine a role for free NO as an intermediate in the denitrification sequence in *Pseudomonas aeruginosa*. These workers viewed NO reduction to N_2O as a separate pathway, not connected to nitrite reduction to N_2O, because they were unable to trap ^{15}N-labelled NO in an exogenous pool of ^{14}N-labelled NO during dissimilation of ^{15}N-labelled nitrite. Furthermore, they could not demonstrate scrambling of ^{15}N and ^{14}N in the final denitrification product when nitrite or exogenous NO were differentially labelled with either ^{15}N or ^{14}N. Scrambling of the labels should have resulted had the denitrification sequences involving nitrite and NO had obligatory intermediates in common.

Contrasting results were obtained by Firestone *et al.* (1979b) for bacteria representative of the *Pseudomonas fluorescens* biotype E. The ^{13}N label from ^{13}N-labelled nitrite was found to exchange readily with pools of added, nonlabelled NO with as much as 54% of the ^{13}N from the labelled nitrite appearing in the pool of NO in *Pseudomonas aureofaciens*. These results are consistent with the concept that NO may be an intermediate in the denitrification sequence; certainly that free NO is in rapid equilibrium with a bound intermediate. They do not support St. John and Hollocher's (1977) outline of separate pathways for reduction of nitrite and NO to N_2O.

Firestone *et al.* (1979b) discuss at length reasons for the disparity between results obtained in their study and that of St. John and Hollocher (1977). They note, for example, that the denitrification rate was considerably higher in St. John and Hollocher's study compared with their own and suggest that the high denitrification rate may have precluded any opportunity of demonstrating appreciable interchange of exogenous NO with membrane bound NO or another intermediate.

Garber and Hollocher (1981) discount the importance of denitrification rate on NO exchange reactions. However, they were able to demonstrate that denitrifiers exhibit considerable diversity in respect to the production and/or utilization of NO during denitrification. For example *Ps. denitrificans* and *Ps. aureofaciens* exhibited enzymatic exchange between the nitrite and NO pools, although to different degrees. On the other hand *Pa. denitrificans*, like *Ps. aeruginosa* did not appear to produce or use NO during denitrification. Another denitrifier *Ps. stutzeri* was anomalous. This denitrifier did not exhibit any exchange of isotope between nitrite and NO, neither was there any unidirectional transfer of isotope between these compounds. However, this species did exhibit scrambling of ^{15}N and ^{14}N when nitrite and NO were differentially labelled. This finding lead Garber and Hollocher (1981) to suggest that the reductive pathways involving nitrite and NO share a common intermediate which is neither free NO nor an enzyme-bound NO in *P. stutzeri*.

The same mechanism could also apply to the other denitrifiers as well. Averill and Tiedje (1982) suggest that NO could arise from the decomposition of a labile ferrous-nitrosyl complex, one of a series of bound intermediates which they suggest could be formed during nitrite reduction to N_2O (Fig. 2.1). In this scheme the initial binding of nitrite is to the axial position of a ferrous porphyrin (I). Protonation and dehydration produce a ferrous-nitrosyl complex (III) which either loses NO, forming the ferric porphyrin (VIII), or undergoes attack by nitrite to form a FeN_2O_3 complex. This complex would yield a coordinated oxyhyponitrite (V) following a two electron reduction. A further two electron reduction, protonation and dehydration would yield a coordinated nitrous oxide compound (VI). Finally, the removal of N_2O would result in the regeneration of the ferrous prophyrin (I).

There are several attractive aspects about this scheme. First, there is a chemical evidence to support the reactions proposed. Second the scheme provides a plausible explanation for the production and utilization of NO during nitrite reduction. Third, the reduction of nitrite to N_2O is described in terms of two electron steps which is consistent with other known biological reduction reactions coupled to phosphorylation or energy conservation.

Zumft and Cardenas (1979) also characterize NO evolution to be an artifact associated with a bound reductase. However, they assume that the reduction of NO to N_2O is undertaken by a cytoplasmic reductase which is not connected to nitrite reduction. The reaction, in their view, serves to remove excess electrons or reducing power. It is doubtful that this is the case. Firestone *et al.* (1979b) were unable to detect appreciable reduction of exogenous NO in the presence of nitrite. This would suggest that the enzymes associated with NO reduction, in *P. aureofaciens* at least, are closely linked with those responsible for nitrite

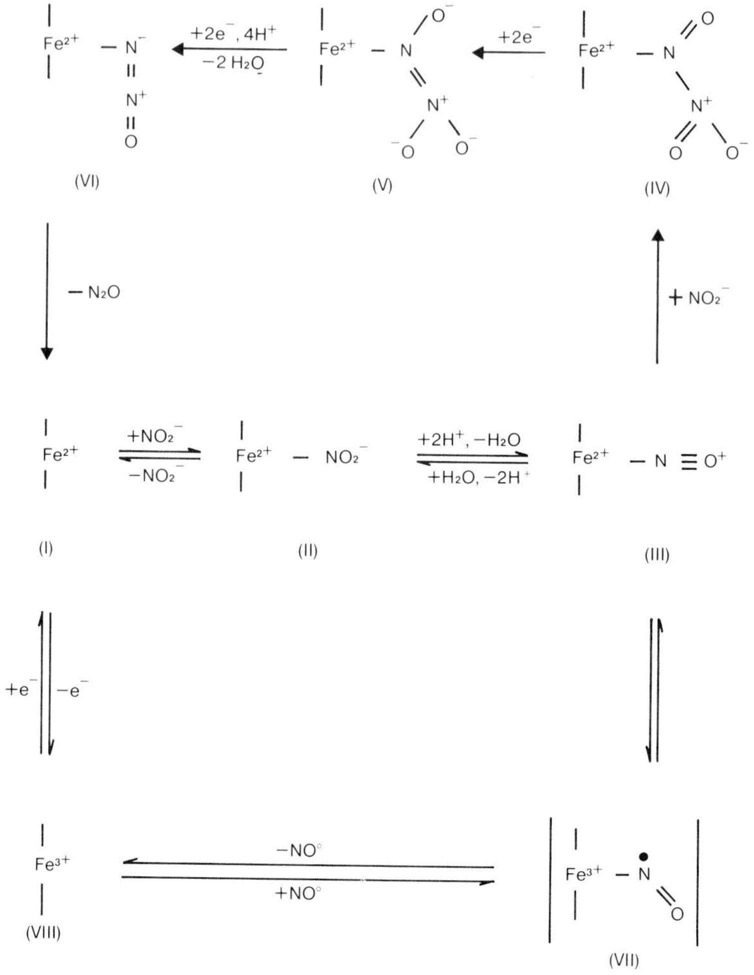

Fig. 2.1. Proposed pathway of nitrite reduction to nitrous oxide (from Averill and Tiedje 1982).

reduction as outlined by Averill and Tiedje (1982).

The role of N_2O as an obligatory intermediate has been more positively demonstrated. The numerous reports of N_2O as a denitrification product, albeit transitory, provide substantial but indirect evidence of the role of N_2O in the reduction sequence of nitrate to N_2. Whereas ^{15}N and ^{13}N studies have proved inconclusive in determining a role for NO, such studies provide definitive proof that N_2O is an obligatory intermediate during denitrification. Nitrogen from labelled nitrite (^{15}N or ^{13}N) has been readily trapped in pools of nonlabelled N_2O, while added labelled N_2O has been shown to be rapidly reduced to N_2 (St. John and Hollocher 1977, Firestone *et al.* 1979a).

These latter reports essentially defuse the issue of whether N_2O is an obligate intermediate in biological denitrification. Also, the earlier findings by Sacks and Barker (1952) and Allen and van Niel (1952) that N_2O is not an intermediate are now no longer considered tenable. For example, the inhibitory effect of nitrate on N_2O reductase accounts for Sacks and Barker's (1952) observed lag in the utilization of N_2O by nitrate-adapted cells. Sacks and Barker (1952) concluded that adaptation to nitrate did not include adaptation to N_2O, ruling out the latter as an obligate intermediate in denitrification.

The finding that nitrite can be reduced to N_2 in the presence of azide and 2,4 dinitrophenol, known inhibitors of N_2O reduction, was also cited by Sacks and Barker (1952) and Allen and van Niel (1952) as evidence against the obligatory involvement of N_2O in the denitrification sequence. However, work by Matsubara and Mori (1968) contradicts these earlier findings. Notably, the latter workers employed gas chromatographic procedures to identify positively the gaseous product of nitrite reduction, shown to be N_2O in the presence of azide or 2,4 dinitrophenol. It is not improbable that the manometric techniques employed earlier by Sacks and Barker (1952) and Allen and Niel (1952) did not discriminate N_2 from N_2O, so invalidating their conclusions.

Results from pure-culture studies involving denitrifying bacteria are also overwhelming in support of N_2O as an intermediate. Cultures of *Ps. denitrificans* (Matsubara and Mori 1968, Miyata *et al.* 1969) *Corynebacterium nephridii* (Renner and Becker 1970) and *Thiobacillus denitrificans* (Baldensperger and Garcia 1975) have been shown to form N_2O during nitrite and NO reduction. In addition *Ps. denitrificans*, *Ps. stutzeri* and *Pa denitrificans* have been grown anaerobically using N_2O as an electron acceptor instead of oxygen, nitrate or nitrite (Allen and van Niel 1952, Pichinoty and D'Ornano 1961, Matsubara 1971). A N_2O reducing system has been isolated in *Ps. denitrificans*, (Matsubara 1971). This was induced by N_2O and repressed by oxygen, an indication that it serves a distinct biological function in the reduction of N oxides to N_2. The quantitative accumulation of N_2O following the specific inhibition of N_2O reduction by acetylene (Balderston *et al.* 1976, Yoshinari and Knowles 1976) is further evidence that N_2O is an intermediate. Perhaps the only remaining question in regard to the role of N_2O in denitrification, is the extent to which N_2O is the terminal product in denitrifying bacteria.

The issue of the involvement of hyponitrite has been discussed at length in earlier reviews. It is possible that a N oxide intermediate in denitrification could arise by dehydration of a hydrated precursor with the same oxidation-reduction state as the intermediate. Hyponitrite spontaneously decomposes in water to yield N_2O thus making it an attractive precursor to N_2O formation. Although this reaction has been studied in denitrifying systems, the results have been inconclusive. The recent findings of Hollocher *et al.* (1980), however, indicate that free hyponitrite is unlikely to be a physiologically important intermediate in denitrification, at least in *Pa denitrificans*. While small pools of ^{13}N and ^{15}N-labelled hyponitrite were detected during dissimilation of ^{13}N-labelled nitrate and ^{15}N-labelled nitrite, respectively, estimates of the N flux through the presumptive hyponitrite pool showed that the denitrification flux could not be accounted for by considering spontaneous dehydration of the detected hyponitrite. Hollocher *et al.* (1980) concluded that hyponitrite cannot be an intermediate, at least in the absence of a hyponitrite dehydrase which can catalyze hyponitrite dehydration by a factor of about one thousand. No such enzyme was identified in *Pa denitrificans*.

In summary, the pathway of denitrification clearly proceeds by way of nitrite and N_2O as shown,

$$NO_3^- \rightarrow NO_2^- \rightarrow X \rightarrow N_2O \rightarrow N_2$$

Evidence for the existence of an intermediate (X) between nitrite and N_2O is mixed. Most recent experimental data does not support the inclusion of NO as intermediate, as previously suggested by Payne (1973). Instead, NO reduction to N_2O is viewed as separate pathway not linked to nitrite reduction i.e.

$$NO_2^- \rightarrow N_2O \rightarrow N_2$$
$$\uparrow$$
$$NO$$

or is closely linked to nitrite reduction via an exchange reaction involving a bound intermediate i.e.

$$NO$$
$$\uparrow\downarrow$$
$$NO_2^- \rightarrow X \rightarrow N_2O \rightarrow N_2$$

2.3.2. Dissimilatory nitrate reductase

Stouthamer (1976) comprehensively reviewed the subject of bacterial nitrate reductase from its many aspects. This review will deal only briefly with the properties of nitrate reductase and will not discuss aspects related to the genetics of nitrate reductase formation. For a treatise on these aspects the reader is referred to Stouthamer (1976).

Dissimilatory nitrate reductase (not to be confused with assimilatory nitrate reductase) is membrane bound in *E. coli* (Cole and Wimpenny 1968, Radcliffe and Nicholas 1970), *Ps. stutzeri* (Kodama 1970), *Ps. denitrificans* (Miyata 1971), *Bacillus stearothermophilus* (Kiszkiss and Downey 1972) and *B. licheniformis* (van 't Riet *et al.* 1979). The dissimilatory nitrate reductases from several bacteria have been solubilized from the membranes, purified and some of their properties determined. These enzymes are for the most part composed of two subunits having molecular weights of about 150,000 (α-subunit) and about 60,000 (β-subunit). The enzymes are rich in iron, sulphur and molybdenum, with the molybdenum attached to the low molecular weight peptide (Mo-cofactor) in *B. licheniformis* (van't Riet *et al.* 1979). In this respect nitrate reductase is similar to other molybdenum containing enzymes.

Electron flow apparently proceeds from cytochrome(s) to the molybdenum cofactor to iron and finally nitrate (Forget and Dervartanian 1972). Formate, lactate, pyruvate and NADH are effective donors for nitrate reduction in particulate preparations from *E. coli* and *Ps. denitrificans* (Cole and Wimpenny 1968, Radcliffe and Nicholas 1970). The presence of an active formate dehydrogenase system in cells grown anaerobically on nitrate and the much lower level of activity in cells grown on oxygen suggest that this system is an integral part of the nitrate respiration system and that formate is of physio-logical importance as an electron donor for nitrate reduction in *E. coli* and *Ps. denitrificans* (Radcliffe and Nicholas 1970).

The electron carriers linked to nitrate respiration in *E. coli* are shown in Fig. 2.2. The branch point from oxygen terminated electron transport occurs at the cytochrome b level. Electrons are transferred either directly to nitrate reductase or through a second b-type and a c-type cytochrome (Payne 1973). Most evidence is in support of direct transfer of electrons from cytochrome b_{556} to nitrate reductase as depicted.

Although it is now widely appreciated that nitrate reductase is membrane bound, the orientation or location of the enzyme on the bacterial membrane, and the site of nitrate reduction are not well understood. Several possible

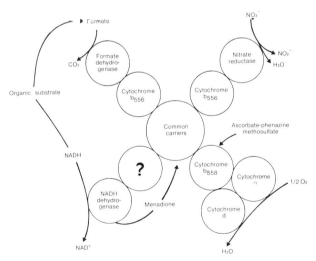

Fig. 2.2. Electron carriers associated with the nitrate reductase and cytochrome oxidase in *E. coli* (from Sanchez *et al.* 1979).

42

configurations have been delineated. In *Klebsiella aerogenes* nitrate reductase appears to have a transmembranous orientation (Wientjes *et al.* 1979). Substrate binding and nitrate reduction are envisaged to occur on the periplasmic or outer side of the membrane and the protons transported across the membrane via the nitrate reductase. On the other hand in the denitrifying bacteria, *Pa denitrificans,* (Kristjansson *et al.* 1978) and *Bacillus licheniformis,* (Wientjes *et al.* 1979) the two subunits of nitrate reductase appear to be attached exclusively to the inner or cytoplasmic side of the bacterial membrane. In this case nitrate binding and reduction are considered to occur on the inner side of the plasma membrane.

The picture is more confused for *E. coli.* Garland *et al.* (1975) depict nitrate reductase to be transmembranous, with nitrate binding and reduction occurring externally and the reductase behaving as a proton pump (Fig. 2.3.). MacGregor and Christopher (1978) and Kristjansson and Hollocher (1979) on the other hand contend that nitrate binding and the site of nitrate reduction lie on the cytoplasmic side of the membrane. Kristjansson and Hollocher based their argument on studies which showed intact cells of *E. coli* to exhibit the same permeability barrier to the reduction of chlorate, relative to nitrate, noted previously for denitrifying bacteria (John 1977). This fact more clearly

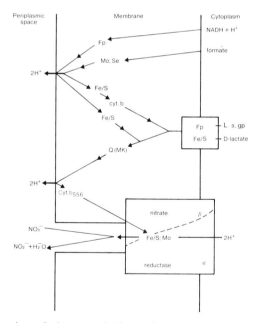

Fig. 2.3. Electron carriers of nitrate respiration and their membrane topography in *E. coli.* Fp, flavoprotein; MoSe, formate dehydrogenase; Fe/S, iron-sulfur protein; L, α-gp, L-α glycerol-phosphate; Q (MK), ubiquinone or menaquinone; α, β, subunits of nitrate reductase (from Haddock and Jones 1977).

establishes that the binding site of the nitrate reductase is located on the inner aspect of the bacterial membrane in *E. coli*, not externally as suggested by Garland *et al.* (1975).

The transmembrane orientation of the nitrate reductase is at least consistent with the chemiosmotic hypothesis (Michell 1966) which predicts a trans-membranous location of electron carriers near phosphorylation sites to drive proton translocation. Wientjes *et al.* (1979) note that oxidative phosphorylation occurs concomitantly with nitrate reduction to nitrite in *K. aerogenes* and *E. coli*, which are considered to have nitrate reductase in a transmembrane orientation. On the other hand the nitrate reductase in *B. licheniformis* and *Pa denitrificans* is considered to serve a different physiological role. Wientjes *et al.* (1979) cite reports which substantiate that oxidative phosphorylation does not occur at the level between cytochrome b_{556} and nitrate reductase, at least in these bacteria. How well this relationship holds for other nitrate respirers and denitrifiers is not clear. At the best it appears a tenuous hypothesis, particularly in the light of the likely different orientation of the nitrate reductase in *E. coli;* and even their own admission that nitrate binding could also possibly occur on the inner aspect of the bacterial membrane in *K. aerogenes*.

Kristjansson and Hollocher (1979) postulate that proton translocation during nitrate reduction is linked to membrane ATPases, not the nitrate reductase. Uptake of nitrate (and also nitrite) by *Pa denitrificans* was found to be coupled with the uptake of one proton per anion transported. Nitrate movement was described as facilitated diffusion down a concentration gradient created by the reduction of nitrate to nitrite on the inner side of the membrane (Kristjansson and Hollocher 1979). The likelihood that nitrate is transported across the membrane by a specific transporter (John 1977, Kristjansson and Hollocher 1979) provides one explanation for the vastly lowered level of chlorate reduction in intact cells of denitrifying bacteria compared to mem-brane-disrupted preparations where chlorate is reduced rapidly in the absence of the intact plasma membrane.

2.3.3. Nitrite reductase

Dissimilatory nitrite reductase is found only in a select group of bacteria (Table 2.1). Unlike nitrate reductase, nitrite reductase has been isolated mainly in the soluble cell fraction, suggesting that this enzyme is either located in the cytoplasm or is only weakly associated with the bacterial membrane (Payne 1973).

Two types of nitrite reductase have been isolated. A copper-containing enzyme is found in *Ps. denitrificans* (Radcliffe and Nicholas 1968, Miyata and Mori 1969) and *Achromobacter cycloclastes* (Iwasaki and Matsubara 1972). Purified enzymes from these bacteria reduce nitrite stoichiometrically to NO providing artificial electron carriers such as reduced viologen dyes or free flavins are present. An additional property, unique to the copper-containing nitrite reductase, is the ability to produce N_2O from nitrite and hydroxylamine. Renner and Becker (1970) also describe this activity in resting cells of *Corynebacterium nephridii*, perhaps an indication that the latter bacterium also possesses the copper-containing reductase.

The second nitrite reductase, classified as a haemoprotein, has been isolated in *Alcaligenes faecalis* (Iwasaki and Matsubara 1971), *Ps. aeruginosa* and *Ps. stutzeri* (Kodama 1970) and *Pa denitrificans* (Newton 1969). The haemoprotein from *A. faecalis* has been characterized and found to contain c and d type haem (cd cytochrome) and two atoms of iron per molecule (Iwasaki and Matsubara 1971). Purified cd cytochromes use reduced forms of cytochrome c, phenazine methosulfate and viologen dyes as electron donors in the reduction of nitrite to NO. There is no evidence of N_2O production from nitrite and hydroxylamine in bacteria possessing the cd type cytochrome nitrite reductase.

The physiologically important dehydrogenases linked to nitrite reduction remain to be characterized. Nitrite reduction can be observed with NADH, succinate and lactate via membrane-bound NADH dehydrogenase, succinate dehydrogenase and lactate dehydrogenase, respectively.

The electron carriers linking the dehydrogenase reactions to nitrate reductase also remain to be fully elucidated. Miyata and Mori (1969) outlined the following electron transport scheme for nitrite reduction in *Ps. denitrificans*.

$$NO_2^- \rightarrow NO$$
$$\uparrow \text{ nitrite reductase}$$
Cytochrome c_{553}
or
Cytochrome c_{552}
$$\uparrow$$
FMN or FAD
$$\uparrow \text{ Lactate dehydrogenase}$$
Lactate

Achromobacter cycloclastes (Iwasaki and Matsubara 1972), and *E. coli* (Cole and Wimpenny 1968) also possess cytochrome c_{553}, and *Pa denitrificans* a c-type cytochrome. These have been implicated in electron transport to nitrite, in the respective organisms. However, whether these cytochromes are the physiological electron carriers remains to be demonstrated. Higher rates of nitrite reduction follow the addition of flavins to cell-free extracts from *Ps. denitrificans*, providing circumstantial evidence that this electron carrier is linked to nitrite reductase.

Recent research is of little help in settling the issue of orientation of the nitrite reductase. Several workers (Alefounder and Ferguson 1980) assign the location of the dissimilatory nitrite reductase in *Pa denitrificans* to the periplasmic side of the membrane. On the other hand, Kristjansson *et al.* (1978) were led by their studies on proton movements after the addition to cells of pulses of nitrite to conclude that nitrite is reduced inside the cell. Work with ferritin-labelled antibodies also led Saraste and Kuronen (1978) to suggest that the enzyme is located on the inner surface of the plasma membrane in *Ps. aeruginosa.*

A periplasmic site for nitrite reductase is consistent with cytochrome c being the immediate physiological electron donor, since it appears that bacterial cytochrome c is located on the outer surface of the plasma membrane (Garrard 1972, Prince *et al.* 1975). A periplasmic site for nitrite reduction also implies that nitrite produced in the cell by the reduction of nitrate must be transported out. The possibility that a specific nitrate transporter facilitates nitrate transport to nitrate reductase has already been highlighted. Alefounder and Ferguson (1980) consider it not unlikely that a nitrate-nitrite antiporter may facilitate the movement of nitrate in and nitrite out in denitrifying bacteria.

2.3.4. Nitric oxide and nitrous oxide reductases

The possibility of NO being an intermediate in denitrification was discussed in some depth earlier in this chapter. Considerable uncertainty remains as to whether NO is a physiologically important intermediate in whole cells of denitrifying bacteria. The fact remains, however, that several researchers have successfully isolated preparations that are active in the reduction of NO in denitrifying bacteria (see Fewson and Nicholas 1960, 1961, Chung and Najjar 1956, Miyata *et al.* 1969, Matsubara and Iwasaki 1972).

The NO reductase in *Alcaligenes faecalis* (Matsubara and Iwasaki 1972) and *Ps. denitrificans* (Miyata *et al.* 1969) was found to be largely particle bound (some was in the soluble component) while that in *Ps. stutzeri* and *Ps. aeruginosa* (Fewson and Nicholas 1960, 1961, Chung and Najjar 1956) was isolated exclusively in the soluble fraction. Matsubara and Iwasaki (1972) characterized the soluble component in *A. faecalis* as a cytochrome cd. However, the particulate fraction active in the reduction of NO was not found to be associated with nitrite reductase activity and nor did the preparation possess spectral characteristics associated with cytochrome cd reductase. Matsubara and Iwasaki (1972) further noted that the particulate and soluble NO reductases reacted differently to high NO concentrations, the particulate fraction being inhibited at 20% (v/v NO) while the soluble reductase remained unaffected at gaseous concentrations up to 60% v/v NO. These workers suggest that the soluble nitrite reductase effects the reduction of NO to N_2O when high levels of nitric oxide accumulate, thereby removing a potentially toxic gas.

No studies have been reported showing the orientation of the NO reductases on the bacterial membranes. A close association between the nitrite reductase and the NO reductase might be anticipated, especially since NO is infrequently detected during denitrification, even in the gas phase above cell extracts of *Ps. denitrificans* reducing nitrite (Miyata *et al.* 1969).

The electron transport system linked to NO reductase is poorly understood. Miyata (1971) suggested the following scheme where lactate serves as a physiological electron donor:

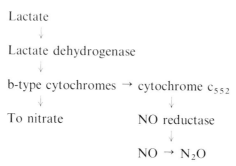

Miyata (1971) based this scheme on results obtained from spectrophotometric and inhibitor studies which showed the involvement of a b-type and a cytochrome c_{552} in the transfer of electrons to the NO reductase in *Ps. denitrificans*. A c-type cytochrome has also been found to be tightly associated with a membrane fraction containing a NO reductase in *P. perfectomarinus* (Payne *et al.* 1971).

Relatively little is known of the N_2O reductases in denitrifying bacteria. The enzyme is located on the bacterial membrane in *Ps. perfectomarinus* and contains metal cofactors (Payne, 1973). Azide, cyanide, 2,4 dinitrophenol (Matsubara and Mori 1968, Kristjansson and Hollocher 1980), acetylene at low concentrations (Federova *et al.* 1973, Balderston *et al.* 1976, Yoshinari and Knowles 1976) and carbon monoxide (Kristjansson and Hollocher 1980) inhibit the enzyme. The quantitative, yet reversible nature of the acetylene inhibition permits the assay of denitrification under atmospheric conditions where the small quantities of N_2 produced during denitrification cannot be distinguished from atmospheric N_2 (Balderston *et al.* 1976, Yoshinari and Knowles 1976).

The electron carrier system linked to N_2O reduction has not been evaluated in any detail. Matsubara (1971) found the amount of a soluble carbon monoxide binding cytochrome c_{553} and a particulate bound cytochrome c_{552} to be appreciably higher in anaerobically grown cells incubated with N_2O instead of nitrate. He suggested that these cytochromes played a role in the nitrous oxide reducing system in *Ps. denitrificans*. In a later, more comprehensive study Matsubara (1975) noted that reduced b and c-type cytochromes were partially oxidized when N_2O was added to intact cells reduced with lactate under anaerobic conditions. The oxidation of these cytochromes was inhibited non-competitively by azide, cyanide, 2,4 dinitrophenol and $CuSO_4$, all of which inhibited N_2O reduction to N_2 when present at high concentrations. Matsubara (1975) was unable to confirm that the c-type cytochrome was either the soluble carbon monoxide binding c_{553} or the particulate bound cytochrome c_{552} identified in an earlier study (Matsubara 1971).

2.3.5. Regulation of dissimilatory reductases

The synthesis and the activity of enzymes and associated electron carriers involved in the reduction of nitrate, nitrite, NO and N_2O, is tightly regulated by oxygen. Payne (1973) and Stouthamer (1976) have extensively reviewed the literature on this subject. Only the salient points need be highlighted in this chapter.

Inhibition by oxygen of the onset of the synthesis of the nitrate respiratory system has been noted in many bacteria, including *E. aerogenes, Proteus mirabilis* and *Aerobacter aerogenes*, while anoxia has been shown to depress nitrate reductase synthesis in *E. coli, Bacillus stearothermophilus, P. mirabilis* and *Haemophilus* sp. (van 't Riet *et al.* 1968, Payne 1973).

In most bacteria, (*B. stearothermophilus, Haemophilus influenzae* are exceptions) nitrate reductase synthesis is induced only when nitrate is present during anoxia. In the case of the noted exceptions, the onset of anoxia alone is sufficient to initiate nitrate reductase synthesis, although the amount of nitrate reductase formed is higher when nitrate is also present (Payne 1973, Stouthamer 1976). Nitrite and azide are also known to induce nitrate reductase formation under anaerobic conditions (Stouthamer 1976).

Transfer from anaerobic to aerobic conditions usually initiates a rapid inhibition of dissimilatory nitrate reduction. There are, however, some reports of cases where oxygen has been found to repress the synthesis of nitrate reductase without immediately affecting the activity of the enzyme already present. Stouthamer (1976) suggests that nitrate respiration could continue as long as the electron transport chain to oxygen remains in a nonfunctional state.

A gradual inactivation of the nitrate reductase has been observed in cultures of *K. aerogenes* (van 't Riet *et al.* 1968) *P. mirabilis* (de Groot and Stouthamer 1970) and *B. stearothermophilus* (Downey *et al.* 1969) following the transfer of anaerobically grown cells to aerobic conditions. The reason for the phenomenon is not well understood. Purified nitrate reductase is not inactivated by exposure to oxygen, although nitrate reductase in cell-free extracts is inactivated (van 't Riet *et al.* 1968). Furthermore de Groot and Stouthamer (1970) were unable to demonstrate inactivation of the reductase in mutants of *P. mirabilis* known not to contain cytochromes associated with aerobic respiration. It would appear that the inactivation of nitrate reductase under aerobic conditions occurs largely when the enzyme is complexed with a functional respiratory chain (Stouthamer 1976).

The synthesis of the nitrite, NO and N_2O reductases is also effected in the absence of oxygen. Nitrate does not appear to inhibit the synthesis of these reductases in *Ps. perfectomarinus* but the activity of the NO reductase is lowered in the presence of nitrate (Payne 1973). In *Ps. stutzeri* (Kodama *et al.* 1969, Kodama 1970) nitrate inhibits both the synthesis and activity of the nitrite reductase. A similar phenomenon appears to occur during nitric oxide reduction in *Ps. denitrificans* (Miyata 1971).

Anaerobic and oxygen limited cells have higher levels of cytochromes b and c and cytochrome cd and lower levels of cytochrome a + a_3 (Sapshead and Wimpenny 1972, Payne 1973). Nitrate, nitrite and N_2O stimulate the synthesis of the cytochromes associated with the respective reductases when cells are grown anaerobically. On the other hand, nitrate can also counteract the stimulatory effect of nitrite on cytochrome synthesis in *E. coli* and *Ps. stutzeri* and the effect of N_2O or synthesis in *Ps. denitrificans* (Cole and Wimpenny 1968, Payne 1973).

The mechanism by which oxygen controls the synthesis and activity of the reductases and associated cytochromes is not well understood. Stouthamer (1976) has reviewed this subject in depth as it pertains to nitrate reductase. He points out that the formation of a reductase for a particular substrate is likely to be prevented when an electron acceptor with higher energy-yielding potential is also present. This effect certainly would be expected, at least from a consideration of thermodynamic principles. However, the nature of the mechanism regulating the synthesis or activity of the reductase is not described.

Showe and De Moss (1968) suggest that the intracellular redox potential is the controlling factor. They consider the intracellular potential to be a function not only of the identity and concentration of the electron acceptors, but also of the capability of electron flow in the cell. Wimpenny and Cole (1976) and Wimpenny (1969), on the other hand, link the extracellular redox to the control of reductase activity and synthesis, and discount that the presence of a specific electron acceptor regulates dissimilatory reductases. Stouthamer (1976) considers that neither explanation (intra or extracellular redox control) is consistent with recent results from redox studies involving transfer of cells from aerobic to anaerobic conditions. Instead Stouthamer proposes that the factor regulating synthesis of reductases is the oxidation-reduction state of components of the respiratory chain. Stouthamer (1976) further comments 'that it is possible that nitrate reductase (and other reductases) cannot be complexed in an active membrane-bound form with other electron-transport components when electron flow to the enzymes is impeded'.

Support for Showe and De Moss' (1968) and Stouthamer's (1976) schemes is evident in the results obtained by Downey *et al.* (1969) and van 't Riet *et al.*

(1968). These workers noted that the rate of synthesis of nitrate reductase was rapid immediately after the onset of anaerobiosis, but declined in time, to a lower steady-state level. They considered that the biphasic kinetics reflected a shift from what they termed 'gratuitous' to 'nongratuitous' induction, and that the lower steady-state level probably reflected the stabilization of the intracellular redox potential following nitrate reduction. The oxidized nitrate reductase could also act as its own repressor, thereby effecting a suspension of the synthesis of nitrate reductase, according to the mechanism described by Stouthamer (1976).

Another novel explanation of the oxygen regulatory mechanism is outlined by Alefounder and Ferguson (1980). These workers suggest that 'the control by oxygen of nitrate reduction is likely to be exerted as a restriction on the accessibility of nitrate to its reductase rather than as a control on the relative flow of electrons to oxygen or nitrate'. The evidence on which this scheme was developed is indirect and circumstantial, notably that nitrate reduction in *Pa denitrificans* was possible in the presence of oxygen where cells were treated with the detergent Triton X 100. The same treatment was also found to substantially increase the rate of chlorate reduction in intact cells, an indication that the detergent had reduced the effectiveness of the plasma membrane to exclude chlorate. John (1977) was also able to demonstrate simultaneous oxygen and nitrate reduction in cell preparations containing 'inside-out vesicles' formed after lysosome treatment. Interestingly, the same preparations rapidly reduced chlorate, a characteristic not normally associated with intact cells with a physically intact plasma membrane.

Alefounder and Ferguson (1980) made no attempt to describe the manner in which oxygen inhibits the activity of nitrite reductase. By their own account the nitrite reductase in *Pa. denitrificans* is located on the periplasmic side of the membrane. Oxygen restriction of nitrite transport might explain the inhibition where nitrite is produced intracellularly during nitrate reduction, but it is less likely to account for the inhibition of nitrite reduction where nitrite is supplied exogenously.

2.4. Biological denitrification in soil

There is an extensive literature dealing with aspects of denitrification in soil and the factors that influence the process in soil. Only the more pertinent findings will be discussed in this chapter. Additional information can be obtained from the reviews by Broadbent and Clark (1965) and Allison (1966) and Focht and Verstraete (1977).

Most literature on denitrification in soil pertains to laboratory studies where denitrification is evaluated under controlled environments, generally in the absence of oxygen or under flooded conditions. Interpretation of the results from these studies is often complicated by the addition of carbonaceous substrate and/or air drying of soil as a pretreatment. The enhancement of denitrification following addition of carbonaceous substrate or rewetting of previously air-dried soil is well documented. However, it is not clear what effect these soil pretreatments have on the number and diversity of nitrate respirers and denitrifiers. Focht and Verstraete (1977) cite reports which appear to indicate that organic matter content is a major factor determining whether nitrate-respiring or denitrifying bacteria dominate in soil.

Air drying of soil may alter the structure of the microbial population, so inducing changes not only in the rate of denitrification, but also in the nature of the denitrification products when the soil is rewet. This problem is likely to be particularly severe where soil has been stored air-dried for long periods before use (Pattern et al. 1980).

Such effects should be kept in mind when evaluating denitrification research. At best, laboratory studies of denitrification in soil serve to set limits on the process. Thus direct extrapolation of data from the laboratory to the plant-soil environment should be done with caution.

The increased awareness of the potential for loss of fertilizer N via denitrification and the concern that N_2O produced during denitrification may be involved in the destruction of the stratospheric ozone layer, has stimulated much research on denitrification during the last decade. An abbreviated account is given in the following sections of the major factors known to influence denitrification in soil.

2.4.1. Aeration status

The importance of oxygen in repressing the synthesis and activity of dissimilatory reductases has already been highlighted. There are numerous reports demonstrating the lack of denitrification in aerobic media, including soil, and of rapid denitrification in soil in the absence of oxygen or where oxygen availability is sharply reduced by flooding. There are also reports of denitrification in otherwise aerobic soil (Broadbent and Stojanovic 1952, Allison et al. 1960, Cady and Bartholomew 1961, Greenland 1962) and supposedly well-aerated liquid culture (Marshall et al. 1953, Collins 1966, Meiklejohn 1946). The concept of 'aerobic' denitrification is no longer tenable. It is now widely accepted that denitrification is restricted to anaerobic microsites within the otherwise aerobic media. As noted by Focht and Verstraete (1977), the critical factor determining whether aerobic or anaerobic metabolism occurs at a particular point, is the soluble oxygen concentration not the average gaseous oxygen content of the surrounding media. Anaerobic microsites can arise, even in supposedly well-aerated liquid media, where the biological demand for oxygen is great and the transport of oxygen too slow to maintain complete aeration.

Polarographic measurements of dissolved oxygen in vigorously stirred soil suspensions showed that denitrification was not induced until very low O_2 concentrations (Km: 2.7×10^{-6}M) were reached (Greenwood 1962). Similar values have been reported for pure cultures of denitrifying bacteria (Skerman and MacRae 1957) although there appear to be notable exceptions where nitrite is used in preference to nitrate as an electron acceptor (Skerman *et al.* 1958). Greenwood (1963) used a critical value of 4.0×10^{-6}M oxygen, to differentiate between aerobic and anerobic processes. He assumed that denitrification could take place at a constant rate, irrespective of oxygen, providing the concentration was below the critical value, but did not occur at higher oxygen concentrations. Thus aerobic and anaerobic processes can be viewed as being separated by sharply defined boundaries, permitting the effect of aeration on denitrification to be described in terms of the relative proportions of aerobic and anaerobic zones in soil (Greenwood 1963).

The aeration status of soil is difficult to determine, given the marked heterogeneity of soil (Fluhler *et al.* 1976). Several attempts have been made to develop models which describe aeration status in terms of soil respiratory activity, oxygen diffusion rate and soil geometry (Greenwood 1963, Smith 1978, Leffelaar 1979). The development of anoxic microsites (Fig. 2.4) is markedly affected by oxygen consumption rate (or soil respiratory activity), the oxygen diffusion rate, which is primarily affected by the soil moisture status, and finally the geometry of the soil, in particular aggregate size.

The model developed by Greenwood (1963) predicted the development of anaerobic microsites where soil was further than 2 mm from a gas/water interface, providing respiration activity in soil was high. Thus denitrification could proceed in the centre of aggregates providing the aggregates are water-saturated and the soil as a whole has a high respiratory activity. Greenwood (1963) suggested that these prequisites could be met during rewetting of soil after periods of drying or where additions of organic matter are made. Smith (1978) stressed the importance of soil moisture content on the development of anaerobic microsites in soil aggregates. He noted that a ten-fold change in the diffusion coefficient for oxygen within aggregates had a more appreciable effect on the extent of anaerobiosis than a comparable change in mere aggregate size, although the latter remained an important variable in his model.

Thus it is not surprising to find that denitrification is strongly influenced by soil moisture content, under oxygen atmospheres. Where soil is incubated in the absence of oxygen soil moisture content does not affect denitrification in soil, at least over the range of saturation to 'field capacity' (or about $-1/3$ bar of tension). It is widely quoted that denitrification ceases when soil becomes drier than about $-1/3$ bar tension (Focht and Verstraete 1977) when incubated

52

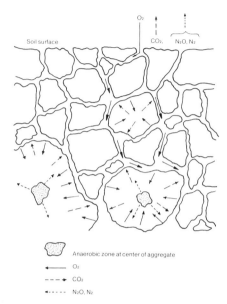

Fig. 2.4. Schematic diagram showing the position of possible anaerobic microsites in otherwise aerobic soil. Key parameters determining the volume of anoxic microsites are soil respiration rate or sink strength for O_2, oxygen diffusion rate in aggregates and aggregate diameter (from Smith, 1978).

under air. However, it would appear from the work of Pilot and Patrick (1972) that the cut-off point may be at much lower soil water tensions. They were unable to detect loss of nitrate when soil was maintained between 20–40 cm water tension, although denitrification was rapid at lower soil moisture tensions. The effect of moisture tension was in governing the air filled porosity of soil. Wesseling and van Wijk (1957) conclude that oxygen diffusion in soil becomes critical at about 85% water saturation of the total pore space.

Aeration status in soil is also affected by the activity of plant roots in soil. The potential for denitrification to occur in the rhizosphere has been demonstrated. Several factors are involved, most notably depression in oxygen availability as a result of root respiration in aggregated soil environments (Woldendorp 1962, 1968, Stefanson 1972).

Few studies have evaluated the effect of oxygen level on the products of denitrification. A tendency for increased N_2O production with improved aeration status is apparent in the results from earlier studies, (Nömmik 1956, Wijler and Delwiche 1954, Cady and Bartholomew 1961) suggesting that N_2O reduction is slowed in soil in the presence of low oxygen levels. Later work by Firestone *et al.* (1979a) using ^{13}N shows that the inhibitory effect of oxygen on N_2O reduction was both direct and manifested at very low oxygen levels. Competition for electrons, preferential inhibition of N_2O reductase, or a general slowdown in the denitrification process in the presence of oxygen thereby allowing N_2O to move away from the site of reduction are all possible explanations for the higher production of N_2O under these circumstances. Analyses of soil atmospheres in the field show highest concentrations of N_2O in the -50 to -200 mbar of tension range, with concentrations generally much lower under saturated conditions (Dowdell and Smith 1974).

2.4.2. Organic matter content

The requirement of oxidisable substrate for the growth of denitrifying bacteria was highlighted earlier in this chapter. In soil, the oxidisable substrate is derived largely from the soil organic matter and denitrification rate is strongly correlated with the amount of readily available organic matter (Burford and Bremner 1975, Stanford et al. 1975).

Soil treatments, including air drying of soil (Pattern et al. 1980) freezing and thawing (McGarity 1962) and the addition of carbonaceous materials (Bremner and Shaw 1958) greatly enhance the rate of denitrification by increasing the amount of soluble or readily available organic matter.

Denitrification is also enhanced in the presence of plants at least in upland soils (Stefanson 1972, Woldendorp 1962, 1968). While part of this effect can be attributed to decreased oxygen availability, discussed in the previous section, the release of exudates in the rhizosphere may contribute significantly to an enhanced rate of soil respiration by heterotrophic soil bacteria and concomitantly higher rates of denitrification where anaerobiosis develops. A positive rhizosphere effect on denitrification has also been found in flooded rice plant-soil systems at least in regard to N_2O reduction (Garcia 1975). However, the higher recovery of ^{15}N found in the presence of rice plants, and the significantly lower recovery from fallow systems (Table 2.2) suggests that rhizosphere denitrification is of little consequence in flooded rice systems where nitrate and not N_2O is the primary substrate for denitrification. Indeed the plant may also serve to reduce denitrification by depleting the inorganic-N pool, thereby conserving nitrogen as appears to be the case shown in Table 2.2.

2.4.3. Temperature

The effects of temperature on the rate of denitrification are reasonably well understood. Denitrification generally shows an exponential increase between 15°C and 30°C but this relationship does not hold below 12°C. Stanford et al. (1975) report that denitrification can occur at temperatures close to freezing, although not unexpectedly, the rate is low.

The optimum temperature for denitrification in soils is surprisingly high. Bremner and Shaw (1958) found the rate to be unchanged between 35°C and

Table 2.2. Recovery of ^{15}N from a silt loam after incorporation of urea, as influenced by water regime and presence of plants (from Fillery and Vlek 1982)

Water regime	Treatment	Total recovery of ^{15}N labelled urea (%)	Plant recovery of ^{15}N labelled urea (%)
Continuously flooded	planted	92	49
	fallow	61	—
Intermittently flooded	planted	91	42
	fallow	53	—

60° C, while Stanford et al. (1975) reported the optimum to be at 35° C, but did not evaluate denitrification rate at higher temperatures. Nömmik (1956) and Keeney et al. (1979) report an optimum about 65° C. Keeney et al. (1979) postulate that the high optimum for denitrification probably reflects chemo-denitrification reactions involving nitrite produced by thermophilic nitrate respirers. Considerably more N_2 was recovered at 50 to 67° C than was initially present in soil systems as nitrate N. This result could best be explained by considering that nitrite, produced during nitrate respiration, reacted with oxidized N functional groups in the soil to form nitrogenous gases, notably NO and N_2O, which were then subject to biological reduction to N_2.

Temperature also influences the products of denitrification, with N_2O production dominating at lower temperatures (Bailey 1976, Keeney et al. 1979) and particularly high temperatures $\sim 70°$ C (Keeney et al. 1979). As soil temperature increases between 15 and 65° C, N_2O persists for shorter periods; although the $N_2O/(N_2 + N_2O)$ ratio is similar immediately after the onset of anaerobiosis. This apparently reflects either a higher rate of induction of N_2O reductase activity with increasing temperature, or a more rapid rate of depletion of more oxidized N acceptors and a correspondingly shorter period of inhibition of N_2O reductase activity.

Keeney et al. (1979) concluded that while the rate of denitrification was low at $< 15°$ C the amount of N_2O evolved could at least be equivalent to that evolved at 25° C. Thus denitrification during the late autumn and early spring in temperate climatic zones could account for a significant portion of the N_2O released over the year, particularly as soils often tend to be saturated or at least have high moisture contents and consequently lower oxygen levels, over these periods as well.

2.4.4. Effect of nitrate

In an earlier review on denitrification, Broadbent and Clark (1965) concluded that there was general agreement among several workers that denitrification rate was unaffected by a fairly wide range of nitrate concentrations. More recent research on denitrification in soil has shown that a refinement of the response of denitrification to nitrate concentration is necessary. It is now recognized that denitrification follows first-order kinetics in respect to nitrate when oxidisable substrate is not limiting and nitrate levels are lower than 40 mg/1 Stanford *et al.* 1975). Denitrification follows zero-order kinetics largely when carbonaceous substrate is limiting or when nitrate is present at concentrations above 40 mg/1 (Starr and Parlange 1975).

Nitrate concentration has a profound effect on the products of denitrification. Nömmik (1956) and Wijler and Delwiche (1954) are credited with the first accounts of nitrate inhibition of N_2O reductase activity. The importance of nitrate level on the production of N_2O during denitrification was re-emphasized in the work by Blackmer and Bremner (1978). Recent work by Firestone *et al.* (1979a), Fillery (1979) and Chao and Sakdinan (1978), essentially verifies the initial reports and the results of Blackmer and Bremner (1978).

The effects of nitrate level on the products of denitrification in non air-dried unamended (except for nitrate) Plano soil are shown in Table 2.3. The inhibition of N_2O reduction at the lower soil pH at all nitrate levels, and the accumulation of NO at the lower pH and high nitrate levels are clearly discernible. The effect of soil pH will be discussed in more detail in the following section.

At the higher soil pH the nitrate inhibition is temporary, although N_2O remains as a significant product for a longer period at the higher nitrate levels. The onset of N_2O reduction cannot be explained in terms of the depletion of the nitrate pool. Indeed the results presented here and by others do not agree with the finding of Nömmik (1956) that N_2O reduction proceeds only in the absence of nitrate. It is plausible that the synthesis of N_2O reductase could be delayed initially until a 'threshold' N_2O level is reached thereby inducing synthesis (Letey *et al.* 1980). However, this explanation does not account for the pH effects shown in Table 2.3 and in other studies.

The pH effect can be accounted for if nitrite and not nitrate is the factor responsible for the inhibition of N_2O reductase, and presumably NO reductase during denitrification. Firestone *et al.* (1979a) report that the inhibitory effect of nitrite on N_2O reductase in soil is much stronger than that of nitrate, with as little as 2.0 ppm nitrite-N causing N_2O to be the predominant product. Nitrite-N concentrations of this order of magnitude were detected by Fillery (1979) during nitrate reduction (Table 2.3). Interestingly, NO and N_2O were the dominant denitrification products over time where nitrite-N exceeded 1.0 ppm. Nitrate is also known to lower the activity of the NO reductase (Miyata 1971, Payne 1973) but the results reported in Table 2.3 suggest that the pH-linked inhibition is more important.

Table 2.3. Effect of initial nitrate concentration and soil pH on the concentrations of nitrogenous products during biological denitrification in a silt loam incubated under helium

Product	Sampling time (days)	pH 4.8			pH 5.8			pH 6.8		
		Initial nitrate concentration ($\mu g/g$)								
		45	125	525	35	115	515	40	120	522
		Concentration of product (μg N/g soil)								
NO_2^-	0.5	0.3	1.0	1.8	Tr	Tr	0.8	Tr	0.5	0.3
	2.0	0.1	2.5	4.5	0.1	Tr	0.1	Tr	Tr	0.3
	16.0	nd	nd	nd	nd	nd	nd	nd	nd	nd
NO	0.5	1.5	1.6	1.7	0.5	0.6	0.9	0.4	0.5	0.8
	1.0	3.3	4.0	6.5	0.6	0.6	1.3	0.3	0.6	1.7
	2.0	4.1	6.5	8.8	0.1	0.3	1.1	Tr	0.1	0.5
	16.0	—	0.3	44.1	—	Tr	Tr	—	nd	nd
N_2O	0.5	2.7	2.3	1.6	2.4	2.7	2.8	1.9	2.4	3.1
	1.0	6.7	5.9	4.1	6.1	6.7	6.5	3.9	6.0	7.9
	2.0	14.6	15.1	12.7	6.3	11.0	12.7	3.0	7.7	15.1
	16.0	—	59.6	86.9	—	7.6	32.7	—	0.7	28.2
N_2	0.5	0.6	0.4	0.4	1.8	0.9	0.5	2.3	2.1	0.9
	1.0	0.8	0.9	0.7	3.1	2.1	0.9	6.5	4.7	2.0
	2.0	1.2	2.6	1.5	8.1	4.3	1.7	14.3	9.8	6.0
	16.0	—	35.3	32.6	—	59.0	34.4	—	82.8	37.8

— = not analyzed; nd = not detected; Tr = <0.1 ppm. Fillery (unpublished data).

2.4.5. Soil pH

By several accounts, the rate of denitrification is strongly affected by pH, being low in acid soil and generally rapid at slightly alkaline pHs (Bremner and Shaw 1958, Nömmik 1956). Indeed Jansson and Clark (1952) report that an alkaline reaction was required for any extensive denitrification. These results are surprising since the denitrifying bacteria are represented in many genera; and that the activity of the soil flora remains largely unaffected over a wide range of pHs. Perhaps a reasonable explanation for the results reported by Bremner and Shaw (1958) and Nömmik (1956) is that the alkali added to amend acid soils to higher pHs, also temporarily solubilised organic material, thereby increasing the level of readily available organic matter, and thus the rate of nitrate respiration under the imposed anaerobic conditions.

Cooper and Smith (1963) in a survey of soils for denitrification activity, were unable to demonstrate a pH effect on denitrification when the soils were grouped according to pH; denitrification rates being more variable within each pH grouping. Likewise, Fillery (1979) was unable to demonstrate any pH effect when soil collected from a long-term lime trial, and ranging in pH from 4.8 to 6.8, was subjected to anoxia. In the latter case, the rate of denitrification more closely followed the soil respiratory activity as determined by production of carbon dioxide, which itself shows no consistent relationship with soil pH.

Soil pH does, however, profoundly affect the products of denitrification. For example N_2O reduction is appreciably reduced at lower soil pHs. Nömmik (1956) found N_2O to accumulate at pHs below 7.0 while Wijler and Delwiche (1954) report N_2O reduction to be impeded below pH 6.0. Blackmer and Bremner (1978) suggested that the inhibitory effect of pH was due to the intervention of nitrate. For example, they were able to show rapid reduction of exogenous N_2O in soils of low pH, in the absence of nitrate. Presumably, the inhibitory effect of nitrate, discussed in an earlier section, is enhanced in acidic media. Exactly how this occurred was not elaborated on by Blackmer and Bremner (1978).

Another explanation for the inhibitory effect of pH is the increased accumulation of nitrite during denitrification in the more acidic soil (Fillery 1979). Nitrite certainly would not have been present in Blackmer and Bremner's study where nitrate was removed prior to incubation of the soil with N_2O, thereby eliminating an important inhibitor of N_2O reductase (Firestone et al. 1979a).

Exactly how soil pH might affect nitrite reductase activity is not clear. A periplasmic location of the enzyme may increase its sensitivity to soil pH, or to soil constituents which are made increasingly available at lower pHs. In this context Bollag and Barabasz (1979) report that heavy metals affect the activity of both the nitrite and N_2O reductases, but the levels which proved inhibitory

58

Table 2.4. Effect of nitrite concentration on the production of nitric oxide in a steam-sterilized acid silt loam (pH 4.8) incubated under helium

Nitrite addition (μgN/g)	Sampling time (days)		
	0.5	1	4
	Nitric oxide produced (μgN/g)		
2	0.2	0.2	Tr
10	2.9	2.7	1.9
20	6.2	5.7	4.7
40	11.5	10.4	9.1

Soil was sterilized 3 times over a period of a week to eliminate biological activity (Fillery, unpublished data)

were considerably above the concentrations typically detected in soil.

Nitric oxide has been found to be a dominant gaseous denitrification product at pHs below 5.0 (Nömmik 1956, Wijler and Delwiche 1954, Fillery 1979). As previously discussed there remains considerable uncertainty as to the source of the NO. As a general rule, nitric oxide production in acidic media is attributed to chemodenitrification reactions, which will be the subject of discussion in the following chapter. The rate of NO production in autoclaved soil is strongly concentration dependent and increases linearly with increasing nitrite added (Table 2.4). However, rates of NO production comparable to those found in biologically active soil (Table 2.3) could only be obtained with the addition of nitrite at levels substantially above that detected during biological denitrification. This finding led Fillery (1979) to conclude that NO found during nitrate dissimilation in acid soil might also arise through the inhibition of NO reductase, presumably by nitrite. Alternatively, it is possible that the ferrous-nitrosyl complex suggested by Averill and Tiedje (1982) to be pivotal in the denitrification reaction, may be increasingly unstable at lower pHs, resulting in increased NO production.

The toxic effect of nitrite on microbial activity in acid soil has been examined by Bancroft *et al.* (1979). These workers point to the need for further evaluation of the potential ecological effect of NO_x (including nitrite) on soil microbiology. Detailed studies of the effect of nitrite on the denitrification process are lacking. Given that nitrite appears to play an important role, it is important that research be directed towards this problem.

2.5. References

Alefounder, P.R. and Ferguson, S.J. 1980 The location of dissimilatory nitrite reductase and the control of dissimilatory nitrate reductase by oxygen in *Paracoccus denitrificans*. Biochem. J. 192, 231–240.

Allen, M.B. and van Niel, C.B. 1952 Experiments on bacterial denitrification. J. Bacteriol 64, 397–412.

Allison, F.E. 1955 The enigma of soil nitrogen balance sheets. Adv. Agron. 7, 213–250.

Allison, F.E. 1966 The fate of nitrogen applied to soils. Adv. Agron. 18, 219–258.

Allison, F.E., Carter, J.N. and Sterling, L.D. 1960 The effect of partial pressure of oxygen on denitrification in soil. Soil Sci. Soc. Am. Proc. 24, 283–285.

Averill, B.A. and Tiedje J.M. 1982 Hypothesis: The chemical mechanism of microbial denitrification. FEBS Letters (in press).

Bailey, L.D. 1976 Effects of temperature and root on denitrification in a soil. Can J. Soil Sci. 56, 79–87.

Baldensperger, J. and Garcia, J.L. 1975 Reduction of oxidized inorganic nitrogen compounds by a new strain of *Thiobacillus denitrificans*. Arch. Microbiol. 103, 31–36.

Balderston, W.L., Sherr, B. and Payne, W.J. 1976 Blockage by acetylene of nitrous oxide reduction in *Pseudomonas perfectomarinus*. Appl. Environ. Microbiol. 31, 504–508.

Bancroft, K., Grant, I.F. and Alexander, M. 1979 Toxicity of NO_2: Effect of nitrite on microbial activity in a acid soil. Appl. Environ. Microbiol. 38, 940–944.

Bollag, J.M. and Barabasz, W. 1979 Effect of heavy metals on the denitrification process in soil. J. Environ. Qual. 8, 196–201.

Blackmer, A.M. and Bremner, J.M. 1978 Inhibitory effect of nitrate on reduction of N_2O to N_2 by soil microorganisms. Soil Biol. Biochem. 10, 187–191.

Bremner, J.M. and Shaw, K. 1958 Denitrification in soil. II. Factors affecting denitrification. J. Agric. Sci. 51, 40–52.

Broadbent, F.E. and Clark, F. 1965. Denitrification. In: Bartholomew, W.V and Clark, F.E. (eds.), Soil Nitrogen. Agronomy 10, pp. 344–359. American Society of Agronomy, Madison.

Broadbent, F.E. and Stojanovic, B.J. 1952 The effect of partial pressure of oxygen on some soil nitrogen transformations. Soil Sci. Soc. Am. Proc. 16, 359–363.

Burford, J.R. and Bremner, J.M. 1975 Relationships between the denitrification capacities of soils and total, water soluble and readily decomposable soil organic matter. Soil Biol. Biochem 7, 389–394.

Cady, F.B. and Bartholomew, W.V. 1960 Sequential products of anaerobic denitrification in Norfolk soil material. Soil Sci. Soc. Am. Proc. 24, 477–482.

Cady, F.B. and Bartholomew, W.V. 1961 Influence of low pO_2 on denitrification processes and products. Soil Sci. Soc. Am. Proc. 25, 362–365.

Cady, F.B. and Bartholomew, W.V. 1963 Investigations of nitric oxide reactions in soils. Soil Sci. Soc. Am. Proc. 27, 546–549.

Chao, C.M. and Sakdinan, L. 1978 Mass spectrometric investigation on denitrification. Can. J. Soil Sci. 58, 443–457.

Chung, C.W. and Najjar, V.A. 1956 Co-factor requirements of enzymatic denitrification. I. Nitrite reductase. J. Biol. Chem. 218, 617–632.

Cole, J.A. and Wimpenny, J.W.T. 1968 Metabolic pathways for nitrate reduction in *Escherichia coli*. Biochim. Biophys. Acta 162, 39–48.

Collins, F.M. 1955 Effect of aeration on the formation of nitrate reducing enzymes by *P. aeruginosa*. Nature (Lond.) 175, 173–174.

Cooper, G.S. and Smith, R.L. 1963 Sequence of products formed during denitrification in some diverse western soils. Soil Sci. Soc. Am. Proc. 27, 659–662.

Daniel, R.M., Smith, I.M., Phillip, J.A.D., Radcliff, H.D., Drozd, J.W. and Bull, A.T. 1980 Anaerobic growth and denitrification by *Rhizobium japonicum* and other rhizobia. J. Gen. Microbiol. 120, 517–521.

de Groot, G.N. and Stouthamer, A.H. 1970 Regulation of reductase formation in *Proteus mirabilis*. II. Influence of growth with azide and haem deficiency on nitrate reductase formation. Biochim. Biophys. Acta 208, 414–427.

Dowdell, R.J. and Smith, K.A. 1974 Field studies of the soil atmosphere. II. Occurrence of nitrous oxide. J. Soil Sci. 25, 231–238.

De Datta, S.K. 1981 Principles and Practices of Rice Production. Chapter 4, pp. 89–145. John Wiley & Sons, Chichester.

Downey, R.J., Kiszkiss, D.F. and Nuner, J.H. 1969 Influence of oxygen on development of nitrate respiration in *Bacillus stearothermophilus*. J. Bacteriol. 98, 1056–1062.

Federova, R.I., Milekhina, E.I. and Ilyukhina, N.I. 1973 Possibility of using the 'gas exchange' method to detect extraterrestrial life: identification of nitrogen-fixing organisms. Izv. Akad. Nauk SSSR. Ser. Biol. 6, 797–806.

Fewson, C.A. and Nicholas, D.J.D. 1960 Utilization of nitric oxide by microorganisms and higher plants. Nature (Lond.) 188, 794–796.

Fewson, C.A. and Nicholas, D.J.D. 1961 Nitrate reductase from *Pseudomonas aeruginosa*. Biochim. Biophys. Acta 49, 335–349.

Fillery, I.R.P. 1979 Denitrification in soils under low oxygen or anaerobic environments. Diss. Abstr. Int. B40, 1529.

Fillery, I.R.P. and Vlek, P.L.G. 1982 The significance of denitrification of applied nitrogen in fallow and cropped rice soils under different flooding regimes. Plant and Soil 65, 153–169.

Firestone, M.K., Smith, M.S., Firestone, R.B. and Tiedje, J.M. 1979a The influence of nitrate, nitrite, and oxygen on the composition of the gaseous products of denitrification in soil. Soil Sci. Soc. Am. J. 43, 1140–1144.

Firestone, M.K., Firestone, R.B. and Tiedje, J.M. 1979b Nitric oxide as an intermediate in denitrification: Evidence from nitrogen-13 isotope exchange. Biochem. Biophys. Res. Commun. 91, 10–16.

Fluhler, H., Stolzy, L.H. and Ardakani, M.S. 1976 A statistical approach to define soil aeration with respect to denitrification. Soil Sci. 122, 115–123.

Focht, D.D. and Verstraete, W. 1977 Biochemical ecology of nitrification and denitrification. Ann. Rev. Microbiol. Ecol. 1, 135–214.

Forget, P. and Dervartanian, D.V. 1972 The bacterial nitrate reductases. E.P.R. studies on nitrate reductase A from *Micrococcus denitrificans*. Biochim. Biophys. Acta 256, 600–606.

Galbally, I.E. and Roy, C.R. 1978 Loss of fixed nitrogen from soils by nitric oxide exhalation. Nature (Lond.) 275, 734–735.

Gamble, T.N., Betlach, M.R. and Tiedje, J.M. 1977 Numerically dominant denitrifying bacteria from world soils. Appl. Environ. Microbiol. 33, 926–939.

Garber, E.A.E. and Hollocher, T.C. 1981 ^{15}N tracer studies on the role of NO in denitrification. J. Biol. Chem. 256, 5459–5465.

Garcia, J.L. 1975 Evaluation de la dénitrification dans les rizieres par la methode de reduction de N_2O. Soil Biol. Biochem. 7, 251–256.

Garcia, J.L. 1977 Analyse de différents groups composant la microflore dénitrifiante des sols de rizière de Sénégal. Ann. Microbiol. (Paris) 128A, 433–446.

Garland, P.B., Downie, J.A. and Haddock, B.A. 1975 Proton translocation and respiratory nitrate reductase of *Escherichia coli*. Biochem. J. 152, 547–559.

Garrard, W.T. 1972 Synthesis, assembly and localization of periplasmic cytochrome c. J. Biol. Chem. 247, 5935–5943.

Greenland, D.J. 1962 Denitrification in some tropical soils. J. Agr. Sci. 58, 227–233.

Greenwood, D.J. 1962 Nitrification and nitrate dissimilation in soil. II. Effect of oxygen concentration. Plant Soil 17, 378–391.

Greenwood, D.J. 1963 Nitrogen transformations and the distribution of oxygen in soil. Chem Ind. (Lond.) 799–803.

Haddock, B.A. and Jones, C.W. 1977 Bacterial respiration. Bacteriol. Rev. 41, 47–99.

Hauck, R.D. 1971 Quantitative estimates of nitrogen cycle processes: concept and review. In: Nitrogen-15 in Soil-Plant Studies. IAEA-PL-341/6 Vienna, Austria.

Hollocher, T.C., Garbers, E., Cooper, A.J.L. and Reiman, R.F. 1980 ^{13}N, ^{15}N isotope and kinetic evidence against hyponitrite as an intermediate in denitrification. J. Biol. Chem. 255, 5027–5030.

Iwasaki, H. and Matsubara, T. 1971 Cytochrome c557 (551) and cytochrome cd of *Alcaligenes faecalis*. J. Biochem. (Tokyo) 69, 847–857.

Iwasaki, H. and Matsubara, T. 1972 A nitrite reductase from *Achromobacter cycloclastes*. J. Biochem. (Tokyo) 71, 645–652.

Jansson, S.L. and Clark, F.E. 1952 Losses of nitrogen during decomposition of plant material in the presence of inorganic nitrogen. Soil Sci. Soc. Am. Proc. 16, 330–334.

John, P. 1977 Aerobic and anaerobic bacterial respiration monitored by electrodes. J. Gen. Microbiol. 98, 231–238.

Keeney, D.R., Fillery, I.R. and Marx, G.P. 1979 Effect of temperature on gaseous N products of denitrification in soil. Soil Sci. Soc. Am. J. 43, 1124–1128.

Kessel, J.F. van 1976 Influence of denitrification in aquatic sediments on the nitrogen content of natural waters. Agric. Res. Rep. (Wageningen) No. 858.

Kiszkiss, D.F. and Downey, R.J. 1972 Localization and solubilization of the respiratory nitrate reductase of *Bacillus stearothermophilus*. J. Bacteriol. 109, 803–810.

Kluyver, A.J. and Verhoeven, W. 1954 Studies on true dissimilatory nitrate reduction. II. The mechanisms of denitrification. Antonie van Leeuwenhoek J. Microbiol. Serol. 20, 241–262.

Kodama, T. 1970 Effects of growth conditions on formation of cytochrome system of a denitrifying bacterium *Pseudomonas stutzeri*. Plant Cell Physiol. 11, 231–239.

Kodama, T., Shimada, K. and Mori, T. 1969 Studies on anaerobic biphasic growth of a denitrifying bacterium *Pseudomonas stutzeri*. Plant Cell Physiol. 10, 855–865.

Kristjansson, J.K., Walter, B. and Hollocher, T.C. 1978 Respiration-dependent proton translocation and the transport of nitrate and nitrite in *Paracoccus denitrificans* and other denitrifying bacteria. Biochemistry 17, 5014–5019.

Kristjansson, J.K. and Hollocher, T.C. 1979 Substrate binding site for nitrate reductase of *Escherichia coli* is on the inner aspect of the membrane. J. Bacteriol. 137, 1227–1233.

Kristjansson, J.K. and Hollocher, T.C. 1980 First practical assay for soluble nitrous oxide reductase of denitrifying bacteria and a partial kinetic characterization. J. Biol. Chem. 255, 704–707.

Leffelaar, P.A. 1979 Simulation of partial anaerobiosis in a model soil in respect to denitrification. Soil Sci. 128, 110–120.

Letey, J., Haddas, A., Vailoras, N. and Focht, D.D. 1980 Effect of preincubation treatments on the ratio of N_2O/N_2 evolution. J. Environ. Qual. 9, 232–235.

MacGregor, C.H. and Christopher, A.R. 1978 Asymmetric distribution of nitrate reductase subunits in the cytoplasmic membrane of *Escherichia coli*: Evidence derived from surface labeling studies with transglutiminase. Arch. Biochem. Biophys. 185, 204–213.

Marshall, R.O., Dishburger, H.J., MacVicar, R. and Hallmark, G.D. 1953 Studies on the effect of aeration on nitrate reduction by *Pseudomonas* species using ^{15}N. J. Bacteriol. 66, 254–258.

Matsubara, T. 1971 Studies on denitrification. XIII. Some properties of the N_2O-anaerobically grown cell. J. Biochem. (Tokyo) 69, 991–1001.

Matsubara, T. 1975 The participation of cytochromes in the reduction of N_2O to N_2 by a denitrifying bacterium. J. Biochem. (Tokyo) 77, 627–632.

Matsubara, T. and Iwasaki, H. 1971 Enzymatic steps of dissimilatory nitrite reduction in *Alcaligenes faecalis*. J. Biochem. (Tokyo) 69, 859–868.

Matsubara, T. and Iwasaki, H. 1972 Nitric oxide-reducing activity of *Alcaligenes faecalis* cytochrome cd. J. Biochem. (Tokyo) 72, 57–64.

Matsubara, T. and Mori, T. 1968 Studies on denitrification. IX. Nitrous oxide, its production and reduction to nitrogen. J. Biochem. (Tokyo) 64, 863–871.

McGarity, J.W. 1962 Effect of freezing of soil on denitrification. Nature (Lond.) 196, 1342–1343.

Meiklejohn, J. 1946 Aerobic denitrification. Ann. Appl. Biol. 27, 558–573.

Mitchell, P. 1966 Chemiosmotic coupling in oxidative and photosynthetic phosphorylation. Glyn Research Ltd., Bodmin, Cornwall.

Miyata, M. 1971 Studies on denitrification. XIV. The electron donating system in the reduction of nitric oxide and nitrate. J. Biochem. (Tokyo) 70, 205–213.

Miyata, M. and Mori, T. 1969 Studies on denitrification. X. The denitrifying enzyme as a nitrite reductase and electron donating system for denitrification. J. Biochem. (Tokyo) 66, 463–471.

Miyata, M., Matsubara, T. and Mori, T. 1969 Studies on denitrification. XI. Some properties of nitric oxide reductase. J. Biochem. (Tokyo) 66, 759–765.

Newton, N. 1969 The two-haem nitrite reductase of *Micrococcus denitrificans*. Biochim. Biophys. Acta 185, 316–331.

Neyra, C.A., Dobereiner, J., Lalande, R. and Knowles, R. 1977 Denitrification by N_2-fixing *Spirillum lipoferum*. Can. J. Microbiol. 23, 300–305.

Neyra, C.A. and van Berkum, P. 1977 Nitrate reduction and nitrogenase activity in *Spirillum lipoferum*. Can. J. Microbiol. 23, 306–310.

Nömmik, H. 1956 Investigations on denitrification in soil. Acta Agric. Scand. VI 2, 195–228.

Pattern, D.K., Bremner, J.M. and Blackmer, A.M. 1980 Effects of drying and air-dry storage of soils on their capacity for denitrification of nitrate. Soil Sci. Soc. Am. J. 44, 67–70.

Payne, W.J. 1973 Reduction of nitrogenous oxides by microorganisms. Bacteriol. Rev. 37, 409–452.

Payne, W.J., Riley, P.S. and Cox, C.D. 1971 Separate nitrite, nitric oxide, and nitrous oxide reducing fractions from *Pseudomonas perfectomarinus*. J. Bacteriol. 106, 356–361.

Pichinoty, F. and D'Orano, L. 1961 Inhibition by oxygen of biosynthesis and activity of nitrate-reductase in *Aerobacter aerogenes*. Nature (Lond.) 191, 879–881.

Pilot, L. and Patrick, W.H., Jr. 1972 Nitrate reduction in soils. Effect of soil moisture. Soil Sci. 114, 312–316.

Prince, R.C., Baccarini-Melandri, A., Hauska, G.A., Melandri, B.A. and Crofts, A.R. 1975 Asymmetry of an energy transducing membrane; the location of cytochrome c_2 in *Rhodopseudomonas spheroides* and *Rhodopseudomonas capsulata*. Biochim. Biophys. Acta 387, 212–227.

Radcliffe, B.C. and Nicholas, D.J.D. 1968 Some properties of a nitrite reductase from *Pseudomonas denitrificans*. Biochim. Biophys. Acta 153, 545–554.

Radcliffe, B.C. and Nicholas, D.J.D. 1970 Some properties of a nitrate reductase from *Pseudomonas denitrificans*. Biochim. Biophys. Acta 205, 273–287.

Renner, E.D. and Becker, G.L. 1970 Production of nitric oxide and nitrous oxide during denitrification by *Corynebacterium nephridii*. J. Bacteriol. 101, 821–826.

Rigaud, J., Bergersen, F.J., Turner, G.L. and Daniel, R.M. 1973 Nitrate dependent acetylene-reduction and nitrogen fixation by soybean bacteroids. J. Gen Microbiol. 77, 137–144.

Sacks, L.E. and Barker, H.A. 1952 Substrate oxidation and nitrous oxide utilization in denitrification. J. Bacteriol. 64, 247–252.

Sanchez, C., Jose, A., Dubourdieu, M. and Chippaux, M. 1979 Localization and characterization of cytochromes from membrane vesicles of *Escherichia coli* K-12 grown in anaerobiosis with nitrate. Biochim. Biophys. Acta 547, 198–210.

Sapshead, L.M. and Wimpenny, J.W.T. 1972 The influence of oxygen and nitrate on the formation of the cytochrome pigments of the aerobic and anaerobic respiratory chain of *Micrococcus denitrificans*. Biochim. Biophys. Acta 267, 388–397.

Saraste, M. and Kuronen, T. 1978 Interaction of *Pseudomonas* cytochrome cd with the cytoplasmic membrane. Biochim. Biophys. Acta 513, 117–131.

Schwartzbeck, R.A., MacGregor, J.M. and Schmidt, E.L. 1961 Gaseous nitrogen losses from nitrogen fertilized soils measured with infra-red and mass spectroscopy. Soil Sci. Soc. Am. Proc. 25, 186–189.

Showe, M.K. and De Moss, J.A. 1968 Localization and regulation of synthesis of nitrate reductase in *Escherichia coli*. J. Bacteriol. 95, 1305–1313.

Skerman, V.B.D. and MacRae, I.C. 1957 The influence of oxygen on the reduction of nitrate by adapted cells of *Pseudomonas denitrificans*. Can. J. Microbiol. 3, 215–230.

Skerman, V.B.D., Carey, B.J. and MacRae, I.C. 1958 The influence of oxygen on the reduction of nitrite by washed suspensions of adapted cells of *Achromobacter liquefaciens*. Can. J. Microbiol. 4, 243–256.

Smith, K.A. 1978 A model of the extent of anaerobic zones in aggregated soils and its application to estimates of denitrification. Abstracts 11th Congress Int. Soc. Soil Sci. Edmonton, Canada, p. 304.

Stanford, G., Vanderpol, R.A. and Dzienia, S. 1975 Denitrification rates in relation to total and extractable soil carbon. Soil Sci. Soc. Am. Proc. 39, 284–289.

Starr, J.L. and Parlange, J.Y. 1975 Nonlinear denitrification kinetics with continuous flow in soil columns. Soil Sci. Soc. Am. Proc. 39, 875–880.

Stefanson, R.C. 1972 Soil denitrification in sealed soil-plant systems. I. Effect of plants; soil water content and soil organic matter content. Plant Soil 33, 113–127.

St. John, R.T. and Hollocher, T.C. 1977 Nitrogen 15 tracer studies on the pathway of denitrification in *Pseudomonas aeruginosa*. J. Biol. Chem. 252, 212–218.

Stouthamer, A.H. 1976 Biochemistry and genetics of nitrate reductase in bacteria. Adv. Microb. Physiol. 14, 315–375.

Terman, G.L. 1979 Volatilization losses of nitrogen as ammonia from surface fertilizers, organic amendments and crop residues. Adv. Agron. 31, 189–220.

Valera, C.L. and Alexander, M. 1961 Nutrition and physiology of denitrifying bacteria. Plant Soil 15, 268–280.

van 't Riet, J., Stouthamer, A.H. and Planta, R.J. 1968 Regulation of nitrate assimilation and nitrate respiration in *Aerobacter aerogenes*. J. Bacteriol. 96, 1455–1464.

van 't Riet, J., Wientjes, F.B., van Doorn, J. and Planta, R.J. 1979 Purification and characterization of the respiratory nitrate reductase of *Bacillus licheniformis*. Biochim. Biophys. Acta 576, 347–360.

Vives, J. and Parés, R. 1975 Enumeration y caracterizacion de la flora desnitricante quimioorganotrota en una pradera experimental. Microbiol. Esp. 28, 43.

Vlek, P.L.G. and Craswell, E.T. 1981 Ammonia volatilization from flooded soils. Fert. Res. 2, 227–245.

Wesseling, J. and van Wijk, W.R. 1957 Land drainage in relation to soils and crops. I. Soil physical conditions in relation to drain depth. In: Luthin, L.N. (ed.), Drainage of Agricultural Lands, pp. 461–504. American Society of Agronomy, Madison.

Wetselaar, R. and Farquhar, G.D. 1980 Nitrogen losses from tops of plants. Adv. Agron. 33, 263–302.

Wientjes, I.B., Kolk, A.H.J., Nanninga, N. and van 't Riet, J. 1979 Respiratory nitrate reductase: its localization in the cytoplasmic membrane of *Klebsiella aerogenes* and *Bacillus licheniformis*. Eur. J. Biochem. 95, 61–68.

Wijler, J. and Delwiche, C.C. 1954 Investigations on the denitrifying process in soil. Plant Soil 5, 155–169.

Wimpenny, J.W.T. 1969 Oxygen and carbon dioxide as regulators of microbial growth and metabolisms. Symp. Soc. Gen. Microbiol. 19, 161–197.

Wimpenny, J.W.T. and Cole, J.A. 1967 The regulation of metabolism in facultative bacteria III. Biochim. Biophys. Acta 148, 233–242.

Woldendorp, J.W. 1962 The quantitative influence of the rhizospere on denitrification. Plant Soil 17, 267–270.

Woldendorp, J.W. 1968 Losses of soil nitrogen. Stikstof 12, 32–46.

Yoshinari, T. and Knowles, R. 1976 Acetylene inhibition of nitrous oxide reduction by denitrifying bacteria. Biochim. Biophys. Res. Commun. 69, 705–710.

Zablotowicz, R.M., Eskew, D.L. and Focht, D.D. 178 Denitrification in *Rhizobium*. Can. J. Microbiol. 24, 757–760.

Zumft, W.G. and Cardenas, J. 1979 The inorganic biochemistry of nitrogen bioenergetic processes. Naturwissenschaften 66, 81–88.

3. Chemodenitrification

P.M. CHALK and C.J. SMITH

3.1. Introduction

It has long been suspected that the poor efficiency of N fertilizers was due in part to conversion to gaseous forms of N which were then evolved from the soil. It was generally assumed that the important pathways of loss were volatilization of NH_3, and reduction of nitrate to N_2O and N_2 by soil microorganisms (biological denitrification). There was little direct evidence to support the assumption of significant losses via these pathways, despite intensive study of the factors affecting the processes. However, recent advances in the methodology for the collection of samples and measurement of gaseous forms of N, have stimulated research on gaseous-N losses from soil. Investigations have shown that several gaseous forms of N may be evolved from soils which are actively nitrifying. The nature of the processes which cause these losses is currently the subject of much investigation and speculation. There is evidence that nitrite produced by nitrifying or denitrifying microorganisms may react chemically to form gaseous N compounds (chemical denitrification).

3.2. Role of nitrite in gaseous nitrogen loss from soil

Clark (1962) proposed that gaseous N losses could occur during nitrification via chemical reactions involving nitrite. Vine (1962) also suggested that gaseous losses of N may occur from 'sidetracking' of nitrification, but did not suggest a specific pathway. Gaseous loss of N associated with nitrite instability was termed 'chemo-denitrification' (Clark 1962). The pathway was proposed to explain observed N-deficits in soils which accumulated nitrite following addition of urea (Soulides and Clark 1958, Clark et al. 1960). Organic compounds were shown to accelerate nitrite dismutation (Clark and Beard 1960). The following diagram illustrates the proposed pathway (Broadbent and Stevenson 1966).

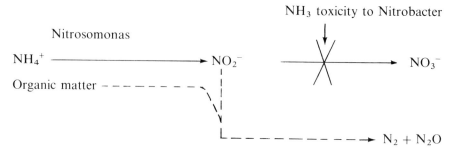

In well-aerated, unfertilized soil, oxidation of nitrite to nitrate by Nitrobacter proceeds at a faster rate than the conversion of ammonium to nitrite by Nitrosomonas. Consequently, nitrite is not normally present in amounts greater than 1 μg N/g of soil. High concentrations may be found, however, when N fertilizers which form alkaline solutions upon hydrolysis are band-applied to soil. Urea, ammonium carbonate, diammonium phosphate, urea ammonium phosphate and anhydrous ammonia hydrolyse to produce an alkaline environment. Nitrite accumulation has also been reported in soils in which biological denitrification is occurring (Cady and Bartholomew 1960, Cooper and Smith 1963).

Initially, at the centre of an alkaline-hydrolysing fertilizer band, soil pH may be as high as 10 (Blue and Eno 1954, Nömmik and Nilsson 1963, Parr and Papendick 1966) and the N concentration may be several thousand μg/g of soil (Chalk *et al.* 1975). Both Nitrosomonas and Nitrobacter are adversely affected at high pH and salt concentration, but Nitrobacter is more sensitive to such environmental stresses. The inhibitory effect at high pH is considered to be due to toxicity of free NH_3, the concentration of which is governed by pH. The optimum soil pH for nitrification is between 7 and 9, with Nitrobacter having a lower optimum range than Nitrosomonas (Alexander 1965). The activity of Nitrobacter is temporarily inhibited to a greater extent than that of Nitrosomonas, and nitrite accumulates. Several hundred μg $NO_2^- - N$/g soil have been measured in fertilizer bands in the field (Chalk *et al.* 1975).

Nitrite may accumulate in both alkaline and acid soils following application of alkaline-hydrolysing fertilizers (Broadbent *et al.* 1957, Nömmik and Nilsson 1963, Hauck and Stephenson 1965, Jones and Hedlin 1970a, Pang et al. 1973, 1975a, b). Nitrite may also accumulate in alkaline soils treated with acid-hydrolysing ammonium fertilizers such as ammonium sulfate (Justice and Smith 1962, Bezdicek *et al.* 1971). Low temperature and moisture (Tyler *et al.* 1959, Justice and Smith 1962, Chalk *et al.* 1975) and low soil buffer capacity (Hauck and Stephenson 1965) have been associated with nitrite accumulation. The type of fertilizer, granule size, rate of application and distribution all affect soil pH, rate of nitrification, and the concentration and distribution of the nitrification products (Hauck and Stephenson 1965, Bezdicek *et al.* 1971, Wetselaar *et al.* 1972).

The change in soil pH which occurs after fertilizer application is an important factor influencing both the accumulation and reactivity of nitrite. The initial increase in soil pH observed when NH_3 or NH_3-producing fertilizers are applied is due to the hydrolysis:

$$NH_3 + H_2O \rightleftharpoons NH_4^+ + OH^-.$$

As nitrification proceeds the pH decreases because the oxidation of ammonium to nitrite by Nitrosomonas is an acid producing reaction, e.g.

$$2NH_4^+ + 3O_2 \rightarrow 2NO_2^- + H_2O + 4H^+.$$

By summation of the above equations, it can be seen that the net effect is the production of 1 mole of acid (H^+) for every mole of NH_3 which is nitrified. Theoretically, when one half of the NH_3 fertilizer has been nitrified, the soil should be at its original pH overall. Oxidation of the remaining fertilizer will produce net acidity.

Nitrite is particularly reactive under acidic conditions, although nitrite decomposition will occur when the measured soil pH is alkaline (Smith and Chalk 1980a). Acid conditions may be encountered around the periphery of a fertilizer band where nitrification of the fertilizer is complete, whereas at the centre of the band where fertilizer N remains, the pH is alkaline and nitrite may be present (Nömmik and Nilsson 1963). Nitrite accumulation has been reported to occur between pH 7 and 8 (Wetselaar *et al.* 1972) and between pH 7.5 and 8.5 (Nömmik and Nilsson 1963). Nitrite formed in the inner alkaline zone could diffuse into the acid peripheral zone. Hauck and Stephenson (1965) suggested that nitrite accumulating at the alkaline fertilizer granule site may be unstable in the acid portion of the soil environment, leading to gaseous loss of N. It has also been suggested that chemodenitrification reactions may take place at organic matter or clay mineral surfaces where pH is lower than the measured soil pH (Porter 1969, Nelson and Bremner 1970b).

3.3. Gaseous products of chemodenitrification

Several gaseous forms of N are produced in sterilized soil treated with nitrite. Dinitrogen (N_2) and nitrous oxide (N_2O) have been identified as products of nitrite decomposition in soils (Smith and Clark 1960, Reuss and Smith 1965, Nelson and Bremner 1970b, Smith and Chalk 1980 a, b). Nitric oxide (NO) and nitrogen dioxide (NO_2) have been individually or collectively measured as nitrite decomposition products (Gerretsen and de Hoop 1957, Reuss and Smith 1965, Nelson and Bremner 1970b, Jones and Hedlin 1970b, Smith and Chalk 1980 a, b). There is indirect evidence that methyl nitrite (CH_3ONO) may be produced from nitrite reactions in soils, since this gas has been identified following the reaction of nitrous acid with lignin preparations (Stevenson and Kirkman 1964, Stevenson and Swaby 1964, Edwards and Bremner 1966).

Nitric oxide has not been positively identified as a product of chemodenitrification in soils under aerobic conditions. It is generally assumed that it is rapidly oxidized by atmospheric O_2 to NO_2, and is therefore commonly determined together with any NO_2 evolved from soil by absorption in acid or alkaline permanganate solutions (Nelson and Bremner 1970b, Smith and Chalk 1980a, b), with determination of the nitrate formed by steam distillation (Bundy and Bremner 1973, Smith and Chalk 1979a). However, the oxidation of NO to NO_2 is concentration dependent (termolecular reaction). At concentrations of 100 ppm or greater, the half life for oxidation in air is one hour or less, whereas at low concentrations (0.01 ppm) it is of the order of 10^4 hours (Galbally and Roy 1978). Failure to detect NO during chemodenitrification (Nelson and Bremner 1970b) may have been due to the use of a gas chromatographic technique lacking sufficient sensitivity.

Nitric oxide evolution has been measured in nitrite treated soils under conditions of low O_2 concentration (Smith and Clark 1960, Reuss and Smith 1965, Keeney et al. 1979). Nitric oxide has also been detected under anaerobic conditions following nitrate addition to soil (Cady and Bartholomew 1960, Cooper and Smith 1963, Keeney et al. 1979). Nitric oxide is not considered to be an intermediate in biological denitrification, but to arise from chemical decomposition of nitrite. The decomposition of nitrous acid has been proposed as the pathway, since significant anaerobic production of NO was only found in acid media (Cady and Bartholomew 1961, 1963). However, Keeney et al. (1979) have measured high concentrations of NO in a nitrite treated sterile soil of pH 6.8, and have proposed that nitrosation reactions may account for the greater than quantitative recovery of nitrite-N as $NO + N_2 + N_2O$ at pH near neutrality.

Some investigators have been unable to detect NO_2 in the aerobic atmosphere above acidic soils treated with nitrite, although N_2 and N_2O evolution occurred under similar experimental conditions (Smith and Clark 1960, Tyler and Broadbent 1960, Meek and MacKenzie 1965). This has been attributed to the use of closed incubation systems that promoted sorption of this gas by moist soil and conversion of NO_2 to nitrate (Reuss and Smith 1965, Nelson and Bremner 1970b). It is generally assumed that NO_2 is very soluble in water. This can be represented as:

$$3NO_2 + H_2O \rightleftharpoons 2HNO_3 + NO$$

or

$$2NO_2 + H_2O \rightleftharpoons HNO_3 + HNO_2.$$

However, at low concentrations in air, NO_2 is very poorly absorbed by water unless reactive solutes such as arsenite or phenoxide ions are present (Nash 1970). It has been established that NO_2 evolved from soil will be reabsorbed if a permanganate trap is not included in a closed incubation vessel (Nelson and Bremner 1970b), or if an inefficient trap (low solution: soil surface area ratio) is used (Smith and Chalk 1979b, 1980a). Smith and Chalk (1979b) demonstrated that the amount of N recovered in an acid permanganate trap, included in a closed incubation vessel containing an acid soil treated with nitrite, depended on the conditions of incubation. Increased gas absorption in the trap was obtained by increasing the temperature, and decreasing the soil water content and depth of soil. It is evident that the trap does not measure the total quantity of $NO + NO_2$ produced by chemodenitrification, but a net amount which is dependent on the experimental conditions.

3.4. Mechanisms of nitrite decomposition in soil

Several mechanisms of chemodenitrification have been proposed to explain the loss of several gaseous forms of N which occurs when nitrite is added to soil. These mechanisms have been reviewed previously by several authors (Allison 1966, Broadbent and Clark 1965, Broadbent and Stevenson 1966, Nelson 1967, Hauck 1968).

3.4.1. Decomposition of nitrous acid

Self-decomposition reactions involve the nitrous acid molecule, rather than the nitrite ion, and are therefore highly pH dependent, since

$$HNO_2 \rightleftharpoons H^+ + NO_2^-.$$

The pK_a for this equilibrium is 3.29 at 25° C (Aylward and Findlay 1971). In a system at pH 6 only 0.2% of the nitrite-N exists as undissociated nitrous acid, while at pH 5, pH 4 and pH 3, the proportions of nitrous acid are 1.9, 16 and 74% respectively. Nitrite is reasonably stable in solutions of pH 5.5 or greater (Allison and Doetsch 1950, Gerretsen and de Hoop 1957, Allison 1963, Porter 1969).

The classical nitrous acid self-decomposition reaction is given as

$$3HNO_2 \rightleftharpoons HNO_3 + 2NO + H_2O.$$

Nelson and Bremner (1970b) obtained evidence that the self-decomposition reaction was better represented by the equation

$$2HNO_2 \rightleftharpoons NO + NO_2 + H_2O.$$

However, the spontaneity of different nitrous acid self-decomposition reactions was calculated for different sets of activities of the participating components (Van Cleemput and Baert 1976). These calculations showed that nitrous acid preferentially decomposes spontaneously to NO and nitrate instead of self-decomposition to NO and NO_2 or N_2O_4. The higher the pH the larger the region of nitrous acid stability. Theoretical evidence for the reaction of nitrous acid with nitric acid with formation of NO_2 was obtained, e.g.

$$HNO_2 + HNO_3 \rightleftharpoons 2NO_2 + H_2O.$$

Calculations of Gibbs free energy changes indicated formation of NO_2 rather than N_2O_4.

The consequences of self-decomposition of nitrous acid are the chemical oxidation of nitrite to nitrate, and the formation of the gases NO and NO_2. Nitric oxide may be oxidized to NO_2 by atmospheric O_2, and the gases may escape to the atmosphere, or be absorbed by the soil (Mortland 1965, Miyamoto et al. 1974, Prather and Miyamoto 1974, Prather et al. 1973a, b).

Nelson and Bremner (1970b) reported that substantial amounts of N_2 and NO_2 and small amounts of N_2O were evolved on treatment of neutral and acidic soils with nitrite. The rate of NO_2 production has been shown to be inversely related to soil pH (Reuss and Smith 1965, Nelson and Bremner 1970b, Smith and Chalk 1980a,b), but Nelson and Bremner (1970b) reported significant evolution of NO_2 from soils where measured pH was above 7. Smith and Chalk (1980a, b) also reported that substantial amounts of N_2 and NO_2 and lesser amounts of N_2O were evolved from 3 soils (pH 5.2 to 7.5) treated with nitrite. Little NO_2 was evolved from a calcareous soil (pH 8.2) although significant and equal amounts of N_2 and N_2O were evolved (Smith and Chalk 1980a).

Nelson and Bremner (1970b) obtained evidence that most of the NO_2 evolved on treatment of soil with nitrite was formed by self-decomposition of nitrous acid, and by oxidation of NO produced in this way. Further evidence for production of NO + NO_2 through self-decomposition of nitrous acid was obtained through addition of ^{15}N-labelled nitrite to γ-irradiated soils (Smith and Chalk 1980b). In 3 soils (pH 6.2 to 7.5) the ^{15}N-enrichment of evolved NO + NO_2 was only slightly less than the ^{15}N-enrichment of added nitrite. The absence of isotopic dilution proved that the gas originated almost entirely from the added nitrite. The production of significant amounts of NO + NO_2 in soils where the measured pH is greater than 5.5, indicates that self-decomposition occurs at colloid surfaces where the pH is considerably lower than the measured soil pH (Porter 1969, Nelson and Bremner 1970b). This hypothesis is supported by evidence that air-drying of soils treated with nitrite increases the rate of nitrite decomposition. The polarizing effect of exhangeable cations on their associated water molecules is concentrated as the soil dries, so that the degree of dissociation and hence the acidity of the surface water increases (White 1979).

3.4.2. Reaction of nitrous acid with amino compounds

Under suitable conditions, compounds which contain free amino groups (e.g. amino acids, urea, amines) will react with nitrous acid to produce N_2, e.g.

$$R - NH_2 + HNO_2 \rightleftarrows R - OH + H_2O + N_2.$$

This reaction is commonly known as the Van Slyke reaction, although the classical Van Slyke reaction involves only α-amino acids (Allison 1966). The reaction involves the nitrous acid molecule and is therefore dependent on pH, as previously discussed. In the laboratory, the reaction proceeds rapidly and quantitatively in the Van Slyke apparatus, under acidic conditions and an atmosphere of NO, which prevents self-deomposition of nitrous acid (Allison 1966). The N_2 formed is derived in equal proportions from the nitrous acid and the amino compound.

Many investigators have concluded, from studies of the reactions of nitrite with amino acids and urea in buffer solutions, that the Van Slyke reaction is not likely to occur to any significant extent in soils (Allison and Doetsch 1950, Allison *et al.* 1952, Smith and Clark 1960, Allison 1963, 1966, Sabbe and Reed 1964, Broadbent and Clark 1965, Bremner and Nelson 1968). These conclusions were based on consideration of the instability of nitrous acid in the acidic environment necessary for the Van Slyke reaction, and the low concentration of free amino compounds in soils. However, Nelson (1967) has shown that nitrous acid will react readily with the amino sugars, glucosamine and galactosamine in pH 5 buffer. It was also shown that glucosamine promoted nitrite decomposition during incubation of acidic and neutral soils, and during air-drying of neutral and alkaline soils. It is known that 5–10% of the total N in most soils is in the form of amino sugars, but it is not known if the amino groups are free or bound. Reuss and Smith (1965) proposed that N_2 evolution in nitrite-treated acid soils could result from a 'Van Slyke-type' reaction involving labile amino groups in the soil organic matter. Porter (1969) also suggested that some aromatic amines in soil organic matter might react with nitrous acid to give a typical Van Slyke-type reaction. This suggestion was based on the observation that the aromatic amine, anthranilic acid, reacted with nitrite at 50° C in pH 5 and 6 buffer to produce N_2 and some NO. Similarly, Smith and Chalk (1980b) suggested that a Van Slyke-type reaction may have partially contributed to the N_2 evolved on treatment of three γ-irradiated soils with nitrite. The ^{15}N-enrichment of evolved N_2 was approximately one-half of the ^{15}N-enrichment of added nitrite, so that approximately equal amounts of nitrite-N and indigenous-N reacted to form N_2.

Arnold (1954) suggested that gaseous losses of N in soil may occur through reaction of nitrite and hydroxylamine formed during nitrification.

$$NH_2OH + HNO_2 \rightleftarrows N_2O + 2H_2O$$

Porter (1969) has shown that N_2O is produced rapidly when nitrite and hydroxylamine are added to buffer solutions of pH 5 or 6. Incubation or air-drying of soils treated with nitrite-hydroxylamine solution led to rapid decomposition of nitrite when measured soil pH was 5.1 to 7.7 (Nelson 1967). Nitrite decomposition did not occur, or was much reduced, in the absence of hydroxylamine. Several workers have shown that hydroxylamine undergoes rapid chemical decomposition in soils with formation of N_2O and N_2 (Arnold 1954, Nelson 1978, Bremner et al. 1980). Bremner et al. (1980) have shown that addition of hydroxylamine to sterilized soils (pH 5.6 to 7.8) treated with nitrite had little effect on the rate of N_2O evolution over 24 hours, compared to soils where only hydroxylamine was added. They concluded that if N_2O is produced in soil by chemical decomposition of hydroxylamine, then reaction of hydroxylamine with nitrous acid is not the pathway.

Results presented by Bremner et al. (1980) apparently conflict with those of Nelson (1967). However, the experiment reported by Bremner et al. (1980) did not show conclusively that nitrite and hydroxylamine were not reacting to form N_2O. Residual nitrite was not measured and the origin of the 2 atoms of N per molecule of N_2O evolved was not established using $^{15}NO_2^-$ or $^{15}NH_2OH$. If measurements had been taken after 48 hours, they might have shown substantially more production of N_2O from the hydroxylamine + nitrite treatment than from the hydroxylamine treatment. Although it appears that hydroxylamine and nitrite can react readily in acidic and neutral soils to produce N_2O, there is no evidence that this reaction is important in chemo-denitrification in soil, or that other hydroxylamine decomposition reactions occur in soil to produce N_2O and N_2. Hydroxylamine has not been detected in research on N transformations in soils, and although it has been established that it is an intermediate in the autotrophic oxidation of ammonium to nitrite, extracellular release of hydroxylamine has not been demonstrated (Bremner et al. 1980).

3.4.3. Reaction of nitrous acid with ammonium

Allison (1963) suggested that significant gaseous loss of N could occur from soils by reaction of ammonium with nitrite to form N_2. Ewing and Bauer (1966) have represented the reaction as

$$HNO_2 + NH_4^+ \rightleftharpoons N_2 + 2H_2O + H^+.$$

The reaction is often described as ammonium nitrite decomposition. Several authors (Gerretsen and de Hoop 1957, Clark 1962, Sabbe and Reed 1964) have noted the similarity between this reaction and the Van Slyke reaction. However, Allison (1963) felt that there were distinct differences, which warranted a different name.

The reaction of nitrite and ammonium has been studied in buffer solutions under a normal atmosphere (Gerretsen and de Hoop 1957, Nelson 1967) or in unbuffered solutions under an atmosphere of NO (Smith and Clark 1960). The NO atmosphere was used to prevent self-decomposition of nitrous acid. Data on recovery of nitrite after 2 days showed little loss at pH 6 or greater, but losses occurred at pH 4 and 5 (Nelson 1967). It was not possible to determine the loss mechanism since data on ammonium recovery and gaseous N losses were not presented. Rapid evolution of N_2 was measured in acid solutions (pH 4 to 5.6) containing ammonium and nitrite, under an atmosphere of NO, but the rate of loss at pH 7 or greater was very slow (Smith and Clark 1960). Increasing loss of NO was reported in solutions containing ammonium and nitrite as pH decreased from 5.5 to 4 (Gerretsen and de Hoop 1957). Self-decomposition of nitrous acid was the probable origin of the NO, although Allison (1965) has suggested that NO could arise from ammonium nitrite decomposition, e.g.

$$3NH_4NO_2 \rightleftharpoons NH_4NO_3 + 2NO + 2NH_3 + H_2O.$$

It therefore appears that reaction of nitrite and ammonium is most likely to occur in soil under acidic conditions (Allison 1963, Ewing and Bauer 1966), but there is little evidence to support its occurrence. Nelson (1967) could not detect any reaction between ammonium and nitrite when moist, acidic or alkaline soils were incubated for several days. Data obtained by Smith and Clark (1960) also suggested that loss of N_2 in an acid soil treated with ammonium and nitrite did not arise from ammonium nitrite decomposition, but from reduction of nitrite by some other agent. Similarly, tracer studies reported by Jones (1951) and Nömmik (1956) showed that N_2 evolved from soils treated with ^{15}N-labelled ammonium and nitrite was not formed by reaction of ammonium and nitrite.

Ewing and Bauer (1966) concluded that N_2 could be formed in alkaline soils through ammonium nitrite deomposition, provided the concentrations of ammonium and nitrite were high enough. In an alkaline-hydrolysing fertilizer band, both ions may be present in high concentrations (Chalk *et al.* 1975). Consideration of reaction kinetics indicated that losses would be greater in dry soil compared to moist soil for equivalent amounts of added ammonium and nitrite and that ammonium nitrite decomposition could occur in dry soil up to pH 8 or 9 (Ewing and Bauer 1966). There is good evidence that ammonium nitrite decomposition can occur in neutral or alkaline soils which are air dried following ammonium and nitrite addition (Wahhab and Uddin 1954, Nelson 1967). Nelson (1967) confirmed ammonium nitrite decomposition through measurement of initial and residual ammonium and nitrite, and measurement of the ^{15}N-enrichment of evolved N_2 when labelled ammonium and unlabelled nitrite were added. Decomposition of ammonium nitrite during air drying was promoted by increase in soil pH and reactant concentration, being most evident in light textured alkaline soils (Nelson 1967). Decomposition during air-drying did not occur at pH values less than 7, but was quite extensive at pH 7 or greater, and occurred when reactant concentration was only 10 ppm nitrite N and 50 ppm ammonium N.

3.4.4. Reaction of nitrous acid with soil organic matter

Organic compounds other than amino compounds may participate in reactions with nitrite which result in gaseous-N evolution. Bremner (1957) found that N_2 and/or N_2O were formed when humic acid and lignin preparations were treated with nitrite in acid medium, and that nitrite N became organically bound or 'fixed'. Clark and Beard (1960) showed that removal of soil organic matter with hydrogen peroxide reduced loss of added nitrite, while addition of organic materials increased loss. Smith and Clark (1960) also noted that loss of N_2 was higher in a soil high in organic matter compared to a soil low in organic matter. Direct evidence for the reaction of nitrite and soil organic matter was obtained by Bremner and Fuhr (1966). When ^{15}N-labelled sodium nitrite was added to acid and neutral soils, part of the N was fixed by the organic matter (10–28%), and part was converted to gaseous forms of N (33–79%). Smith and Chalk (1980b) found a significant positive correlation between ^{15}N fixed and ^{15}N not recovered in 5 soils treated with nitrite and sampled after 1, 2 and 3 days of incubation. The ^{15}N not recovered was presumed to be lost through evolution of N_2 and N_2O, and possibly CH_3ONO.

Studies of factors affecting chemical transformations of nitrite in soils (Bremner and Fuhr 1966, Nelson and Bremner 1969) have shown that the extent of decomposition and fixation of nitrite was inversely related to pH, but some fixation did occur in neutral and alkaline soils. With increasing soil organic matter content, nitrite fixation and decomposition increased when soil pH was between 5 and 7, and for soil pH less than 5, fixation increased but the amount of N volatilized decreased (Nelson and Bremner 1969). The rate of decomposition and fixation of nitrite increased with increasing nitrite concentration, and air-drying increased decomposition but did not markedly affect fixation (Nelson and Bremner 1969).

The nitrite which is fixed by soil organic matter is resistant to decomposition by soil micro-organisms (Bremner 1957, Bremner and Fuhr 1966, Smith and Chalk 1979c). Indirect evidence for resistance was provided by acid hydrolysis studies (Bremner 1957, Bremner and Fuhr 1966). Only 50 to 60% of nitrite N fixed by soils was released by boiling with 6N hydrochloric acid for 12 hours. Tracer studies have provided direct evidence for resistance to decomposition of fixed nitrite N (Smith and Chalk 1979c). Only 20% of fixed N was recovered as inorganic-N after 70 days of incubation, or was recovered in the roots and 3 harvests of ryegrass grown for 197 days. Fixed N was more available than indigenous N ($<10\%$ recovered), but with increasing time fixed N became more resistant to mineralization. Bremner (1968) has suggested that the application of high rates of NH_3 or NH_3-yielding fertilizers could lead to accumulation of fixed nitrite N in soil, because of resistance to mineralization.

Studies of the mechanism of nitrite fixation in soils have been limited by inadequate knowledge of the structure of soil organic matter (Bremner and Fuhr 1966). Experiments using model substances, which are believed to exist in soil organic matter, have been performed to determine their fixation capacity and ability to decompose nitrite (Bremner 1957, Stevenson and Swaby 1964, Bremner and Fuhr 1966, Nelson 1967, Porter 1969, Stevenson *et al.* 1970). Studies have been conducted in buffered media, usually under acidic conditions at pH 5 or 6 (Nelson 1967, Porter 1969, Stevenson *et al.* 1970). Bremner (1968) concluded that aromatic compounds with phenolic hydroxyl groups were largely, if not entirely, responsible for fixation of nitrite and formation of N_2 and N_2O. Plant lignin and soil organic matter contain aromatic compounds with phenolic hydroxyl groups (Bremner and Fuhr 1966, Bremner 1968). Humic acids contain lignin-like building units, and are able to fix nitrite (Bremner 1957).

The gases N_2, N_2O and NO have all been detected following the reaction of nitrite with soil organic matter (Bremner 1957, Reuss and Smith 1965, Bremner and Fuhr 1966, Nelson and Bremner 1970b, Steen and Stojanovic 1971, Cawse and Cornfield 1972, Van Cleemput *et al.* 1976). Another nitrogenous gas, methyl nitrite (CH_3ONO), has also been identified after the reaction of nitrous acid and aromatic methoxyl groups in lignin preparations (Stevenson and Kirkman 1964, Stevenson and Swaby 1964, Edwards and Bremner 1966). The formation of CH_3ONO provides an explanation to the finding that the reactions of lignin and humic acid preparations with nitrite lead to a reduction of the methoxyl group content of the preparations (Bremner 1957, Steen and Stojanovic 1971). The following reaction of nitrous acid and phenolic ethers is probably involved in CH_3ONO formation (Sobolev 1961).

Methyl nitrite evolution from soil has not been reported. Stevenson *et al.* (1970) also failed to detect CH_3ONO evolution from the reaction of nitrite with humic acids, fulvic acids, lignin and aromatic substances when the pH was 6 or greater, although Steen and Stojanovic (1971) reported a decrease in the methoxyl group content of lignin and humic acid preparations over the pH range 6.5 to 8. Methyl nitrite is unstable and undergoes atmospheric photo-dissociation. The lifetime of CH_3ONO is nearly independent of altitude and is approximately 2 minutes (Taylor *et al.* 1980). The major products of the photolysis of methyl nitrite are CH_2O, N_2O and H_2O, but N_2 may also be formed (Wiebe and Heicklen 1973).

The mechanisms involved in the formation of N_2, N_2O and NO by reaction of nitrous acid with phenolic constituents are only partially understood (Stevenson *et al.* 1970). The reactions are known as nitrosation reactions. Nitrosation denotes the addition of the nitroso group ($-N=0$) to an organic molecule by reaction with nitrous acid and with other compounds which form species of the type $O=N-X$ (Austin 1961). Nitrosation leads to the formation of nitroso ($C-N=0$) and oximino ($C=NOH$) compounds, e.g.

The oximino compounds subsequently react with excess nitrous acid to form N_2O (Austin 1961), e.g.

$$\diagdown C = NOH + HNO_2 \rightarrow \diagdown C = O + N_2O + H_2O.$$

There is also evidence that diazonium ($-N=N^+$) compounds are formed from the nitroso derivatives, and that diazonium compounds decompose at room temperature with the formation of N_2 (Philpot and Small 1938), e.g.

Evidence has been obtained that nitrous acid may oxidize aromatic rings, which results in ring cleavage, production of gaseous N and oxidation of some ring carbon to CO_2 (Austin 1961, Stevenson and Swaby 1964, Nelson 1967, Porter 1969, Stevenson *et al.* 1970), e.g.

Several mechanisms have been proposed for the formation of NO by nitrosation reactions. Oxidation of aromatic rings may lead to formation of unsaturated hydroxy compounds (reductones), which can react with nitrous acid (Austin 1961), e.g.

$$\begin{array}{ccc} \text{HO} & \text{OH} & \quad\quad \text{O} \quad\;\; \text{O} \\ | & | & \quad\quad || \quad\;\; || \\ -\,\text{C} = \text{C} + 2\;\text{HNO}_2 & \rightarrow & -\,\text{C} - \text{C} - \; + 2\;\text{H}_2\text{O} + 2\;\text{NO} \end{array}$$

Under certain conditions, hyponitrous acid ($\text{HO} - \text{N} = \text{N} - \text{OH}$) produced by the action of nitrous acid on oximes can be oxidized to NO (Austin 1961).

It is probable that several mechanisms are involved in the production of gaseous N through reaction of nitrous acid and soil organic matter. However, it should be noted that the only reactions which have been proposed to account for nitrite fixation are those which involve formation of nitroso or oximino compounds. Smith and Chalk (1980b) found that the ratio of nitrite N fixed: nitrite N lost was approximately $1:2$ for 4 soils, but the ratio was appreciably higher for another soil. This suggested that different mechanisms were operating. Also, the ^{15}N enrichment of evolved N_2 in all soils was approximately one-half of the ^{15}N enrichment of the added nitrite. These data do not support the hypothesis of N_2 formation through decomposition of nitrosophenols formed by reaction of added nitrous acid and phenolic constituents, which would involve no isotope dilution. Isotope dilution would only occur if added ^{15}N-labelled nitrous acid reacted with indigenous nitrosophenols or other compounds containing N. The approximately equal contribution of labelled and unlabelled sources of N to the N_2 evolved, suggested that a reaction of the Van Slyke-type may have been partly responsible (Smith and Chalk 1980b). Stevenson et al. (1970) also suggested that some N_2 was generated from the soil organic matter through the reaction of nitrite with amino compounds, because a higher proportion of the gas recovered occurred as N_2 with soils than with lignins and model phenolic compounds.

3.4.5. Reduction of nitrite by metallic ions

Wullstein and Gilmour (1964, 1966) reported that transition metals such as Fe, Cu and Mn in the reduced state are important in the non-enzymatic reduction of nitrite and the formation of NO and N_2. Evidence in support of this hypothesis was based on the observation that less NO was produced when a clay or soil was extracted with sodium chloride before addition of nitrite. The sodium chloride extracts from soils containing metal ions also promoted nitrite decomposition. A general equation was proposed (Wullstein 1967):

$$\text{M}^{++} + \text{NO}_2^- + 2\text{H}^+ \rightleftharpoons \text{M}^{+++} + \text{NO} + \text{H}_2\text{O}.$$

Nelson and Bremner (1970a) showed that only ferrous, cuprous and stannous ions promoted nitrite decomposition in acid media, but concluded that well-aerated soils do not contain sufficient amounts of these cations to be significant in chemodenitrification. These results confirmed calculations made by Chao and Kroontje (1963) of standard electrode potentials of various cation-nitrous acid couples, which indicated that ferrous, cuprous and stannous ions can reduce nitrite in acid media to form NO and N_2O, but not manganous ion as proposed by Wullstein and Gilmour (1964).

Under anaerobic conditions, reduction of nitrite by reduced metal ions may occur due to higher concentrations. Van Cleemput et al. (1976) have suggested that reaction of manganous and ferrous ions with nitrite may be responsible for N_2, N_2O and NO formation under anaerobic conditions. Chao and Kroontje (1963, 1966) found that nitrite could be readily reduced by ferrous ion to NO at pH 5, but at pH 6 or higher, the reaction stopped. A small amount of N_2O was also produced by the reaction. Moraghan and Buresh (1977) reported that chemical decomposition of nitrite occurred anaerobically in a high ferrous ion environment. Nitrous oxide, N_2 and ammonium were produced, N_2O being the principal product. Nitrous oxide was stable in the ferrous ion medium at pH 6, but at pH 8, N_2O was reduced to N_2 in the presence of cuprous ion.

3.5 Gaseous nitrogen loss during nitrification

Numerous papers (Gerretsen and de Hoop 1957, Broadbent et al. 1958, Soulides and Clark 1958, Wagner and Smith 1958, Carter and Allison 1960, 1961, Clark and Beard 1960, Clark et al. 1960, Cady and Bartholomew 1961, Schwartzbeck et al. 1961, Meek and McKenzie 1965, Hauck and Stephenson 1965, Steen and Stojanovic 1971, Bundy and Bremner 1974) have reported losses of N following the nitrification of NH_3 or NH_3-producing fertilizers added to soils in the laboratory. Nitrogen deficits exceeding 25% of the applied N were noted in poorly buffered soils (Soulides and Clark 1958, Wagner and Smith 1958, Clark et al. 1960, Carter and Allison 1961). In these experiments NH_3 or NH_3-producing fertilizers were applied at concentrations equivalent to those measured in fertilizer bands. Nitrite was found to accumulate during nitrification. Soulides and Clark (1958) and Clark et al. (1960) observed that the largest N-deficit occurred on initially alkaline soils which became acidic during nitrification. The deficit was attributed to chemodenitrification.

Previous work indicated that urea was particularly susceptible to loss in soils that tended to accumulate nitrite under aerobic conditions (Soulides and Clark 1958, Clark et al. 1960, Hauck and Stephenson 1965). Bundy and Bremner (1974) reported significant N deficits in soils that accumulated nitrite after incubation with urea. However the N deficits were not reduced by the addition of the nitrification inhibitor, nitrapyrin, even though nitrite accumulation was prevented. The conclusion drawn from these observations was that the main N deficit observed was not due to gaseous loss of urea N through chemodenitrification, which should have been markedly reduced in the presence of nitrapyrin (Bremner 1977). It was shown that part of the deficit was due to volatilization of NH_3 and fixation of ammonium formed by urea hydrolysis. The results illustrate the need for direct measurments of gaseous losses when evaluating the significance of chemical denitrification.

There have been few attempts to determine the amounts of the various gases evolved during nitrification in laboratory samples. Usually only one or two N gases have been determined quantitatively in any one study. Small losses of NO and/or NO_2 have been measured (Wagner and Smith 1958, Bundy and Bremner 1974). Losses amounted to less than 1.5% of urea applied at 400–500 μg N/g of soil. Similarly, Steen and Stojanovic (1971) measured NO losses of 5.3% of urea and 1.3% of ammonium sulfate applied to calcareous soils at 300 μg N/g soil. Evolution of N_2 and N_2O occurred during nitrification of ammonium sulfate added to calcareous soil (Meek and MacKenzie 1965). The total loss was 1.3% of the applied-N, N_2 being the major component. Small amounts of ^{15}N-labelled N_2 (less than 0.3% of applied $^{15}NH_4Cl$ or $^{15}NH_4NO_3$) and traces of $NO + NO_2$ were detected during incubation of soils at field capacity (Schwartzbeck *et al.* 1961), while Wagner and Smith (1958) reported qualitative evidence of N_2O evolution during incubation of urea-treated soils. Laboratory studies reported by Bremner and Blackmer (1978) showed small losses of N_2O (<1% of added N) in soils fertilized with urea or ammonium sulfate.

Recently, Smith and Chalk (1980a) measured significant gaseous losses of N during nitrification in soils treated with NH_3 fertilizer. Evolution of N_2, N_2O and $NO + NO_2$ occurred in 3 soils, when measured soil pH was greater than 7.5. The highest loss amounted to 16.5% of the applied N in a calcareous soil incubated for 28 days at $30°C$. Dinitrogen was the major gaseous form of N evolved in all soils, but a substantial amount of N_2O (6% of applied N) was evolved from the calcareous soil. Only small amounts of $NO + NO_2$ were recovered. Smith and Chalk (1980a) concluded that the gaseous N losses resulted from chemodenitrification. Evidence for loss via chemical pathways involving nitrite was obtained from studies of the effect of nitrapyrin on losses. Addition of the nitrification inhibitor, nitrapyrin, with the NH_3 fertilizer prevented loss of N_2O and markedly reduced losses of N_2 and $NO + NO_2$. The addition of nitrapyrin completely prevented accumulation of nitrite, but did not prevent nitrification as shown by the decrease in exchangeable ammonium and increase in nitrate during incubation. Bremner and Blackmer (1978) have also shown that nitrapyrin is effective in reducing emission of N_2O from well-aerated soils treated with ammonium sulfate or urea.

Further evidence for gaseous N losses by chemodenitrification (Smith and

Chalk 1980a) was obtained by comparison of the pattern of loss from NH_3-treated soils with the pattern of loss from nitrite treated soils. Although the rates of loss were less in the NH_3 fertilized soils, the patterns of loss were very similar. In a calcareous soil, N_2O was a major component of the N evolved after both nitrite and NH_3 treatments. In other similarly treated soils, N_2 was the major component; $NO + NO_2$ was the minor component for all soils. Studies on the effect of pH on gaseous N evolution from soils treated with nitrite showed that loss of N_2 and $NO + NO_2$ was inversely related to pH, but N_2 and N_2O were evolved at appreciable rates under alkaline conditions (Smith and Chalk 1980a).

Acid soils have been used in many studies of nitrite decomposition (Smith and Clark 1960, Reuss and Smith 1965, Nelson and Bremner 1970b). Because of the effect of pH on gaseous N loss, results of such studies may give a misleading indication of possible gaseous loss during nitrification under alkaline conditions. For example, Nelson and Bremner (1970b) found substantial loss of nitrite, NO_2 and N_2, but negligible loss of N_2O, in several acid soils (NO_2 loss equalled or exceeded N_2 loss, but the proportion of N_2 increased as the soil pH increased). Because of the apparent importance of NO_2 as a product of nitrite reactions, Bundy and Bremner (1974) used it as an indication of chemodenitrification during urea transformations in soil. When little NO_2 loss occurred, it was concluded that very little loss of urea-N occurred via chemodenitrification. The work of Smith and Chalk (1980a) demonstrated that it is essential to measure all forms of gaseous N evolved before valid conclusions can be drawn.

Loss of N_2O has been reported in field studies involving unfertilized soils (Freney et al. 1978, Denmead et al. 1979) and soils fertilized with ammonium nitrate (McKenney et al. 1978) or ammonium sulfate and urea (Ryden et al. 1978, Breitenbeck et al. 1980) or NH_3 (Hutchinson and Mosier 1979, Bremner et al. 1981, Cochran et al. 1981, Mosier and Hutchinson 1981) when soil moisture was at or below field capacity. Emission of NO has also been detected in unfertilized field soil (Galbally and Roy 1978). Denmead et al. (1979) measured N_2O emissions in a pasture at very low soil moisture. The production of N_2O was accompanied by a net increase in soil nitrate. Emission of N_2O at low soil moisture has been confirmed in laboratory experiments (Freney et al. 1978, Freney et al. 1979). Addition of the nitrification inhibitor, carbon disulphide, completely blocked the oxidation of ammonium to nitrate, and reduced the rate of N_2O production by half (Freney et al. 1979). These results suggested that an appreciable amount of N_2O was being produced during nitrification. Bremner and Blackmer (1979) also reported that addition of acetylene, which inhibits nitrification and the microbial reduction of N_2O to N_2, prevented emissions of N_2O in ammonium fertilized soils near field capacity. They concluded that most, if not all, of the N_2O evolved was produced by nitrifying organisms. However, Freney *et al.* (1979) concluded that although a considerable part of the N_2O was produced by the oxidation of ammonia, biological denitrification could not be ruled out. For example, addition of azide, which inhibits nitrification and the microbial reduction of N_2O to N_2, enhanced emission of N_2O.

Losses of N_2O from soils fertilized with urea or $(NH_4)_2SO_4$ in the field have generally been small. Ryden *et al.* (1978) reported that loss of N_2O-N during 76 hours amounted to 0.6% of the N applied as ammonium sulfate. Breitenbeck *et al.* (1980) determined that fertilizer-induced emissions of N_2O-N observed in 96 days from plots treated with 250 kg N/ha as ammonium sulfate or urea represented <0.2% of the applied-N. Emission of N_2O from plots treated with calcium nitrate were no higher than emissions from unfertilized plots, indicating that the N_2O was not derived from biological denitrification. McKenney *et al.* (1978) estimated that the loss of N_2O-N from band-applied ammonium nitrate may have amounted to only 0.05% of the fertilizer-N applied at 224 kg N/ha. Recently, Bremner *et al.* (1981) measured significant losses of N_2O-N in 96 days (4 to 6.8% of applied-N) in plots fertilized with 250 kg anhydrous NH_3-N/ha. Bremner *et al.* (1981) drew attention to the much higher losses of N_2O from NH_3-fertilized soil compared to similar studies (Breitenbeck *et al.* 1980) involving soils fertilized with ammonium sulfate or urea. It is significant that the urea and ammonium sulfate were broadcast over plots by dissolving in water and sprinkling on, whereas the NH_3 was injected at a single point at 20 cm depth in the centre of small, 50 cm-diameter plots. Other workers have reported smaller losses of N_2O from soils when anhydrous NH_3 was band-applied. Mosier and Hutchinson (1981) measured a 1.3% loss over 4 months (200 kg NH_3-N/ha), while Cochran *et al.* (1981) found that less than 0.1% of the applied-N (220 kg NH_3-N/ha) was emitted as N_2O over 35 days.

Several authors have suggested that N_2O losses measured in unfertilized soils or soils fertilized with ammonium or ammonium-producing fertilizers may have been produced by the nitrifying organisms during oxidation of ammonium to nitrate (Freney *et al.* 1978, 1979, Bremner and Blackmer 1979, Breitenbeck *et al.* 1980, Bremner *et al.* 1981). Evolution of N_2O has been measured under certain conditions with intact cells or cell-free extracts of *Nitrosomonas europaea* (Hooper 1968, Yoshida and Alexander 1970, Ritchie and Nicholas 1972). However, it has not been suggested that losses of N_2 and $NO + NO_2$ during nitrification are produced by the nitrifying bacteria. There is evidence that N_2 and $NO + NO_2$ produced in N-fertilized soils during nitrification is the result of chemodenitrification reactions, and that N_2O produced could also be a product of such reactions.

3.6. Conclusions

A considerable number of laboratory experiments have shown that nitrite added to sterile soils may undergo rapid chemical reaction. Part of the added nitrite becomes organically bound or fixed, while some N is concurrently converted to N_2, N_2O and $NO + NO_2$. There is indirect evidence that another nitrogenous gas, CH_3ONO, may be formed. The gaseous N products, N_2 and N_2O, are readily evolved from soil, and there is evidence that small concentrations of NO and NO_2 may also be evolved. There is evidently a need for direct measurement of CH_3ONO evolution, and for studies on the reactions of CH_3ONO in soil.

Chemical reactions involving nitrite are dependent on pH and concentration. The extent of nitrite decomposition and fixation is inversely related to soil pH, but several studies have demonstrated that nitrite is reactive in alkaline soils. With increasing nitrite concentration, the amounts fixed and decomposed increase, but the percentages of the added N which are fixed and decomposed decrease (Nelson and Bremner 1969). The available data suggest that nitrite is reactive even at low concentrations.

Several laboratory studies have shown that losses of N_2, N_2O and $NO + NO_2$ occur during nitrification in soils amended with alkaline-hydrolysing fertilizers such as urea and ammonia, or in calcareous soils fertilized with ammonium sulfate. Losses have usually been small ($<2\%$ of applied-N), but recently Smith and Chalk (1980a) measured significant losses of N_2 (10% of applied NH_3-N) and N_2O (6% of applied NH_3-N) in a calcareous soil. There were marked differences between soils in the amounts and proportions of N_2, N_2O and $NO + NO_2$ evolved, but losses occurred when the measured soil pH was 7.5 or higher. It is clearly desirable that studies be carried out to identify the factors which influence these gaseous N losses.

The results of laboratory incubation studies indicate that significant gaseous N losses can occur by chemodenitrification when the competitive biological oxidation of nitrite by Nitrobacter is inhibited by an alkaline-hydrolysing fertilizer. A result of the inhibition of Nitrobacter is nitrite accumulation, which occurs when soil pH is between 7 and 8.5. Nitrite decomposition may occur in soil over this pH range, particularly at high concentrations of nitrite. It has generally been assumed or inferred that nitrite accumulation is a necessary prerequisite for chemodenitrification (Meek and MacKenzie 1965, Bundy and Bremner 1974). Soulides and Clark (1958) found a direct relationship between nitrite accumulation and the magnitude of the N deficit, while Wagner and Smith (1958) and Steen and Stojanovic (1971) found that the period of rapid $NO + NO_2$ evolution coincided with the period of maximum nitrite accumulation. However results obtained by Smith and Chalk (1980a) showed that substantial loss of N_2 could occur without nitrite accumulation.

Although a considerable amount of research has been carried out into the nature of nitrate reactions in solutions using model compounds, there is no general agreement on the mechanisms of chemodenitrification in soils. The complexity of chemical denitrification processes is emphasized by the array of gaseous N products, which suggests that a series of reactions may take place. All of the reactions which have been proposed involve nitrous acid. Chemo-denitrification is therefore favoured by acid conditions in soil. Since chemo-denitrification is also known to occur in soil when the measured pH is alkaline, the reactions must take place at sites where the pH is lower than that measured.

Several authors have concluded that NO is formed in soil through the self-decomposition of nitrous acid. However there is also evidence that NO could be formed by nitrosation reactions, and that it could be produced in anaerobic environments by reaction of nitrite with reduced metallic ions. It appears that nitrosation reactions may be important in N_2O formation. Formation of N_2 may also result from nitrosation reactions, but there is evidence that N_2 may also be formed by reaction of nitrous acid with amino compounds and ammonium. Further progress in evaluating the relative importance of chemo-denitrification mechanisms could be made by ^{15}N tracer studies. Measurement of the ^{15}N abundance of evolved nitrogenous gases in $^{15}NO_2^-$ treated soils is required, and studies are needed to identify the factors which may influence the labelling of the gaseous products.

Although laboratory experiments have shown that significant gaseous N losses may occur in small samples treated with nitrite or NH_3 fertilizer, the significance of gaseous N losses during nitrification of band-applied N fertilizers in the field, has not been established. Measurement of N_2O, NO and NO_x losses from N-fertilized soils in the field are required, and studies are needed to identify the factors which may influence losses. Measurement of N_2 loss in the field would be technically very difficult, but desirable, since laboratory studies have shown that N_2 losses exceed N_2O and $NO + NO_2$ losses. Field studies should be complemented by laboratory studies on gaseous N losses. Studies should also be conducted with larger samples in pots or cylinders to confirm results obtained in both laboratory and field experiments. Losses which were measured directly in laboratory experiments (Smith and Chalk 1980a), were not measured by N balance in a glasshouse pot experiment (Smith and Chalk 1980c). In addition to gaseous N losses, the low availability of nitrite fixed by soil organic matter may further reduce the efficiency of NH_3 or NH_3-forming fertilizers. Further studies are required to define more precisely the relationships between nitrite fixation and gaseous N loss, and whether fixed nitrite influences gaseous N loss from soil.

There is evidence that nitrification inhibitors may significantly reduce gaseous N losses during nitrification, by reducing the rate of oxidation of ammonium to nitrite and preventing nitrite accumulation. This is a new

84

practical aspect of the use of nitrification inhibitors. Previously they have been proposed as a means of inhibiting nitrate formation, in order to reduce leaching and biological denitrification losses. The prevention of nitrite accumulation would also prevent possible nitrite phytotoxicity problems (Court *et al.* 1962), but the inhibitor itself may be phytotoxic. Hauck and Bremner (1969) have drawn attention to this and other possible undesirable effects associated with the use of nitrification inhibitors. Increased ammonium fixation and NH_3 volatilization and fixation may occur. It is obvious that more research is needed on the use of nitrification inhibitors to improve N fertilizer efficiency.

3.7 References

Alexander, M. 1965 Nitrification. In: Bartholomew, W.V. and Clark, F.E. (eds.), Soil Nitrogen, pp. 307–343. American Society of Agronomy, Madison.

Allison, F.E. 1963 Losses of gaseous nitrogen from soils by chemical mechanisms involving nitrous acid and nitrites. Soil Sci. 96, 404–409.

Allison, F.E. 1965 Evaluation of incoming and outgoing processes that affect soil nitrogen. In: Bartholomew, W.V. and Clark, F.E. (eds.), Soil Nitrogen, pp. 573–606. American Society of Agronomy, Madison.

Allison, F.E. 1966 Fate of nitrogen applied to soils. Adv. Agron. 18, 219–258.

Allison, F.E. and Doetsch, J. 1950 Nitrogen gas production by the reaction of nitrites with amino acids in slightly acidic media. Soil Sci. Soc. Am. Proc. 15, 163–167.

Allison, F.E., Doetsch, J.H. and Sterling, L.D. 1952 Nitrogen gas formation by interaction of nitrites and amino acids. Soil Sci. 74, 311–314.

Arnold, P.W. 1954 Losses of nitrous oxide from soils. J. Soil Sci. 5, 116–126.

Austin, A.T. 1961 Nitrosation in organic chemistry. Sci. Prog. XLIX, 619–640.

Aylward, G.H. and Findlay, T.J.V. 1971 S.I. Chemical Data. John Wiley and Sons, Sydney. 112 pp.

Bezdicek, D.F., MacGregor, J.M. and Martin, W.P. 1971 The influence of soil-fertilizer geometry on nitrification and nitrite accumulation. Soil Sci. Soc. Am. Proc. 35, 997–1000.

Blue, W.G. and Eno, C.F. 1954 Distribution and retention of anhydrous ammonia in sandy soil. Soil Sci. Soc. Am. Proc. 18, 420–424.

Breitenbeck, G.A., Blackmer, A.M. and Bremner, J.M. 1980 Effects of different nitrogen fertilizers on emission of nitrous oxide from soil. Geophys. Res. Lett. 7, 85–88.

Bremner, J.M. 1957 Studies on soil humic acids. II. Observations on the estimation of free amino groups. Reactions of humic acid and lignin preparations with nitrous acid. J. Agric. Sci. 48, 352–360.

Bremner, J.M. 1968 The nitrogenous constituents of soil organic matter and their role in soil fertility. In: Organic Matter and Soil Fertility, pp 143–185. Pontificia Academia Scientarium, Rome.

Bremner, J.M. 1977 Role of organic matter in volatilization of sulphur and nitrogen from soils. In: Soil Organic Matter Studies, Vol. 2, pp. 229–239. IAEA, Vienna.

Bremner, J.M. and Blackmer, A.M. 1978 Nitrous oxide: Emission from soils during nitrification of fertilizer nitrogen. Science (Wash. D.C.) 199, 295–296.

Bremner, J.M. and Blackmer, A.M. 1979 Effects of acetylene and soil water on emissions of nitrous oxide from soils. Nature (Lond.) 280, 380–381.

Bremner, J.M. and Fuhr, F. 1966. Tracer studies of the reaction of soil organic matter with nitrite. In: The Use of Isotopes in Soil Organic Matter Studies, pp. 337–346. Pergamon Press, Oxford.

Bremner, J.M. and Nelson, D.R. 1968 Chemical decomposition of nitrite in soils. Trans. 9th Int. Congr. Soil Sci. 2, 495–503.

Bremner, J.M., Blackmer, A.M. and Waring, S.A. 1980 Formation of nitrous oxide and dinitrogen by chemical decomposition of hydroxylamine in soils. Soil Biol. Biochem. 12, 263–269.

Bremner, J.M., Breitenbeck, G.A. and Blackmer, A.M. 1981 Effect of anhydrous ammonia fertilization and emission of nitrous oxide from soils. J. Environ. Qual. 10, 77–80.

Broadbent, F.E. and Clark, F. 1965 Denitrification. In: Bartholomew, W.V. and Clark, F. (eds.), Soil Nitrogen, pp. 344–359. American Society of Agronomy, Madison.

Broadbent, F.E. and Stevenson, F.J. 1966 Organic matter interactions. In: McVickar, M.H., Martin, W.P., Miles, I.E. and Tucker, H.H. (eds.), Agricultural Anhydrous Ammonia, pp. 169–187. Agricultural Ammonia Institute, Memphis.

Broadbent, F.E., Hill, G.N. and Tyler, K.B. 1958 Transformation and movement of urea in soils. Soil Sci. Soc. Am. Proc. 22, 303–307.

Broadbent, F.E., Tyler, K.B. and Hill, G.N. 1957 Nitrification of ammoniacal fertilizers in some California soils. Hilgardia 27, 247–267.

Bundy, L.G. and Bremner, J.M. 1973 Determination of ammonium-N and nitrate-N in acid permanganate solution used to absorb ammonia, nitric oxide and nitrogen dioxide evolved from soils. Commun. Soil Sci. Plant. Anal. 4, 179–184.

Bundy, L.G. and Bremner, J.M. 1974 Effects of nitrification inhibitors on transformations of urea nitrogen in soils. Soil Biol. Biochem. 6, 369–375.

Cady, F.B. and Bartholomew, W.V. 1960 Sequential products of anaerobic denitrification in Norfolk soil material. Soil Sci. Soc. Am. Proc. 24, 477–482.

Cady, F.B. and Bartholomew, W.V. 1961 Influence of low pO_2 on denitrification processes and products. Soil Sci. Soc. Am. Proc. 25, 362–365.

Cady, F.B. and Bartholomew, W.V. 1963 Investigations of nitric oxide reactions in soils. Soil Sci. Soc. Am. Proc. 27, 546–549.

Carter, J.N. and Allison, F.E. 1960 Investigations on denitrification in well-aerated soils. Soil Sci. 90, 173–177.

Carter, J.N. and Allison, F.E. 1961 The effect of rates of application of ammonium sulphate on gaseous losses of nitrogen from soils. Soil Sci. Soc. Am. Proc. 25, 484–486.

Cawse, P.A. and Cornfield, A.H. 1972 Biological and chemical reduction of nitrate to nitrite in γ-irradiated soils and factors leading to eventual loss of nitrite. Soil Biol. Biochem. 4, 497–511.

Chalk, P.M., Keeney, D.R. and Walsh, L.M. 1975 Crop recovery and nitrification of fall and spring applied anhydrous ammonia. Agron. J. 67, 33–37.

Chao, T.T. and Kroontje, W. 1963 Inorganic nitrogen oxidation in relation to associated changes in free energy. Soil Sci. Soc. Am. Proc. 27, 44–47.

Chao, T.T. and Kroontje, W. 1966 Inorganic nitrogen transformations through the oxidation and reduction of iron. Soil Sci. Soc. Am. Proc. 30, 193–196.

Clark, F.E. 1962 Losses of nitrogen accompanying nitrification. Trans. Int. Soc. Soil Sci., Comm. IV and V, pp. 173–176.

Clark, F.E. and Beard, W.E. 1960 Influence of organic matter on volatile loss of nitrogen from soil. Trans. 7th Int. Congr. Soil Sci. 2, 501–508.

Clark, F.E., Beard, W.E. and Smith, D.H. 1960 Dissimilar nitrifying capacities of soils in relation to losses of applied nitrogen. Soil Sci. Soc. Am. Proc. 24, 50–54.

Cochran, V.L., Elliott, L.F. and Papendick, R.I. 1981 Nitrous oxide emissions from a fallow field fertilized with anhydrous ammonia. Soil Sci. Soc. Am. J. 45: 307–310.

Cooper, G.S. and Smith, R.L. 1963 Sequence of products formed during denitrification in some diverse western soils. Soil Sci. Soc. Am. Proc. 27, 659–662.

Court, M.N., Stephen, R.C. and Waid, J.S. 1962 Nitrite toxicity arising from the use of urea as a fertilizer. Nature (Lond.) 194, 1263–1265.

Denmead, O.T., Freney, J.R. and Simpson, J.R. 1979 Studies of nitrous oxide emission from a grass sward. Soil Sci. Soc. Am. J. 43, 726–728.

Edwards, A.P. and Bremner, J.M. 1966 Formation of methyl nitrite in reaction of lignin with nitrous acid. In: The Use of Isotopes in Soil Organic Matter Studies, pp. 347–348. Pergamon Press, Oxford.

Ewing, G.J. and Bauer, N. 1966 An evaluation of nitrogen losses from the soil due to the reaction of ammonium ions with nitrous acid. Soil Sci. 102, 64–68.

Freney, J.R., Denmead, O.T. and Simpson, J.R. 1978 Soil as a source or sink for atmospheric nitrous oxide. Nature (Lond.) 273, 530–532.

Freney, J.R., Denmead, O.T. and Simpson, J.R. 1979 Nitrous oxide emission from soils at low moisture contents. Soil Biol. Biochem. 11, 167–173.

Galbally, I.E. and Roy, C.R. 1978 Loss of fixed nitrogen from soils by nitric oxide exhalation. Nature (Lond.) 275, 734–735.

Gerretsen, F.C. and de Hoop, H. 1957 Nitrogen losses during nitrification in solutions and in acid sandy soils. Can. J. Microbiol. 3, 359–380.

Hauck, R.D. 1968 Soil and fertilizer nitrogen – a review of recent work and commentary. Trans. 9th Int. Congr. Soil Sci. 2, 475–486.

Hauck, R.D. and Bremner, J.M. 1969 Significance of the nitrification reaction in nitrogen balances. In: The Biology and Ecology of Nitrogen, pp. 31–39. National Academy of Science, Washington, D.C.

Hauck, R.D. and Stephenson, H.F. 1965 Nitrification of nitrogen fertilizers. Effect of nitrogen source, size and pH of the granule, and concentration. J. Agric. Food Chem. 13, 486–492.

Hooper, A.B. 1968 A nitrite-reducing enzyme from *Nitrosomonas europaea*: preliminary characterization with hydroxylamine as an electron donor. Biochim. Biophys. Acta 162, 49–65.

Hutchinson, G.L. and Mosier, A.R. 1979 Nitrous oxide emissions from an irrigated cornfield. Science (Wash. D.C.) 205, 1125–1126.

Jones, E.J. 1951 Loss of elemental nitrogen from soils under anaerobic conditions. Soil Sci. 71, 193–196.

Jones, R.W. and Hedlin, R.A. 1970a Ammonium, nitrite and nitrate accumulation in three Manitoba soils as influenced by added ammonium sulphate and urea. Can. J. Soil Sci. 50, 331–338.

Jones, R.W. and Hedlin, R.A. 1970b Nitrite instability in three Manitoba soils. Can. J. Soil Sci. 50, 339–345.

Justice, J.K. and Smith, R.L. 1962 Nitrification of ammonium sulphate in a calcareous soil as influenced by combinations of moisture, temperature, and levels of added nitrogen. Soil Sci. Soc. Am. Proc. 26, 246–250.

Keeney, D.R., Fillery, I.R. and Marx, G.P. 1979 Effect of temperature on the gaseous nitrogen products of denitrification in a silt loam soil. Soil Sci. Soc. Am. J. 43, 1124–1128.

McKenney, D.J., Wade, D.L. and Findlay, W.I. 1978 Rates of N_2O evolution from N-fertilized soil. Geophys. Res. Lett. 5, 777–780.

Meek, B.D. and MacKenzie, A.J. 1965 The effect of nitrite and organic matter on aerobic gaseous losses of nitrogen from a calcareous soil. Soil. Sci. Soc. Am. Proc. 29, 176–178.

Miyamoto, S., Prather, R.J. and Bohn, H.L. 1974 Nitric oxide sorption by calcareous soils. II. Effect of moisture on capacity, rate, and sorption products. Soil Sci. Soc. Am. Proc. 38, 71–74.

Moraghan, J.T. and Buresh, R.J. 1977 Chemical reduction of nitrite and nitrous oxide by ferrous ion. Soil Sci. Soc. Am. J. 41, 47–50.

Mortland, M.M. 1965 Nitric oxide adsorption by clay minerals. Soil Sci. Soc. Am. Proc. 29, 514–519.

Mosier, A.R. and Hutchinson, G.L. 1981 Nitrous oxide emissions from cropped fields. J. Environ. Qual. 10: 169–173.

Nash, T. 1970 Absorption of nitrogen dioxide by aqueous solutions. J. Chem. Soc. (Lond.) A, 3023–3024.

Nelson, D.W. 1967 Chemical transformations of nitrite in soils. Ph.D. Thesis, Iowa State University, Ames. 150 pp.

Nelson, D.W. 1978 Transformations of hydroxylamine in soils. Proc. Indiana Acad. Sci. 87, 409–413.

Nelson, D.W. and Bremner, J.M. 1969 Factors affecting chemical transformations of nitrite in soils. Soil Biol. Biochem. 1, 229–239.

Nelson, D.W. and Bremner, J.M. 1970a Role of soil minerals and metallic cations in nitrite decomposition and chemodenitrification in soils. Soil Biol. Biochem. 2, 1–8.

Nelson, D.W. and Bremner, J.M. 1970b Gaseous products of nitrite decomposition in soils. Soil Biol. Biochem. 2, 203–214.

Nömmik, H. 1956 Investigations on denitrification in soils. Acta Agric. Scand. 6, 195–228.

Nömmik, H. and Nilsson, K.O. 1963 Nitrification and movement of anhydrous ammonia in soil. Acta Agric. Scand. 13, 205–219.

Pang, P.C., Cho, C.M. and Hedlin, R.A. 1975a Effect of pH and nitrifier population on nitrification of band-applied and homogeneously mixed urea nitrogen in soils. Can. J. Soil Sci. 55, 15–21.

Pang, P.C., Cho, C.M. and Hedlin, R.A. 1975b Effect of nitrogen concentration on the transformation of band-applied nitrogen fertilizers. Can. J. Soil Sci. 55, 23–27.

Pang, P.C., Hedlin, R.A. and Cho, C.M. 1973 Transformation and movement of band-applied urea, ammonium sulphate and ammonium hydroxide during incubation in several Manitoba soils. Can. J. Soil Sci. 53, 331–341.

Parr, J.F. and Papendick, R.I. 1966 Retention of anhydrous ammonia by soil. II. Effect of ammonia concentration and soil moisture. Soil Sci. 101, 109–119.

Philpot, J.S. and Small, P.A. 1938 The action of nitrous acid on phenols. Biochem. J. 32, 534–541.

Porter, L.K. 1969 Gaseous products produced by anaerobic reaction of sodium nitrite with oxime compounds and oximes synthesized from organic matter. Soil Sci. Soc. Am. Proc. 33, 696–702.

Prather, R.J. and Miyamoto, S. 1974 Nitric oxide sorption by calcareous soils: III. Effect of temperature and lack of oxygen on capacity and rate. Soil Sci. Soc. Am. Proc. 38, 582–585.

Prather, R.J., Miyamoto, S. and Bohn, H.L. 1973a Sorption of nitrogen dioxide by calcerous soils. Soil Sci. Soc. Am. Proc. 37, 860–863.

Prather, R.J., Miyamoto, S. and Bohn, H.L. 1973b Nitric oxide sorption by calcareous soils: I. Capacity, rate, and sorption products in air dry soil. Soil Sci. Soc. Am. Proc. 37, 877–879.

Reuss, J.O. and Smith, R.L. 1965 Chemical reactions of nitrites in acid soils. Soil Sci. Soc. Am. Proc. 29, 267–270.

Ritchie, G.A.F. and Nicholas, D.J.D. 1972 Identification of the sources of nitrous oxide produced by oxidative and reductive processes in *Nitrosomonas europaea*. Biochem. J. 126, 1181–1191.

Ryden, J.C., Lund, L.J. and Focht, D.D. 1978 Direct in-field measurement of nitrous oxide flux from soils. Soil Sci. Soc. Am. J. 42, 731–737.

Sabbe, W.E. and Reed, W.L. 1964 Investigations concerning nitrogen loss through chemical reactions involving urea and nitrite. Soil Sci. Soc. Am. Proc. 28, 478–481.

Schwartzbeck, R.A., MacGregor, J.A. and Schmidt, E.L. 1961 Gaseous nitrogen losses from nitrogen fertilized soils measured with infrared and mass spectroscopy. Soil Sci. Soc. Am. Proc. 25, 186–189.

Smith, C.J. and Chalk, P.M. 1979a Determination of nitrogenous gases evolved from soils in closed systems. Analyst (Lond.) 104, 538–544.

Smith, C.J. and Chalk, P.M. 1979b Factors affecting the determination of nitric oxide and nitrogen

88

dioxide evolution from soil. Soil Sci. 128, 327–330.

Smith, C.J. and Chalk, P.M. 1979c Mineralization of nitrite fixed by soil organic matter. Soil Biol. Biochem. 11, 515–519.

Smith, C.J. and Chalk, P.M. 1980a Gaseous nitrogen evolution during nitrification of ammonia fertilizer and nitrite transformations in soils. Soil Sci. Soc. Am. J. 44, 277–282.

Smith, C.J. and Chalk, P.M. 1980b Fixation and loss of nitrogen during transformations of nitrite in soils. Soil Sci. Soc. Am. J. 44, 288–291.

Smith, C.J. and Chalk, P.M. 1980c Comparison of the efficiency of urea, aqueous ammonia and ammonium sulphate as nitrogen fertilizers. Plant Soil 55, 333–337.

Smith, D.H. and Clark, F.E. 1960 Volatile losses of nitrogen from acid or neutral soils or solutions containing nitrite and ammonium ions. Soil Sci. 90, 86–90.

Sobolev, I. 1961 Lignin model compounds. Nitric acid oxidation of 4-methylguaiacol. J. Org. Chem. 26, 5080–5085.

Soulides, D.A. and Clark, F.E. 1958 Nitrification in grassland soils. Soil Sci. Soc. Am. Proc. 22, 308–311.

Steen, W.C. and Stojanovic, B.J. 1971 Nitric oxide volatilization from a calcareous soil and model aqueous solutions. Soil Sci. Soc. Am. Proc. 35, 277–283.

Stevenson, F.J. and Kirkman, M.A. 1964 Identification of methyl nitrite in the reaction product of nitrous acid and lignin. Nature (Lond.) 201, 107.

Stevenson, F.J. and Swaby, R.J. 1964 Nitrosation of soil organic matter. I. Nature of gases evolved during nitrous acid treatment of lignins and humic substances. Soil Sci. Soc. Am. Proc. 28, 773–778.

Stevenson, F.J., Harrison, R.M., Wetselaar, R. and Leeper, R.A. 1970 Nitrosation of soil organic matter. III. Nature of gases produced by reaction of nitrite with lignins, humic substances, and phenolic constituents under neutral and slightly acidic conditions. Soil Sci. Soc. Am. Proc. 34, 430–435.

Taylor, W.D., Allston, T.D., Moscato, M.J., Fazekas, G.B., Kozlowski, R. and Takacs, G.A. 1980 Atmospheric photodissociation lifetimes for nitromethane, methyl nitrite, and methyl nitrate. Int. J. Chem. Kinetics XIII: 231–240.

Tyler, K.B. and Broadbent, F.E. 1960 Nitrite transformations in California soils. Soil Sci. Soc. Am. Proc. 24, 279–282.

Tyler, K.B., Broadbent, F.E. and Hill, G.N. 1959 Low temperature effects on nitrification in four California soils. Soil Sci. 87, 123–129.

Van Cleemput, O. and Baert, L. 1976 Theoretical considerations of nitrite self-decomposition reactions in soils. Soil Sci. Soc. Am. J. 40, 322–324.

Van Cleemput, O., Patrick Jr., W.H. and McIlhenny, R.C. 1976 Nitrite decomposition in flooded soil under different pH and redox potential conditions. Soil Sci. Soc. Am. J. 40, 55–60.

Vine, H. 1962 Some measurements of release and fixation of nitrogen in soil of natural structure. Plant Soil 17, 109–130.

Wagner, G.H. and Smith, G.E. 1958 Nitrogen losses from soils fertilized with different nitrogen carriers. Soil Sci. 85, 125–129.

Wahhab, A. and Uddin, F. 1954 Loss of nitrogen through reaction of ammonium and nitrite ions. Soil Sci. 78, 119–126.

Wetselaar, R., Passioura, J.B. and Singh, B.R. 1972 Consequences of banding nitrogen fertilizers in soil. I. Effect on nitrification. Plant Soil 36, 159–175.

White, R.E. 1979 Introduction to the Principles and Practice of Soil Science. Blackwell Scientific Publications, Oxford. 198 pp.

Wiebe, H.A. and Heicklen, J. 1973 Photolysis of methyl nitrite. J. Am. Chem. Soc. 95: 1–7.

Wullstein, L.H. 1967 Soil nitrogen volatilization. Agric. Sci. Rev. Coop. State Res. Ser. U.S. Dep. Agric. 5, 8–12.

Wullstein, L.H. and Gilmour, C.M. 1964 Non-enzymatic gaseous loss of nitrite from clay and soil systems. Soil Sci. 97, 428–430.

Wullstein, L.H. and Gilmour, C.M. 1966 Non-enzymatic formation of nitrogen gas. Nature (Lond.) 210, 1150–1151.

Yoshida, T. and Alexander, M. 1970. Nitrous oxide formation by *Nitrosomonas europaea* and heterotrophic microorganisms. Soil Sci. Soc. Am. Proc. 34, 880–882.

4. The measurement of denitrification

J.C. RYDEN and D.E. ROLSTON

4.1. Introduction

The progress of research on the magnitude of denitrification has been restricted by the difficulties involved in making accurate field measurements. Numerous laboratory studies have provided useful information on the factors affecting denitrification in soils but these findings cannot be used to predict the extent of denitrification in the field. In recent years, increased interest in the efficiency of nitrogen (N) use in agriculture and the impact of fertilizer N on the environment have spurred research on the development of field methods for measurement of denitrification loss (Rolston et al. 1976, 1978, Focht 1978, Ryden et al. 1978, 1979a,b, Denmead 1979, McKenney et al. 1978, Matthias et al. 1980).

Molecular nitrogen (N_2) is one of the principal free products of denitrification and is produced in quantities that are usually insufficient to be measured against its ambient concentration. It is for this reason that denitrification losses from soils in the field are a major unknown in the quantification of the N cycle. The other principal product of denitrification is nitrous oxide (N_2O) which can be readily measured against its low ambient concentration (approximately 300 ppb v/v) using gas chromatography (Wentworth and Freeman 1973, Rassmusen et al. 1976, Mosier and Mack 1980). In some laboratory studies nitric oxide has also been a significant product of denitrification (Nömmik 1956, Bailey and Beauchamp 1973). Oxidation of NO to nitrogen dioxide (NO_2) followed by its dissolution to form nitric and nitrous acids suggests that field losses of NO will be low. Field measurements have indicated NO losses in the range of only 0.2 to 2 kg N/ha/year (Galbally and Roy 1978).

Until recently, field assessment of denitrification has depended on the determination of unaccounted-for N in N balance studies (Allison 1966, Hauck 1971) and on the estimation of denitrification potential of a soil using indices such as available carbon, redox potential, soil water potential and the number of denitrifying bacteria (Focht 1978). In this chapter, recent developments in methods for the measurement of denitrification in soils will be discussed. Particular emphasis will be placed on methods for the direct measurement of denitrification in the field.

4.2. Experimental and analytical techniques

4.2.1. Basic concepts

Two basic approaches have been used to determine the extent of denitrification in soils. The easier relies on measurement of the change in soil nitrate concentration. The more satisfactory is determination of the amount of nitrogenous gases produced in soil.

The determination of denitrification as nitrate disappearance has been adopted in numerous laboratory studies (Pilot and Patrick 1972, Misra *et al.* 1974, Westerman and Tucker 1978, Reddy *et al.* 1980) and measurement of soil nitrate content is also used in indirect field estimates of denitrification loss (Lund *et al.* 1978). Estimates of denitrification based on nitrate disappearance are valid only if assimilatory nitrate reduction by soil microorganisms is negligible during the study. As discussed by Focht (1978), assimilatory nitrate reduction is expected to be significant in laboratory studies only for systems to which organic matter of high C:N ratio is added prior to incubation in the presence of a sustained oxygen concentration. In the field, the estimation of denitrification as 'unaccounted-for N' (Allison 1966) is valid only if net mineralization is equal to net immobilization (Fried *et al.* 1976).

With a few exceptions (e.g. Wijler and Delwiche 1954, Nömmik 1956), it is only in recent studies that denitrification has been measured by direct gas analysis. This approach has been facilitated by the development of improved separatory materials and detectors for use in gas chromatography and to some extent by the wider availability of mass spectrometers for determination of ^{15}N-labelled gases.

The improvements in gas chromatographic techniques now permit direct analysis for N_2, N_2O, and NO during incubation of soils under N_2 free atmospheres (Bailey and Beauchamp 1973, Blackmer and Bremner 1977a, Ryden *et al.* 1979a). Direct measurement of denitrification in both the laboratory (Wijler and Delwiche 1954, Nömmik 1956, Focht *et al.* 1980) and the field (Rolston *et al.* 1976, 1978) has been achieved by mass spectrometric analysis of ^{15}N-labelled N_2 and N_2O after addition of labelled nitrate to soils. More recently, the radioactive isotope, ^{13}N, has been used for the measurement of nitrogenous gas production in short term laboratory studies (Gersberg *et al.* 1976, Tiedje *et al.* 1979).

4.2.2. Measurement of nitrate and nitrite

The extraction of nitrate and nitrite from soils is usually achieved with molar solutions of potassium chloride or potassium sulfate that are also used in extraction of exchangeable ammonium (Bremner 1965a). To avoid changes in soil nitrate and nitrite concentration, soils should not be dried prior to extraction. Field moist soils may be extracted using a procedure similar to the saturation extraction described in Richards (1954). Water may be used as an alternative extractant except for soils containing appreciable amounts of hydrous oxides and amorphous aluminosilicates (Espinoza *et al.* 1975).

Colorimetric determination of nitrate and nitrite based on the Griess-Ilosvay reaction has been adopted by many laboratories equipped with auto-analytical systems (Methods for Chemical Analysis of Waters and Wastes, 1976). Nitrate is determined as the difference in the nitrite concentration of the sample prior to and following reduction in a column of finely divided cadmium. Frequently, however, nitrite concentration is negligible in relation to nitrate and the determination of nitrite prior to reduction is unnecessary, but this must be established for samples of different origin.

Prior to the widespread availability of auto-analytical systems, steam distillation (Bremner 1965a) was the method most frequently used for nitrate determination. Steam distillation is still the only technique for the determination of the distribution of ^{15}N between various inorganic fractions after addition of labelled N to soils. The procedure involves the conversion of ammonium, nitrate and nitrite to NH_3 in three separate reactions, the NH_3 being distilled over and trapped in 2% boric acid for subsequent analysis.

Other methods have been used in the determination of nitrate and nitrite. Nitrate specific electrodes have been successfully applied to solutions of low chloride concentration (Mahendrappa 1969, Keeney *et al.* 1970). Hansen *et al.* (1977) have described an injection-flow system permitting analysis of ninety samples per hour. Ultraviolet spectrophotometry has also been used in analysis of nitrate in waters, effluents and soil extracts (Hoather and Rackham 1959, Cawse 1967, Miles and Espejo 1977). More recently, low and high pressure liquid chromatography has been applied to the determination of nitrate and nitrite (Dick and Tabatabai 1979, Tiedje *et al.* 1979, Edwards *et al.* 1980). The anions may be separated using an anion exchange column and their amounts determined by an appropriate detector such as a conductivity cell or an ultraviolet spectrophotometer.

4.2.3. Gas chromatography

Recent improvements in gas chromatographic equipment now permit the direct identification and quantification of all gases of interest in denitrification studies. The instrumentation and choice of column packing is usually dictated by the different analytical requirements of laboratory and field studies. In laboratory studies, it is usual that determination of several gases is required. These gases include N_2, N_2O, NO, CO_2, O_2 and in some studies C_2H_2, Ar and Ne. As direct measurement of N_2 is not practical in most field studies, analytical requirements are usually for determination of N_2O and in a few cases NO. Payne (1973), however, has demonstrated that for saturated soils and sediments it may be feasible to measure N_2 evolution into a N_2-free atmosphere enclosed at the surface.

No single column packing will effect the separation of all the gases specified above. Packings of the Porapak series (porous beaded polyaromatic compounds; Hollis 1966) are most frequently used in the separation of N_2O, NO, and CO_2 in air (Wentworth and Freeman 1973, Bailey and Beauchamp 1973). Separation of N_2 and O_2 is usually achieved on 5Å molecular sieve at above ambient temperatures (Bailey and Beauchamp 1973, Payne 1973) or on Porapak Q at sub-ambient temperatures (Blackmer and Bremner 1977a). Complete separation of the suite of gases specified from a single sample may be attained by coupling columns of Porapak and molecular sieve in series or parallel (Fig. 4.1).

Burford (1969) coupled three columns of molecular sieve in series, and holding each at a different temperature, separated N_2, N_2O, CO_2, Ar and O_2.

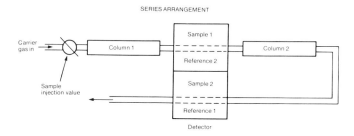

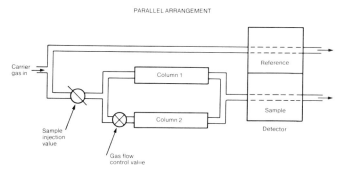

Fig. 4.1. Series and parallel column arrangements for gas chromatographic analysis of N_2, O_2, CO_2, N_2O and NO in a single sample. Column (1) is usually Porapak Q at above ambient temperature and column (2) is 5Å molecular sieve at above ambient temperature or Porapak Q at sub-ambient temperature.

Porter (1969) used a similar arrangement of two columns; Porapak Q followed by molecular sieve. Blackmer *et al.* (1974) used two columns of Porapak Q in series. Nitrogen and O_2 were separated on the second column which was cooled to $-78°$ C. One disadvantage of series arrangements of columns in conjunction with a thermal conductivity detector is that the polarity of the recorder has to be switched as gases pass through the different sides of the detector. This was obviated in a system adapted by Blackmer and Bremner (1977a) from the earlier arrangement described by Blackmer *et al.* (1974) by the use of an ultrasonic detector, each side of which functions independently.

Polarity switching poses no problem in parallel column couplings (Fig. 1). Smith and Dowdell (1973) used separate detectors attached to each column outlet. In contrast, Beard and Guenzi (1976) employed a single detector and attained separation of components by adjustments to column length and carrier gas flow rate.

The choice of detector is also determined by the study to be undertaken. In many laboratory studies, the thermal conductivity (TC) detector has proved sufficiently sensitive for the analysis of most gases of interest (Bailey and Beauchamp 1973, Ryden *et al.* 1979a). Recent improvements in the control of filament current, gas flow and detector temperature now permit the detection of N_2 and N_2O in amounts in the range 10^{-8} to 10^{-7} g N. The linear response of the TC detector is usually four decades and its improved stability eliminates the need for frequent calibration.

The ultrasonic detector used in methods described by Blackmer and Bremner (1977a, 1978) combines greater sensitivity with many of the operational advantages of TC detectors. Blackmer and Bremner (1977) quote a detection limit of 3–9 ng for N_2O. A similar sensitivity has been reported for a helium ionization detector (Delwiche and Rolston 1976). The rather poor stability of this detector has been a detraction to its adoption in denitrification studies.

For each detector discussed above, the response to CO_2 may mask that to N_2O when CO_2 is present in amounts more than 10 times greater than that of N_2O. If necessary, the amount of CO_2 may be decreased by incorporation of a soda lime or Ascarite trap at the inlet port of the column (e.g. Mosier and Mack 1980). Water vapour may also be removed by use of a cold trap (Blackmer and Bremner 1977a) or dessicant such as anhydrous calcium sulphate or magnesium perchlorate.

The detectors discussed above are not sufficiently sensitive for determination of ambient N_2O without prior concentration. This has been achieved by passage of a known volume of air through 5Å molecular sieve on which N_2O is quantitatively retained (Hahn 1972, La Hue *et al.* 1971, Ryden *et al.* 1978). The amount of N_2O passed, however, must not exceed the adsorption capacity of the molecular sieve at the N_2O concentration in the air flow. Nitrous oxide may be conveniently recovered in a smaller volume for GC analysis by addition of water to the molecular sieve in an evacuated flask (Dowdell and Crees 1974, Ryden *et al.* 1978). An alternative approach involves the concentration of N_2O by adsorption on Porapak Q cooled to $-135°$ C (Blackmer and Bremner 1978). At this temperature, Xe is also adsorbed and provides an internal standard during subsequent GC analysis.

Other than mass spectrometric analysis, the use of a [63]Ni electron capture (EC) detector operated at 250 to 350° C offers the only approach to the direct determination of N_2O concentrations near the ambient level. The EC detector was first applied to N_2O analysis by Wentworth *et al.* (1971) and has been used in several subsequent studies (Wentworth and Freeman 1973, Rassmussen *et al.* 1976, Brice *et al.* 1977, Pierotti *et al.* 1978, Cicerone *et al.* 1978, Mosier and Mack 1980, Rolston *et al.* 1982). Several problems are attendant with the use of EC detectors. Appreciable interference is caused by water vapour and freon (Cicerone *et al.* 1978) and by variations in CO_2 concentration (Goldan *et al.* 1978). Cicerone *et al.* (1978) used a column backflush to eliminate these interferences but in so doing increased the time for analysis of each sample. Mosier and Mack (1980) removed water vapour and CO_2 using appropriate adsorbents at the GC inlet. Analysis time was decreased to 4.25 min by separating N_2O and O_2 from other sample gases on a pre-column of Porapak Q which was backflushed while complete separation of N_2O was achieved on a second column.

Pierotti *et al.* (1978) also noted that detector response was affected by fluctuations in laboratory temperature. This effect was minimized by careful control of ambient temperature and frequent checks on detector response by analysis of standard samples. With adequate control of operational procedures, the standard deviation on repeated analyses of the same sample can be reduced to ± 1 to 2 ppb for N_2O concentrations of approximately 320 ppb (Pierotti *et al.* 1978, Mosier and Mack 1980).

4.2.4. ^{15}N-labelled tracers

The stable isotope ^{15}N has been used extensively as a tracer of nitrogen in soils, plants, animals, and water. Several techniques are available to enrich nitrogen from the natural abundance of 0.366 atom % to as high as 99.5 atom % [15]N. Thus, the scope of ^{15}N use in denitrification studies is limited only by cost. Recent reviews (Jansson 1971, Fiedler and Proksch 1975, Hauck and Bremner 1976) have covered all aspects of ^{15}N use and only the basic principles, as they apply in particular to denitrification studies, are outlined below.

The ^{14}N/^{15}N ratio of the sample is generally determined by means of mass or emission spectrometric analysis of ^{15}N-labelled N_2 obtained by direct sampling or by generation from a sample containing inorganic or organic N. The ^{14}N and ^{15}N atoms in the sample of N_2 are paired to form the molecules [28]N_2,

$^{29}N_2$, and $^{30}N_2$. Both mass and emission spectrometric methods provide output signals which are proportional to the number of the three types of molecules. If equilibrium exists between the $^{28}N_2$, $^{29}N_2$, and $^{30}N_2$ molecules, only the $^{28}N_2$ and $^{29}N_2$ signals need to be measured to calculate ^{15}N abundance (Fiedler and Proksch 1975). This equilibrium generally exists for samples of ammonium, nitrite, nitrate, or organic N due to the reduction and digestion procedure that is performed to release N_2 gas. For N_2 gas evolved from ^{15}N fertilizer, however, equilibrium may not exist and all three molecules must be measured.

Determination of denitrification from changes in ^{15}N content with time for intact soil-plant systems requires analysis for $^{14}N/^{15}N$ ratio of N compounds in soil, water and plants. In more simple incubation studies it may be necessary only to determine the ratio for nitrate N. In all such studies, however, inorganic and organic N must be converted to N_2. Nitrogen in samples of water, soil extracts, or plant digests is initially reduced to ammonium, detailed procedures for which have been discussed by Bremner (1965a). After titration for ammonium, the solution is acidified and evaporated to a small volume for determination of the $^{14}N/^{15}N$ ratio. The ammonium in the extract is oxidized by sodium or lithium hypobromite to N_2 which can be admitted to a mass spectrometer for determination of the $^{14}N/^{15}N$ ratio (Rittenberg 1948, Bremner 1965b). An alternative procedure, used by Dowdell and Webster (1980), involves combustion of plant and water samples over CuO to produce N_2 gas (the Dumas procedure). The products of Dumas combustion require careful purification to remove CO_2, CO, H_2O, CH_4 and other gases. Dowdell and Webster (1980) collected N_2 on 5Å molecular sieve at $-196°$ C before its admission to the mass spectrometer.

Gas samples collected from denitrification experiments must be scrubbed to remove gases other than N_2 which affect the measurement of $^{14}N/^{15}N$ ratio. Rolston et al. (1978) accomplished this by injecting gas samples into traps containing ascarite, magnesium perchlorate, and a commercial O_2 scrubber to remove CO_2, H_2O and O_2 from the gas sample. Siegel et al. (1982) used a very sensitive mass spectrometer which required very high purification of gas samples. This was accomplished by passing gas samples successively through liquid N_2 cold traps, and high capacity O_2 traps. Focht et al. (1980) used interfaced gas chromatography and mass spectrometry for measurements of denitrification by direct gas analysis. Nitrogen and N_2O were separated from other gases in a particular sample using columns of 5Å molecular sieve and Porapak Q, respectively, before entry to the mass spectrometer.

Recent applications of mass and emission spectrometry to the determination of denitrification and other N transformations in soils have been described by Focht et al. (1980), Dowdell and Webster (1980), Riga et al. (1980), Lippold and Förster (1980), Siegel et al. (1982), Rolston et al. (1982), Phieffer-Madsen

(1977) and Beer *et al.* (1977). Each laboratory tends to have its own unique operational system developed over an appreciable period and designed to meet the requirements of a particular experimental approach. Consequently only the basic principles of the two instruments are outlined below.

Mass spectrometers comprise five separate units. These are an inlet system for the introduction of N_2 gas, an ion source where N_2 molecules are bombarded with electrons and become charged and accelerated, a magnetic field in which the charged molecules are separated into different paths according to their mass, a collector where the molecules are discharged and the resulting currents amplified, and a recorder which registers the amplified currents as peaks. A more detailed description of the mass spectrometric systems available for $^{14}N/^{15}N$ ratio analysis may be found in the review by Fiedler and Proksch (1975).

Although the principles of mass spectrometry are relatively simple, practical operation generally requires appreciable technical skill. Facilities for efficient N_2 preparation, the maintenance of a high vacuum and electronic stability are particularly important features of any system for routine analyses.

In emission spectrometry, an external energy source of high frequency microwaves is used to bring N_2 molecules in a sample tube to an excited state. When the excited molecules return to the ground state, the energy difference is emitted as electromagnetic radiation of a characteristic wavelength. Small differences exist in the wavelength of the light emitted when the excited $^{28}N_2$, $^{29}N_2$, and $^{30}N_2$ molecules return to the ground state. The emitted light is resolved by a monochromator, and the peaks are recorded in a manner similar to mass spectrometry. Further details of emission spectrometry may be found in Goleb (1970) and Perschke and Proksch (1971) as well as the review by Fiedler and Proksch (1975).

Emission spectrometry is simpler than mass spectrometry since a high vacuum is not required and the instrument can be installed and supervised easily by trained technicians. As the amount of N required for emission spectrometry is about 100 times smaller than that for mass spectrometry, sample preparation requires more care to prevent sample contamination, particularly by atmospheric N_2.

The standard deviation associated with the measurement of ^{15}N abundance using emission spectrometry is $\pm 3\%$, but only $\pm 0.5\%$ using mass spectrometry (Fiedler and Proksch 1975). Thus, for the small amounts of ^{15}N in gases evolved during denitrification studies, particularly those in the field, the emission spectrometer may be less sensitive than the mass spectrometer in the detection of denitrification loss.

4.2.5. Acetylene inhibition of nitrous oxide reduction

The difficulties involved in measurement of total denitrification in soils may be overcome by the observation that small concentrations of acetylene (C_2H_2) inhibit the reduction of N_2O to N_2 during denitrification in pure cultures of denitrifying bacteria (Federova et al. 1973, Balderston et al. 1976, Yoshinari and Knowles 1976) and in soils and sediments (Yoshinari et al. 1977, Klemedtsson et al. 1977, Sørensen 1978, Smith et al. 1978, Yeomans and Beauchamp 1978, Ryden et al. 1979a). Laboratory studies have shown that in the presence of C_2H_2 concentrations between 0.1 and 1% (v/v), the sole product of denitrification is N_2O which appears in amounts, expressed as total N, equivalent to the combined production of N_2O and N_2 in the absence of C_2H_2. The effect is illustrated by data in Table 4.1 for denitrification in a soil slurry incubated in the absence and presence of C_2H_2 at a concentration of 1% (v/v). Although both N_2O and N_2 were produced during the early stages of incubation in the absence of C_2H_2, the sole product in the presence of C_2H_2 was N_2O. Significantly, production of N_2O-N in the presence of C_2H_2 was equivalent to the combined production of N_2-N and N_2O-N in the absence of C_2H_2. Furthermore, C_2H_2 appears to have no appreciable effect on microbial respiration. Production of CO_2 was essentially the same irrespective of C_2H_2 treatment. Consequently, measurement of N_2O production in the presence and absence of C_2H_2 provides a direct measurement of the extent of denitrification loss as well as an indication of the distribution of denitrification products between N_2 and N_2O.

Unequivocal evidence for the effectiveness of C_2H_2 inhibition of N_2O

Table 4.1. Dinitrogen, N_2O and CO_2* produced during incubation of slurries of a loam with different amounts of added nitrate under different atmospheres in the presence and absence of 1.0% (v/v) C_2H_2 (data from Ryden et al. 1979a)

Incubation time	Without C_2H_2			With C_2H_2		
	N_2	N_2O	CO_2	N_2	N_2O	CO_2
87.7 µg NO_3^--N/g; He atmosphere						
36	10.6	27.9	32.0	0	42.3	36.0
108	87.9	5.2	60.2	0	92.1	67.6
200	91.0	0	82.0	0	92.1	86.0
268 µg NO_3^--N/g; 20% O_2 – 80% Ar atmosphere						
40	2.63	13.9	28.8	0	15.2	29.5
90	12.8	18.9	54.2	0	31.3	55.4
200	31.1	15.4	84.1	0	49.7	84.3

* µg N or C/g soil

reduction has been obtained in studies using ^{15}N and ^{13}N tracers. Ryden *et al.* (1979a) found no production of $^{29}N_2$ or $^{30}N_2$ during incubation of two soils with ^{15}N-labelled nitrate (83 μgN/g) in the presence of 1% (v/v) C_2H_2. The detection limit for $^{30}N_2$ was 0.002 μg N/g. For another soil only 1% of the denitrification products appeared as ^{15}N-labelled N_2 in the presence of C_2H_2 in contrast to 100% ^{15}N-labelled N_2 in the absence of C_2H_2. However, Smith *et al.* (1978), using ^{13}N-labelled nitrate, demonstrated that for low nitrate concentrations (2.0 μg N/g) C_2H_2 concentrations of approximately 5% may be required to effect inhibition of N_2O reduction. In contrast, Ryden *et al.* (1979a) observed N_2O reduction, following exhaustion of soil nitrate, only during incubation with C_2H_2 concentrations $\leqslant 0.1\%$. No reduction of N_2O was detected in the presence of 1% C_2H_2 during a period of 100 h following nitrate exhaustion.

Yeomans and Beauchamp (1978) concluded that for some soils, inhibition of N_2O reduction by C_2H_2 may be overcome after exposure to C_2H_2 for more than seven days. They suggested that this may arise from the development of bacteria or enzymes capable of reducing N_2O in the presence of C_2H_2. It is apparent from their data, however, that N_2O reduction occurred when added nitrate had been consumed, their findings therefore being consistent with those discussed above. More recent work (Yeomans and Beauchamp 1982) has indicated that the decline in the efficiency of inhibition of N_2O reduction by C_2H_2 may be related to a 'decrease' in the concentration of C_2H_2 during prolonged incubation. It was also suggested that in some circumstances C_2H_2 may be used as an organic substrate thereby enhancing denitrification loss relative to that in the absence of C_2H_2. Germon (1980) also reported an enhanced loss of nitrate after incubation of soil with C_2H_2 for 4 to 11 days.

Watanabe and de Guzman (1980) reported that C_2H_2 'disappeared' during anaerobic incubation of a fresh rice paddy soil. The rate of disappearance was inhibited by the presence of nitrate and to some extent by small amounts of O_2, but no increase in CO_2 production was observed. It appears that there may be a relationship between diminished effectiveness of C_2H_2 inhibition and C_2H_2 disappearance on exhaustion of the nitrate supply.

It is difficult to assess the significance of the findings discussed above in relation to those in most other studies where no effects of C_2H_2 on denitrification have been observed (e.g. Yoshinari *et al.* 1977, Smith *et al.* 1978, Ryden *et al.* 1979a). It appears that 'disappearance' or 'decomposition' of C_2H_2 and its attendant effects on denitrification loss are associated with low nitrate concentrations, strongly reducing conditions and periods of continuous incubation with C_2H_2 in excess of 4 to 6 days. Watanabe and de Guzman (1980) detected formic, acetic, propionic and butyric acids following incubation of soil with C_2H_2. They concluded that during studies of denitrification, soils must not be treated with C_2H_2 for a 'prolonged period'.

Several workers have shown that C_2H_2 concentrations effective in inhibiting N_2O reduction also inhibit nitrification in microbial cultures and soils (Hynes and Knowles 1978, Walter *et al.* 1979, Mosier 1980). Walter *et al.* (1979) argued that the coupled inhibition of nitrification may lead to an under-estimate of denitrification due to decreased oxygen demand and to elimination of the nitrate supply in systems in which this is dependent on the nitrification of ammonium. Recent unpublished work (Ryden) has demonstrated equivalent rates of denitrification in soils amended with ammonium nitrate and incubated in the presence and absence of C_2H_2 under conditions which permitted concurrent nitrification in the absence of C_2H_2. Only when nitrate supply was exhausted were differences observed in the extent of denitrification in the presence and absence of C_2H_2.

These findings suggest that measurements of denitrification during concurrent nitrification using C_2H_2 inhibition are valid providing the nitrate supply is not limited by the concurrent inhibition of nitrification. To achieve this condition, the C_2H_2-inhibition technique should be applied for only short periods, probably less than one day. With sufficient care in the mode of its application, the technique remains a useful approach to the direct measurement of denitrification in laboratory and, in particular, field studies.

4.2.6. ^{13}N-labelled tracers

Four groups of workers have recently demonstrated the value of the radioactive isotope ^{13}N in studies of denitrification (Gersberg *et al.* 1976, Tiedje *et al.* 1979, Stout and More 1980, Hollocher *et al.* 1980). Until these applications, the use of ^{13}N had been considered inappropriate due to its short half life (9.96 min). Also, its use demands access to a cyclotron which must be in close proximity to laboratory or field experimental areas. These factors still limit the use of ^{13}N tracer techniques.

Millicurie amounts of ^{13}N-labelled nitrate may be produced within 10 to 20 min by focusing a proton beam on a distilled water target. After irradiation Gersberg *et al.* (1976) found that 99.6% of the radiochemical content was $^{13}NO_3^-$. Tiedje *et al.* (1979) achieved a radiochemical purity of 75 to 90% $^{13}NO_3^-$. An alternative technique for production of $^{13}NO_3^-$ involves bombardment of a lithium carbonate pellet with a 2 MeV deuteron beam from a Van de Graff accelerator. McNaughton and More (1979) have used this technique to produce up to 57 μCi $^{13}NO_3^-$ with a purity of 85%. The isotope decays by emission of positrons which on annihilation yield gamma rays.

The $^{13}NO_3^-$ produced in the cyclotron is rapidly purified and then introduced into a mixed soil slurry or a soil core. The gases produced are flushed from the reaction vessel by a continuous flow of He. After drying, ^{13}N-labelled gases are retained in appropriate traps. Gersberg *et al.* (1976) collected

N_2O and N_2 as a single sample on 13X molecular sieve cooled to $-196°$ C in liquid N_2. In the method described by Tiedje *et al.* (1979), N_2O was first removed by condensation in an aluminium coil at $-196°$ C followed by adsorption of N_2 on cooled molecular sieve. The trapped gases are counted using NaI (Tl) crystal gamma ray detectors at intervals of about 1 min and data are processed by computer to account for decay, counting efficiency and background. Tiedje *et al.* (1979) also developed analytical procedures for the determination of specific activity based on GC separation followed by mass and isotope analysis of nitrogenous gases and for the determination of ^{13}N-labelled nitrate, nitrite and ammonium using high pressure liquid chromatography. Further discussion of the preparation and use of ^{13}N in denitrification studies may be found in Tiedje *et al.* (1981).

Despite the inherent disadvantage of its short half life, the use of ^{13}N in denitrification studies has several exclusive advantages. Of particular importance is the high sensitivity of the technique which permits the detection of low rates of N_2 production in any atmosphere. Additions of N as ^{13}N are negligible (usually about 10^{-12} g N) in contrast to the use of ^{15}N which invariably leads to a significant increase in the nitrate concentration. Consequently ^{13}N may be used to monitor denitrification in systems of very low inorganic N content.

Access to a cyclotron and the short half life of ^{13}N appear to eliminate the use of this isotope in field studies. Restricted diffusion in undisturbed soils is likely to result in complete decay of ^{13}N before labelled gases may be collected. The isotope appears to have appreciable potential in evaluating the mechanism of denitrification (Firestone *et al.* 1979a, Hollocher *et al.* 1980) and factors affecting the rate of denitrification and the composition of products (Firestone *et al.* 1979b, 1980).

4.2.7. *Infra-red analysis*

Infra-red (IR) gas analysis has been applied to measurements of N_2O concentration since the pioneering work of Arnold (1954) who first estimated N_2O loss from a soil surface. It has subsequently been applied in laboratory studies (Hauck and Melsted 1956) and in measurements of N_2O dispersion in a crop canopy (Legg 1975). Although IR gas analysis has been displaced to some extent by the advent of improved GC systems, Denmead (1979) has described a useful application which permits continuous monitoring of N_2O concentration in an air stream drawn from the free atmosphere or through a cover placed over the soil surface in the field.

Quantitative IR gas analysis is based on the difference in absorption intensity between the unknown and a standard sample. Nitrous oxide absorbs at 1285 cm^{-1} but small amounts of water vapour, CO_2 and CO interfere with the analysis. Denmead (1979) found that 1 ppm concentrations of water vapour, CO_2 and CO gave apparent N_2O concentrations of 0.03, 1.2 and 34 ppb, respectively. Water vapour may be removed using magnesium perchlorate, CO_2 using Ascarite or soda lime, and CO by oxidation over silver oxide at 250° C followed by absorption of CO_2. Intermittent calibration, careful flow control and removal of interfering components gave a detection limit of 6 ppb N_2O in air drawn from the free atmosphere as compared with that drawn through a soil cover. This system, based on IR gas analysis offers the only practical approach to continuous monitoring of N_2O exchange at the soil surface.

4.3. Laboratory studies

4.3.1. Incubation of soils

By far the least complex assay of denitrification in soils involves the incubation of soil slurries with added nitrate under an atmosphere purged of N_2 and O_2 by evacuation and backfilling with an inert gas such as He or Ar (Bailey and Beauchamp 1973, Kaplan et al. 1977, Blackmer and Bremner 1977b, Ryden et al. 1979a, Cho et al. 1979). Denitrification is measured as the production of N_2, N_2O, and where appropriate, NO by GC analysis or by mass-spectrometry where ^{15}N-labelled substrates are introduced. The effect of O_2 on denitrification may be evaluated by the introduction of varying amounts of O_2 into the back-filling gas. The choice of backfill gas may be determined solely on practical grounds as inert gases including He, Ar, Kr and Ne cannot be distinguished from N_2 in terms of their effect on the rate of denitrification (Blackmer and Bremner 1977b). For backfilling incubation vessels, it is usual to select the same gas as that used as the carrier in the GC system adopted for analysis of denitrification products.

The rate of denitrification observed during incubation of a soil slurry can probably be viewed as a maximum for that soil. Incubation is frequently conducted under conditions of complete anoxia which are unlikely to occur in undisturbed soils; the addition of nitrate eliminates the effects of low nitrate concentrations on denitrification rate, and the addition of exogenous carbon (e.g. Westerman and Tucker 1978) stimulates microbial activity. It should also be noted that the use of air dry soils also increases denitrification rate (McGarity 1962, Patten et al. 1980) probably due to the formation of additional, more-readily decomposable organic matter during drying (Birch 1958, Yamane and Sato 1968). Although the incubation of soil slurries has no direct relevance to the extent of field denitrification losses, such incubations have value in assessing the relative effect of environmental parameters such as pH, temperature, and O_2 concentration on denitrification rate.

Attempts to avoid the limitations of slurry incubation have involved the use of undisturbed soil cores (Myers and McGarity 1972, Focht *et al.* 1980), and the incubation of soil crumbs or dried soils rewetted to realistic moisture contents (Pilot and Patrick 1972, Smith and Tiedje 1979a, Freney *et al.* 1979, Keeney *et al.* 1979). The system described by Focht *et al.* (1980) has attraction in that undisturbed cores from the field may be incubated under different atmospheres and at various temperatures and moisture contents. Focht *et al.* (1980) used ^{15}N-labelled nitrate addition in combination with interfaced GC-mass spectrometric analysis to measure denitrification. The inherent variability in bulk density of small soil cores with attendent variability in soil moisture regime and aeration demands adequate replication in such studies. Furthermore, during incubation of soil cores, the rate of diffusion of gaseous products from the core to headspace must also be taken into account (Letey *et al.* 1980). Several days are required for N_2O produced in a saturated core to diffuse to the surrounding atmosphere. The problems posed by diffusion of gases from incubated soil cores are eliminated if denitrification is measured as nitrate disappearance (Pilot and Patrick 1972), although such studies provide no information on the composition of denitrification products.

A particularly attractive incubation system has been described by Galsworthy and Burford (1978). Soils were incubated in vessels placed in a constant temperature bath. The head space above the soil was flushed using a continuous flow of He at 1 cm^3/min. The system was constructed to permit the introduction of a sample of the He stream onto a pair of columns arranged in series for GC separation and determination of N_2, N_2O, NO and CO_2. A generally similar system has been described by Wickramasinghe *et al.* (1978). An important feature of these systems is that denitrification products are removed as they appear at the soil surface as would be the case in the field, thereby achieving a more reliable estimate of the ratio of N_2O to N_2 production. However, it still remains almost impossible to duplicate field conditions such as O_2 concentration and diffusion due to altered geometry of the soil pore space.

4.3.2. Soil column studies

Column studies offer an approach to the measurement of denitrification which may closely simulate field conditions. In these studies, a plastic cylinder 0.5 to 3 m in length and up to 20 cm in diameter is packed with soil collected from the field site of interest (Lance and Whisler 1972, Doner *et al.* 1974, Rolston *et al.* 1976, Gilliam *et al.* 1978, Gilbert *et al.* 1979) or intact columns may be taken directly from the field (Guthrie and Duxbury 1978). The base of the column is usually fitted with a porous plate so that a constant water potential may be applied. Small porous ceramic cups are built into the column at various depths so that samples of the soil solution may be extracted. In some studies gas sampling ports have also been included (Starr *et al.* 1974, Rolston *et al.* 1976) or a system installed at the column head to collect gases emanating from the soil surface (Guthrie and Duxbury 1978). Columns are then leached to remove native soil nitrate followed by leaching with a solution of constant initial nitrate concentration. The rate, extent and depth at which denitrification occurs can be determined by measuring the difference in the nitrate concentration of the input solution and that at various depths within the column.

Although the rate and depth of denitrification can be inferred from changes in nitrate concentration, direct measurement of gas production is to be preferred in the estimation of total denitrification loss. Rolston *et al.* (1976) calculated fluxes of N_2 and N_2O derived from additions of ^{15}N-labelled fertilizer after measuring the concentration profiles of these gases in laboratory prepared columns. Guthrie and Duxbury (1978) collected N_2O produced at the top of columns by flushing the enclosed headspace with air and absorbing N_2O on 5Å molecular sieve. Nitrous oxide removed in the column leachate was also determined. This constituted 78 to 98% of the N_2O recovered from flooded columns and 11 to 52% of that recovered from drained columns. It should be noted, however, that N_2O in the column headspace was determined only intermittently.

Soil columns provide a basis for more ambitious measurements of denitrification than have so far been attempted. If the column base is effectively sealed with an unbroken head of water, it may be possible to measure N_2 and N_2O evolution directly by flushing the headspace with 80% He–20% O_2 using a system similar to that described by Galsworthy and Burford (1978). Time would have to be allowed for N_2 initially present in the soil pore space to diffuse from the column before addition of fertilizer and initiation of measurements of gas evolution. An alternative approach would be to leach columns with solutions containing dissolved C_2H_2 using a system similar to that described by Sørensen (1978). Determination of N_2O evolution at the column head and in the leachate will provide a direct estimate of denitrification loss in systems in which nitrate supply is not primarily dependent on nitrification. Nevertheless, the column experiments outlined above will only approximate field situations with high water tables due to the prevention of gas transport at the lower boundary of the column.

4.3.3. Sealed growth chamber studies

Ross *et al.* (1964, 1968) and Stefanson (1970) have described an experimental system which permitted the incubation of soils with growing plants in gas-tight growth chambers. Production of N_2 and N_2O was monitored against a circulating N_2-free atmosphere of 80% Ar–20% O_2 using GC analysis. Although this approach requires an elaborate experimental system, particularly with respect to maintaining N_2-free conditions, valuable information on factors affecting denitrification loss has been attained. Stefanson (1972a, b, c, 1973) used sealed growth chambers to assess the effect of plants, soil-water content, soil organic matter, and form of fertilizer N applied on the extent of denitrification and the distribution of products between N_2 and N_2O.

A more simple approach to the determination of denitrification in soils with intact plants has been described by Smith and Tiedje (1979a). In this study, planted soils in pots were incubated in gas bags with and without added C_2H_2. Total denitrification and the distribution of products was estimated from the production of N_2O in the presence and absence of C_2H_2. Although this less complex system is useful for the evaluation of the relative effects of growing plants on the extent and nature of denitrification loss, the rates of denitrification observed have little relevance to the field situation.

4.3.4. Relevance of laboratory studies to the field

Laboratory studies have provided much useful information on the factors affecting the rate and extent of denitrification in soils and on factors influencing the distribution of products between N_2 and N_2O (Nömmik 1956, Bowman and Focht 1974, Starr and Parlange 1975, Firestone et al. 1980). They have also provided information on the mechanism of denitrification (Payne 1973), Smith and Tiedje 1979b, Firestone et al. 1979a). However it is difficult to extrapolate the findings of laboratory studies to the field due to the disturbance of pore space geometry and conditions affecting gas exchange in the soil between its collection from the field and use in the laboratory. Nevertheless, Schultz et al. (1978) have reported that estimates of denitrification based on laboratory incubation of soil with added nitrate agreed closely with the imbalance in the N budget for the field from which the soil had been collected. These laboratory studies indicated that denitrification rate was primarily dependent on the supply of oxidizable organic C and could be estimated from the rate of CO_2 production.

One of the objectives of some laboratory incubation studies has been to measure the ratio of N_2 to N_2O production under conditions which simulate the field as closely as possible, and to use the ratios to extrapolate field measurements of N_2O loss to estimates of net denitrification loss (Burford and Stefanson 1973, Galsworthy and Burford 1978, Letey et al. 1980). This approach has proved to be extremely difficult due to the wide variation in N_2 to N_2O production in time and with soil and plant factors (Wijler and Delwiche 1954, Nömmik 1956, Stefanson 1973, Firestone et al. 1980). This approach has been further complicated by the observation that significant amounts of N_2O

may also be produced during nitrification in soils (Bremner and Blackmer 1978, 1979, Breitenbeck *et al.* 1980). Hence net production of N_2O at the soil surface may represent contributions from both denitrification and nitrification.

In contrast to the conditions existing in many incubation studies, the use of soil cores and columns attempts to simulate the soil physical conditions occurring in the field; in the case of leached soil columns, there is a limited residence time for nitrate in horizons with high denitrification potential. Leached soil columns may be viewed as model systems for the estimation of denitrification in irrigated agriculture and during periods of drainage in humid regions. In this respect it is interesting to note that estimates of N removal by denitrification based on column studies agree with data obtained in actual field operations during land treatment of wastewater at Flushing Meadows, Arizona (Lance and Whisler 1972, Lance *et al.* 1973). Similarly, denitrification rates for the top 16 cm of a ponded sandy loam were in the range 0.027 to 0.056 μg N/g soil/hr (Volz *et al.* 1975) and compared favourably with those observed in column studies using the same soil (Donor *et al.* 1974). Furthermore, rates and patterns of ^{15}N-labelled N_2 and N_2O emission from columns of a Yolo loam were similar to those observed from field plots on the same soil (Rolston *et al.* 1976).

Column studies are not representative of conditions of wetting and drying cycles that frequently occur during periods of soil water deficit. At these times, the lower part of the soil profile remains dry whilst the upper 10 to 30 cm may be wetted as a result of rainfall or irrigation. Laboratory studies of denitrification aimed at simulating such conditions may be conducted using soil cores taken from the upper portion of the profile and incubated under representative conditions. Erich and Duxbury (1979) have outlined such an approach using the C_2H_2-inhibition technique. Small intact cores were removed from the field and incubated with C_2H_2. The rate of N_2O production was used to estimate instantaneous denitrification rate in the field. This approach is similar to that used for field estimates of N_2 fixation (Goh *et al.* 1978, Hoglund and Brock 1978). Several aspects of the techniques used to measure N_2 fixation could be adopted in estimates of denitrification. In particular, appropriate design of an incubation vessel may permit its return to the position from which the core was taken thereby minimizing the effect of disturbance on soil temperature and denitrification rate (e.g., Ball *et al.* 1979). It may also be feasible to measure both denitrification and N_2 fixation in a single operation for cores in which legumes are present as both estimates rely on the use of C_2H_2.

4.4. Field studies

4.4.1. Indirect estimates

The traditional approach to the field measurement of denitrification has been based on the inbalance in the N budget for a lysimeter or drained field plot. The amount of N leached and that entering the harvested parts of plants is determined and compared with the N input. Denitrification loss has been assumed to be that portion of the N input unaccounted for in the budget (Allison 1955, 1966, Hauck 1971) although this will also include any N loss arising from NH_3 volatilisation.

This approach is valid only for agricultural systems assumed to be at equilibrium with respect to reserves of soil N (Fried et al. 1976). Complications arise if a significant fraction of the total N flow is retained in plant roots, immobilized during microbial decomposition of organic matter or enhanced by mineralization of soil N. This can only be resolved by the use of ^{15}N tracers which permit complete partitioning of applied N. Some recent applications of this technique for estimating denitrification have been reported by Kissel et al. (1977), Jolley and Pierre (1977), Kowalenko and Cameron (1978), Olsen et al. (1979), Rolston et al. (1979) and Riga et al. (1980).

Nitrogen balance studies have also been extended to estimate denitrification loss from freely drained fields where direct measurement of the amount of N leached is not feasible (Lund et al. 1978, Pratt et al. 1978). In these studies, chloride is assumed to be a conservative anion (Kimble et al. 1972, Gast et al. 1974) and nitrogen leached is calculated from changes with time in the nitrate to chloride concentration ratio in the soil solution below the depth of plant uptake using the equation:

$$N_1 = \frac{[NO_3]_s \cdot Cl^-_{ex}}{[Cl^-]_s} \tag{1}$$

N_1 is the N leached, [ ]$_s$ designates the anion concentrations in the soil solution, and Cl^-_{ex} is the amount of chloride in rainfall, irrigation water, or applied directly (Cameron et al. 1978) corrected for the amount removed in the crop.

The assumptions involved in equation (1) are that excess chloride is leached from the root zone, that chloride is not adsorbed on or excluded from mineral or organic matter and that chloride is not released by weathering processes. Although in most cases these assumptions are valid, equation (1) can only be applied to sites that have received relatively uniform management for five to ten years. Another complicating factor is the impact of spatial variability on estimates of anion concentration at a particular depth (Nielsen et al. 1973). Statistical analysis of data from preliminary soil sampling programmes have indicated that reliable estimates of N_1 will only be achieved after removing tens of cores from each study area and taking 10 to 20 samples per core (Pratt *et al.* 1976, 1978, Rible *et al.* 1976). The analytical problems involved with such intensive sampling can be eased to some extent by bulking samples from equivalent depths of the different cores (Pratt *et al.* 1976).

4.4.2. *Measurement of nitrous oxide emission*

For several years, efforts have been made to develop methods for direct estimation of denitrification in the field based on measurement of the efflux of nitrogenous gases from the soil surface. Initial work centred on the measurement of N_2O losses as this gas can be easily measured against its ambient level (Arnold 1954, Burford and Millington 1968, Schutz *et al.* 1970, Rolston *et al.* 1976, Burford and Hall 1977, Focht 1978, Ryden *et al.* 1978, McKenney *et al.* 1978, Denmead 1979, Matthias *et al.* 1980). Since N_2O losses can occur both by denitrification and nitrification (Bremner and Blackmer 1978, 1979, Mosier 1980), measurements of N_2O emission will inevitably include contributions from both processes unless measures are taken to inhibit nitrification. Under field conditions the process of major importance in determining N_2O emission may usually be inferred from prevailing soil conditions (Ryden and Lund 1980, Breitenbeck *et al.* 1980). Measurement of N_2O emission is also complicated by the observation that significant loss of N_2O may occur due to its dissolution in water draining from a soil at field capacity (Guthrie and Duxbury 1978, Burford *et al.* 1981). The importance of this pathway of N_2O loss in relation to emission from the soil surface is difficult to assess from the present literature and requires further study.

In many earlier studies, N_2O emission from a soil surface was calculated from measured N_2O concentration gradients in the profile using equations for gaseous diffusion in soils. The determination of gas flux in this way is discussed under section 4.4.3. Although this approach is subject to many uncertainties, it permits the identification of zones of source and sink activity for N_2O within the soil profile (Rolston *et al.* 1976, Delwiche *et al.* 1978).

Micrometeorological techniques are the preferred approach to the measurement of any gas flux from a land surface for reasons outlined in Chapter 5. The application of this approach to measurement of N_2O losses, however, is subject to severe limitations. In an analysis of micrometeorological techniques, Lemon (1978) demonstrated that even for windspeeds as low as 3 m/sec and N_2O fluxes as high day as 0.24 to 0.48 kg N/ha/day, the differences in atmospheric N_2O concentration between heights of 0.5 and 1.0 m would not exceed the ambient N_2O concentration by more than 1%. This is close to the detection limit for N_2O concentration using the hot ^{63}Ni EC detector after GC separation. As the majority of N_2O fluxes reported to date have been below 0.24 to 0.48 kg N/ha/day (Burford and Hall 1977, McKenney *et al.* 1978, 1980, Rolston *et al.*

1978, Ryden *et al.* 1979b, Ryden and Lund 1980, Bremner *et al.* 1980), use of micrometeorological techniques in the measurement of N_2O emission does not seem feasible in most instances. A similar conclusion has been recorded by Denmead (1979). However, Hutchinson and Mosier (1979) used micro-meteorological techniques to estimate N_2O loss from a cornfield. When N_2O flux was in the range 0.1 to 1.0 kg N/ha/day, estimates based on micro-meteorological measurements were in reasonable agreement with those ob-tained using a soil cover.

The analytical limitations imposed on micrometeorological methods have led several workers to conclude that soil covers offer the only practical approach to the measurement of N_2O flux (Kimball 1978, Rolston 1978, Denmead 1979). Two basic designs have been used in measurements of N_2O flux, and may be termed 'closed' and 'continuous-flow'. The former, in which the change in N_2O concentration in the enclosed airspace is measured with time, has been used by Burford and Hall (1977), Findlay and McKenney (1979), Rolston *et al.* (1978, 1982), Conrad and Seiler (1980) and Matthias *et al.* (1980). The N_2O flux density (f) from the enclosed soil surface is calculated from equation (2):

$$f = (V/A) (273/T) (\Delta C/\Delta t) \tag{2}$$

where V is the volume of the cover, A is the area enclosed, ΔC is the change in N_2O concentration (w/v) during the time (Δt) that the cover is closed, and T is the absolute temperature of the air within the cover.

The closed cover design has been criticised on the grounds that it may lead to an underestimation of N_2O flux due to increasing N_2O concentration in the enclosed airspace (Matthias *et al.* 1978, Rolston *et al.* 1978, Denmead 1979, Focht 1978). Matthias *et al.* (1978) calculated that covers with heights of 5 to 30 cm may lead to an underestimation of N_2O flux by between 10 to 55%. In contrast to theory, McKenney *et al.* (1978), Conrad and Seiler (1980) and Matthias *et al.* (1980) observed a linear increase in N_2O concentration with time during field assessment of closed soil covers. They concluded that increasing N_2O concentration in the enclosed airspace had a negligible effect on the estimation of flux provided a tall chamber was used and the chamber remained in place for only short periods (30 to 60 min). Operational details of closed soil covers are discussed under section 4.3 where their use in the determination of total denitrification loss is described.

With adequate precautions in use (Matthias *et al.* 1980), the closed cover should offer a relatively simple and sensitive approach to direct measurement of N_2O flux over short periods; the detection limit using the closed soil cover is approximately 3×10^{-4} kg N/ha/day. However, for each estimate of flux several determinations of N_2O concentration are required (McKenney *et al.* 1978, Matthias *et al.* 1980, Burford *et al.* 1981).

The continuous flow soil cover provides an alternative approach to determination of N_2O flux. This design minimizes the effect of concentration increase below the cover on the measured flux (Matthias *et al.* 1978, Jury *et al.* 1982) and thereby permits measurements to be made over relatively long periods. During operation, the airspace enclosed by a vented cover is swept by a flow of air drawn from the atmosphere external to the cover. The amount of N_2O in the air leaving the cover is measured by adsorption on 5Å molecular sieve (Schutz *et al.* 1970, Ryden *et al.* 1978) or by infrared analysis (Denmead 1979). The N_2O flux density (f) is calculated from the difference between the N_2O concentration in air drawn through the cover and that in an equivalent flow of air external to the cover (equation 3).

$$f = (R\Delta C/A)\,(273/T) \tag{3}$$

where ΔC is the difference in N_2O concentration (w/v) between the two air flows, R is the air flow rate, A is the area of soil covered and T is the absolute temperature of the air within the cover.

As for the closed soil cover, operational conditions are dictated by the magnitude of the N_2O flux density and the accuracy with which ΔC can be measured. This is coupled with the need to keep ΔC at a minimum, which for a particular flux is achieved by increasing the air flow rate. The detection limit for a flow rate of four exchange volumes per hour and using 5Å molecular sieve and GC analysis is in the order of 0.001 kg N/ha/day. The use of an infrared analyser provides potential for more sensitive measurement of ΔC. In the application using an IR gas analyzer described by Denmead (1979) relatively high flow rates (10 to 30 exchange volumes per hour) resulted in a detection limit of 0.0017 kg N/ha/day. Lower flow rates would have given a sensitivity comparable if not greater than that for the closed cover.

The use of covers in the measurement of gas fluxes from soils has been criticised in that they shield surfaces from pressure fluctuations due to air turbulence which may influence gaseous emission from soils (Kimball and Lemon 1971, Lemon 1978). This problem is minimized in the case of continuous-flow covers in that they are 'coupled' to the external atmosphere by the air-inlet port. Consequently even fairly rapid fluctuations in atmospheric pressure will be experienced within the cover. In the case of closed soil covers, Hutchinson and Mosier (1981) incorporated a vent, the design of which allowed equilibration of pressure fluctuations yet minimised contamination of the accumulating gases with ambient air.

Continuous-flow covers have also been criticised in that small pressure deficits induced by drawing air through the cover may induce mass flow of gases from the profile (Pearson *et al.* 1965, Kanemasu *et al.* 1974, Morgansen 1977). Two studies (Ryden *et al.* 1978, Denmead 1979) have shown that this problem can be avoided in the measurement of N_2O flux by balancing the size of the air-inlet port with the desired range in air-flow rates through the cover.

For pressure deficits less than 0.005 mm H_2O, no effect on the N_2O flux is observed. The size of the air-inlet port required to maintain the pressure deficit within this limit for a particular air flow rate can be calculated from Pouseuille's equation. In the use of continuous-flow soil covers, it is also important that sufficient time is allowed for equilibration between the air flow and the rate of N_2O production before commencing the measurement of ΔC (Denmead 1979).

Another potential problem with respect to the use of soil covers is the induction of elevated temperatures at the enclosed soil surface. Provided that the cover is relatively opaque, discrepancies in air and soil temperature between the enclosed and open surface have never been more than ± 2 to $3°C$ (McGarity and Rajaratnam 1973, Denmead 1979, Matthias et al. 1980). Although soil covers inevitably produce some disturbance to the natural transfer of N_2O to the atmosphere, such effects can be minimized and kept within acceptable limits by careful attention to design and operation of soil covers (Ryden et al. 1978, Denmead 1979, Matthias et al. 1980). Soil covers combine good detection sensitivities and practical simplicity thereby providing a basis for routine monitoring of N_2O emission from soils.

4.4.3. Measurement of total denitrification loss: ^{15}N methods

Methods for direct measurement of denitrification using ^{15}N depend on applying highly enriched ^{15}N fertilizer to soil in the field followed by measurement of the isotopic enrichment of evolved gases. This method measures only denitrification from applied fertilizer and cannot measure denitrification from native or residual N. Of the two major gases from denitrification, the isotopic composition of N_2 is most important. In general, the concentration or total mass of N_2O in a gas sample is too small to accurately measure the isotopic composition of the N_2O produced unless the sample is concentrated. Thus, measured N_2O usually includes that derived from both fertilizer and residual N.

The cost of highly enriched ^{15}N fertilizer limits field experiments to relatively small plots. The measurement of evolving gases must be made in the centre of a plot of sufficient area that the sample represents that coming only from the ^{15}N fertilizer and not that from non-labelled nitrogen outside the plot area. Rolston et al. (1976) used a plot of 4.4 m^2 with sampling occurring within the centre 1 m^2 of the plot. For cases where most of the denitrification occurs in the upper few cm of soil, the plot can be isolated from the rest of the soil using barriers of sufficient depth to minimize lateral diffusion of applied fertilizer and evolved gases. Rolston et al. (1978, 1982) used 0.6 m deep wood borders to isolate each 1 m^2 plot. For situations where denitrification occurs much deeper than the 0.3 m profile depth, this method will not result in accurate measurements of denitrification.

It is important to point out that the ^{15}N methods discussed below are generally used to measure the surface flux of the denitrification gases. Thus, dependent on the depth of denitrification activity, a time lag between gas production and the appearance of gas at the soil surface will occur, especially in very wet soils. This time lag becomes especially important in the case of N_2O, as N_2O may be further reduced to N_2 during transport to the soil surface. Thus, measurement of the surface flux of N_2O and N_2 may not be equal to the rate of production of N_2O and N_2 within the soil.

Two approaches have been used to determine the flux of ^{15}N-labelled gases at the soil surface. These are (i) the calculation of flux from gaseous diffusion theory and (ii) the collection and analysis of labelled gases using the closed soil cover technique described under Section 4.4.2.

Calculation of flux from gaseous diffusion theory is based on independent measurements of the soil-gas diffusivity and gas concentration gradients followed by the use of Fick's Law to calculate the flux of gas within the soil profile or at the soil surface. This approach has been used by Burford and Millington (1968) and Burford and Stefanson (1973) for determining the flux of N_2O and by Rolston *et al.* (1976) for measuring the flux of both N_2O and ^{15}N-labelled N_2 from field soils. To determine the surface flux, the concentration gradients of the denitrification gases should be measured very near the soil surface.

The steady flux of gases in soil can be described by Fick's law:

$$\frac{F}{At} = f = - D_p \frac{dC}{dx} \tag{4}$$

where F is the amount of gas diffusing (kg gas), A is the cross sectional area of the soil (m^2 soil), t is time (sec), f is the gas flux density (kg gas/m^2 soil/sec), C is the concentration in the gaseous phase (kg gas/m^3 soil air), x is distance (m soil), and D_p is the soil-gas diffusivity (m^3 soil air/m soil/sec). The total amount, F, diffusing from the soil or across any particular depth in the soil over some time period t_1 can be calculated from

$$F = - AD_p \int_0^{t_1} \frac{dC}{dx} \, dt \tag{5}$$

In order to apply either equation (4) or (5), the soil-gas diffusivity D_p, and the concentration gradient, dC/dx, must be known at the depth for which the gas flux is to be calculated.

Values of D_p can be measured or calculated by the method discussed by Rolston (1982). The concentration gradient is measured by sampling the soil-gas concentration at a minimum of two depths within the soil. These depths should be as close to each other as possible, yet be far enough apart that a measurable difference in concentration is detected. Concentrations must be measured *in situ* by removal of small samples of gas at the depths of interest. Various samplers have been devised (Taylor and Abrahams 1953, Dowdell *et al.* 1972, De Camargo *et al.* 1974, Jorgenson 1974, Rolston 1981) but all must be designed with a small internal gas volume, as the sampler must be purged before a sample is removed for analysis. For denitrification, it is of particular interest to measure the gas flux at the soil surface. Thus, the concentration should be measured at $x = 0$ and $x = 2$ or 5 cm below the soil surface. It is also important to measure gas concentration at a particular depth at several locations in order to determine an 'average' concentration for that depth. The number of samples required will depend on the soil variability. Rolston *et al.* (1979) used 5 samplers at each depth within 1 m^2 plots.

The greatest uncertainties in calculating the flux of denitrification gases according to equation (4) arise from the uncertainties and variability of soil-gas diffusivity and concentration gradients within field soils (Rolston 1978). This uncertainty can be especially large for soils near saturation, the condition at which much of the denitrification activity in soils may occur. Associated with the variability in the concentration gradient is the inability to adequately determine the slope and shape of the concentration gradient near the surface.

The primary advantage of this method is that it does not greatly disturb the soil environment by placement of chambers or other devices over the soil surface. If samples are taken at several depths, the method gives information about gas concentrations within the soil profile and the zones of sources and sinks for gases produced during denitrification. The method requires a large number of samples to adequately determine soil-gas diffusivity and concentration gradients. However, increased sensitivity in measuring a denitrification rate or surface flux is possible due to the fact that higher concentrations of ^{15}N-labelled N$_2$ gas can be obtained within the soil pore space than within chambers placed over the soil due to dilution of ^{15}N-labelled N$_2$ gas by air within the chamber. The detection limit for the method using gas diffusion theory is about 0.1 kg N/ha/day for a fertilizer enrichment of 10% ^{15}N (Rolston *et al.* 1976).

The concept of the closed soil cover technique has been discussed under Section 4.4.2. In its application to the measurement of total denitrification loss in combination with ^{15}N-labelled fertilizer, the surface flux is determined directly, and is not dependent on measurement of individual parameters as is the case in the use of diffusion theory. However, only the flux at the soil surface is determined and no information on flux at deeper depths in the soil profile is provided.

Closed soil covers permit measurement of gas flux for only short periods following placement of the chamber over the soil surface. As soon as a small amount of gas diffuses into the chamber, the concentration at the soil surface increases, the concentration gradient decreases, and the flux must also decrease. In practice, the cover is placed over the soil surface, the increase in the concentration of the gas of interest within the chamber is measured over a short time period, and the cover is removed. This procedure should be repeated several times in order to estimate fluxes during daily or longer intervals. The measured gas concentration is used to calculate the flux from the soil surface (equation 2).

The use of a closed cover is necessitated in measurements of denitrification using ^{15}N as considerable dilution of ^{15}N labelled N_2 occurs when the gas diffuses into a chamber at atmospheric N_2 concentrations. A closed soil cover results in the least amount of dilution. A continuous-flow cover would cause so much dilution that measurement of ^{15}N-labelled N_2 emanating from the soil would be impossible.

Several closed cover designs have been reported. Rolston *et al.* (1978, 1982) used a chamber consisting of a thick sheet of acrylic plastic with rubber tubing on the lower edge to make an airtight seal at the top of a wood border inserted to the 0.6 m soil depth. This design was necessary in order to measure the entire amount of ^{15}N-labelled N_2 gas evolved from a small plot enclosed by the wooden borders. Other chambers have been described by Findlay and McKenney (1979), Matthias *et al.* (1980), and Hutchinson and Mosier (1981). The cover design used by Findlay and McKenney (1979) and Huchinson and Mosier (1981) involves insertion of its base into the soil surface. The design of Matthias *et al.* (1980) allows the cover to rest on the soil surface with a seal formed by a high density polyurethane foam collar. A larger polyethylene film is used to enclose the cover. This film acts as a windbreak to prevent or minimize wind induced movement of air into and out of the chamber. Sample bottles (Matthias *et al.* 1980) and gas-tight syringes (Rolston *et al.* 1978, 1982, Hutchinson and Mosier 1981, Findlay and McKenney 1979) have been used to remove gas samples from the chamber.

As discussed in Section 4.4.2, a potential error associated with closed covers is the underestimation of gas flux due to the increase in gas concentration within the cover. Matthias *et al.* (1978) and Jury *et al.* (1982) conducted a simulation analysis of the closed cover method for measuring nitrous oxide production in the field. Their analyses show that the flux of gas measured using a closed cover can be in error for high soil air contents if the chamber is left on the surface for long periods.

Rolston *et al.* (1978) proposed a technique for correcting the flux due to the concentration increase. This involved sampling the gas concentration at one depth within the soil. Hutchinson and Mosier (1981) developed a correction method which involved sampling the chamber concentration at three times separated by equal intervals but did not require sampling within the soil.

Matthias *et al.* (1978) demonstrated that errors in flux estimates could be reduced to between 1 and 4% by using a cover height of at least 15 cm and an exponential curve fitting technique which permitted an estimation of the initial slope of the relationship between measured gas concentration and time. If the gas concentration within the cover as a function of time is not linear, one of the correction techniques described above must be employed.

The detection limit for ^{15}N-labelled N_2 gas using closed covers is dependent on the enrichment of the applied fertilizer, the volume of the cover, the precision of mass spectroscopic analysis and the time the covers are left in place over the soil. For the design and fertilizer enrichment (55% ^{15}N) used by Rolston *et al.* (1982) the minimum detection limit for ^{15}N-labelled N_2 was 0.1 to 0.2 kg N/ha/day. Siegel *et al.* (1982) used a very sensitive mass spectrometer and high enrichments (64% ^{15}N) to measure fluxes as small as 2.5 g N/ha/day.

4.4.4. Measurement of total denitrification loss: acetylene inhibition

A potentially versatile and widely applicable approach to direct field estimates of denitrification loss lies in the inhibition of N_2O reduction during denitrification by small concentrations of acetylene (section 4.2.5.). Field application of this technique has been described by Patriquin *et al.* (1978) and Ryden *et al.* (1979b), whilst Chan and Knowles (1979) have reported an application to *in situ* measurement of denitrification in freshwater sediments. Patriquin *et al.* (1978) measured the rate of N_2O evolution into the airspace enclosed by soil covers, some of which contained 10% (v/v) C_2H_2. Although approximately twice as much N_2O was measured below the covers containing C_2H_2, it was not possible to ascertain whether the field methods used achieved complete inhibition of N_2O reduction.

Successful application of the C_2H_2-inhibition technique depends on the distribution of C_2H_2 at concentrations effective in inhibiting N_2O reduction over the depth of soil in which denitrification is expected to occur. This is followed by the measurement of N_2O flux using one of the techniques outlined in Section 4.4.2. An adequate C_2H_2 distribution in the soil profile may be achieved by injecting C_2H_2 into the soil through probes inserted to a depth below which denitrification is expected to be very low. Using this technique, Ryden *et al.* (1979b) have shown that sustained C_2H_2 concentrations effective in the inhibition of N_2O reduction, can be established and maintained in the atmosphere of a moderate volume of undisturbed soil. It should be noted that the development of low C_2H_2 concentrations in the soil atmosphere using this technique does not involve mass flow of C_2H_2 through the soil pore space. Acetylene diffuses radially into the soil pore space from the 'column' of high concentration generated along the length of the supply probe.

Data in Table 4.2 indicate that C_2H_2 concentrations effective in inhibition of

Table 4.2. Mean acetylene concentrations in the air-filled pore space at various depths in different soil profiles during and following supply of acetylene through probes inserted into the soil

Description of field site	C$_2$H$_2$ supply	Time after initiating C$_2$H$_2$ supply (h)	C$_2$H$_2$ concentration at the depth specified (% v/v)			
			7.5 cm	15 cm	30 cm	60 cm
Irrigated loam cropped to celery. Air-filled porosity 10% throughout profile	Continuous at 233 ml/min	0.5	0.15	0.33	0.32	0.18
		1.0	0.40	0.69	0.68	0.48
		2.0	0.45	0.85	1.24	0.77
		5.0	0.43	1.02	1.64	0.90
			10 cm	20 cm	40 cm	60 cm
Drained loam under rye-grass. Air-filled porosity 20% throughout profile	800 ml/min for 15 min; then 300 ml/min for an additional hour	1.0	0.28	0.62	0.54	0.42
		2.0	0.69	0.94	0.74	0.61
		4.0	0.66	0.94	0.67	0.68

N_2O reduction may be established within a reasonable time following initiation of C_2H_2 supply. Even for the irrigated soil, in which saturated conditions prevailed in the upper few centimetres of the profile for 12 to 24 hours after irrigation, adequate C_2H_2 concentrations could be induced within one hour. Variability of C_2H_2 concentration within the profile poses no problem as inhibition of N_2O reduction is effective over a wide range in C_2H_2 concentration (Yoshinari et al. 1977, Ryden et al. 1979a). Addition of ^{15}N-labelled nitrate can be used to demonstrate essentially complete inhibition of N_2O reduction in a C_2H_2-treated field plot. Ryden et al. (1979b) observed no [30]N_2 evolution from C_2H_2-treated plots that had received highly enriched ^{15}N-labelled KNO_3.

The time required for C_2H_2 treatment before appropriate concentrations are established in the soil space can be calculated from diffusion theory. The data in Table 4.2 show reasonable agreement with treatment times calculated by Ryden et al. (1979b). It should be noted, however, that these data relate only to C_2H_2 concentration in the macropore space. Nevertheless, calculations indicate that C_2H_2 concentrations effective in inhibition of N_2O reduction are established at the centre of even saturated soil crumbs up to 1 cm in diameter within 10 to 20 min of the external concentration reaching 0.5 to 1% (Ryden et al. 1979b).

It is essential that critical aspects of the C_2H_2-inhibition technique be evaluated before its application to the measurement of denitrification at a particular field site. Of crucial importance is the determination of the minimum C_2H_2 concentration required to effect complete inhibition of N_2O reduction over the range in nitrate concentration likely to be experienced in the soil. The ease with which C_2H_2 can be distributed in the soil profile must also be determined for the range in air-filled porespace expected during the period of study.

Following the establishment of suitable C_2H_2 concentrations in a soil profile, an equilibration period is required for the rate of N_2O evolution at the soil surface to approach a relatively steady state. Unpublished results (Ryden) indicate that for soils in which denitrification occurs in the upper 10 to 20 cm of the profile the rate of N_2O evolution increased by no more than 5 to 8% during a 4 h period following the 1.25 h C_2H_2 treatment described in Table 4.2. The rate at which N_2O evolution reaches an apparent equilibrium following or during C_2H_2 treatment will depend on several site-specific factors. Possibly the most important of these are soil water content and the depth in the soil at which denitrification occurs. This facet of the C_2H_2-inhibition technique must also be evaluated before its routine application at a particular site.

Distribution of C_2H_2 is expected to pose special problems in poorly drained and paddy soils where saturated conditions develop at various times during the year. At such sites prolonged C_2H_2 treatment may be required, although the relatively high solubility of C_2H_2 will facilitate the development of high concentration gradients and thereby diffusion of the gas in the aqueous phase. Chan and Knowles (1979) have described a system for measurement of denitrification in surficial lake sediments which may have application to the measurement of denitrification in flooded paddy soils. In this system, C_2H_2 was added to water circulating through an enclosed chamber placed over the sediment surface to establish a concentration of 5 to 10% C_2H_2 in solution. Production of N_2O in the presence of C_2H_2 was determined from the increase in dissolved N_2O concentration in the confined water volume.

Denitrification measured using the C_2H_2-inhibition technique has been compared with estimates made using the ^{15}N-soil cover method described in section 4.4.3 (Rolston *et al.* 1982); the pattern of denitrification was similar. However, immediately after each irrigation, gas flux measured using the C_2H_2-inhibition technique was frequently greater than that using [15]N. With increasing time after irrigation, the opposite tended to prevail. The differences in gas flux measured by the two methods can probably be attributed to differences in the rates of production and diffusion of N_2O and of $N_2 + N_2O$ in the presence and absence of C_2H_2, respectively. Differences in gas flux could also arise from differences in the cover designs used to collect gases and to the appreciable difference in the detection limit for each method. Many of the measurements using ^{15}N were at the detection limit for this method. Despite the discrepancies between individual flux measurements, estimates of denitrification loss during six weeks for three different irrigation treatments were very similar for both methods.

A further aspect of the technique requiring evaluation is the possibility that conditions favouring the decomposition of C_2H_2 or its utilization as a substrate in denitrification may develop during routine field studies (Watanabe and de Guzman 1980, Germon 1980). Ryden and Dawson (1982) found no evidence for either possibility during incubation of soils collected from field locations that had received twenty 1-h treatments with C_2H_2 during a study of denitrification lasting 37 days.

Walter *et al.* (1979) have criticised field application of the C_2H_2-inhibition method in that C_2H_2 also inhibits nitrification. Their criticism was based on findings in laboratory studies and on the assumption that inhibition of nitrification may significantly alter the oxygen demand within a soil. Although this may pose problems during concurrent nitrification and denitrification in the vicinity of banded ammonical fertilizers, the overall impact on estimates of denitrification will depend on the contribution of nitrification to the oxygen demand of the soil as a whole. Under most circumstances this contribution to oxygen demand is not expected to be large. Furthermore, in a wet soil of low oxygen content, autotropic nitrifiers will be poor competitors for oxygen against the heterotropic population (Focht and Verstraete 1977).

The major potential problem associated with the coupled inhibition of nitrification by C_2H_2 is the exhaustion of nitrate if C_2H_2 is repeatedly applied to the same location in the field (Ryden 1982). Ryden and Dawson (1982), however, have shown that short-term (1 to 2 h) treatment with C_2H_2 of a fallow soil and soils under grass at intervals as frequent as 3 out of 4 days had no significant effect on the distribution of inorganic N between nitrate and ammonium. This observation suggests that nitrification activity in soils in the field recovers more rapidly than that in soils incubated with C_2H_2 in the laboratory. Walter et al. (1979) found that nitrification was affected for 13 to 17 days following exposure of a moist soil to C_2H_2 for 24 h.

If the inhibition of nitrification by C_2H_2 leads to appreciable effects on the amounts of nitrate and ammonium in soils at other sites, this problem will probably be mitigated by limiting C_2H_2 treatments to periods of only one to two days at the same location. Denitrification over longer periods may be assessed from measurements made at different locations within the same study area at different times. Such an approach will permit the recovery of nitrification activity before denitrification is again measured at the same location.

With respect to the effect of coupled inhibition of nitrification on field estimates of denitrification, Lund et al. (1978) and Ryden and Lund (1980) obtained close agreement between indirect estimates of denitrification loss and those measured using C_2H_2 inhibition in irrigated vegetable fields. This was achieved despite the fact that essentially all the fertilizer N was banded in ammoniacal froms and that no special procedures were adopted to account for coupled inhibition of nitrification.

The C_2H_2-inhibition method appears to provide a particularly versatile approach to the direct measurement of denitrification loss. However, the method must be carefully evaluated for any limitations before application to routine measurements at a particular field site. The field equipment and laboratory instrumentation are relatively inexpensive and simple to operate. Furthermore, the detection limit for net denitrification flux is the same as that for the method used in the measurement of N_2O evolution. As fluxes as low as 0.001 kg N/ha/day can be measured using a continous-flow soil cover, the technique is at least two orders of magnitude more sensitive than the ^{15}N method reported by Rolston et al. (1982).

4.4.5. Some direct estimates of denitrification loss from fields

The rate and extent of denitrification in soils is dependent on several interacting factors (Focht and Verstraete 1977). These include nitrate concentration, available carbon, temperature, pH and soil water content and its relationship with aeration. The effect of these factors on denitrification has been evaluated in numerous laboratory studies (Wijler and Delwiche 1954, Bremner and Shaw 1958, Bowman and Focht 1974, Keeney *et al.* 1979). However, there are few field studies from which the impact of these factors can be assessed. In the field, the presence of plants may further complicate the interaction of these factors in that they affect soil-water and O_2 relationships and the pools of available nitrate and carbon.

The pattern of denitrification loss following fertilization and irrigation or rainfall is typified by the relationships in Fig. 4.2. Maximum denitrification loss is usually observed within 24 to 48 hours after the onset of conditions conducive to denitrification (Rolston *et al.* 1976, 1978, 1982, Ryden *et al.* 1979b, Ryden and Lund 1980). The time required to reach a peak denitrification rate depends on the ease of diffusion of gases from the profile (Fig. 4.2a) and in some cases on the time required for the development of zones of O_2 depletion within the soil after it is wetted by rainfall or irrigation (Fig. 4.2b). The decline in denitrification loss reflects the movement of nitrate out of the depth of maximum denitrification potential (the case for the relationship in Fig. 4.2a), the drying of the profile allowing greater aeration (Fig. 4.2b), or the uptake of nitrate by plants thereby limiting the supply available for denitrification. Recent work (Ryden 1981) indicates that plant uptake of nitrate is a major factor limiting denitrification loss from grassland in humid temperate zones. The pattern of denitrification loss is also strongly affected by soil water potential and its relationship with soil aeration. For soils in which denitrifica-

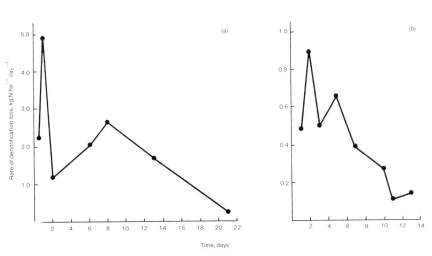

Fig. 4.2. Rates of denitrification loss (a) following fertilizer application to a wet soil under ryegrass maintained at a soil water potential of -14.7 mbar and (b) following N application to and irrigation of an initially dry soil (-400 mbar) cropped to artichokes. Data in (a) were obtained using ^{15}N-labelled fertilizer (Rolston *et al.* 1978) and in (b) using the C_2H_2-inhibition technique (Ryden and Lund 1980).

tion was not limited by nitrate concentration, significant N loss was associated with soil water potential in the range -50 to -200 mbar (Ryden and Lund 1980).

The effects of organic carbon level and temperature on rates of denitrification are indicated by data in Table 4.3. For Yolo loam, the maximum rate of denitrification at $23°$ C for plots cropped to ryegrass was twice that for the uncropped plots. Application of manure to the uncropped plots increased the maximum rate of loss by a factor of approximately 34. Effects of organic matter input for Yolo loam are apparent in Table 4.3 and have also been reported by Ryden et al. (1979b). After incorporation of residues from a celery crop and a salt-elution irrigation, rates of denitrification loss from a Salinas loam were twice those following the final irrigation before harvest.

Lower rates of denitrification from the Yolo loam were observed during winter studies when the soil temperature averaged $8°$ C. With the exception of the manure-treated plots, rates of denitrification loss were $\leqslant 1$ kg N/ha/day, the detection limit of the ^{15}N technique used in these studies. The effect of low winter temperatures on denitrification loss is also indicated by rates for a Wickham loam under ryegrass. These low rates of N loss also reflect a limitation on denitrification imposed by the low soil nitrate concentrations that prevailed during winter.

Rates of denitrification loss from a Salinas loam cropped to vegetables were similar during summer irrigation and winter rainfall. The mean rate of N loss during winter rain, however, was twice that during the period of irrigation due to the maintenance, over longer periods, of a soil water regime conducive to denitrification. Only towards the end of winter did soil nitrate concentration fall to a level expected to limit denitrification loss (Ryden and Lund 1980). Rates of denitrification loss from Arlington sandy loam of low organic C content never exceeded 0.05 kg N/ha/day and resulted in a loss of only 1.5 kg N/ha during two months.

Rates of denitrification loss from a Wickham loam under ryegrass were at a maximum when N fertilizer application coincided with frequent rain. This was particularly apparent following fertilizer application to the soil at field capacity in spring; 13% of the N applied was lost. At the higher rate of fertilization, denitrification losses between 0.3 and 0.7 kg N/ha/day were maintained for two to three weeks following fertilizer application. Denitrification losses declined when soil water content fell below 20% (Ryden 1981). This decline was compounded by the decrease in soil nitrate concentration due to plant uptake.

In the studies summarized in Table 4.3, N_2O losses were also measured. As much of the N_2O loss measured in the studies cited could be attributed to denitrification, it is worth assessing the impact of soil conditions on the proportion of total N loss occuring as N_2O. For the Yolo loam summer experiment, the proportion of N lost as N_2O decreased with increasing organic

Table 4.3. Some field measurements of denitrification loss[a]

Soil type and location	Land use	Soil water potential (m bar)	Soil temperature (°C)	Soil organic C (mg/g)	N applied (kg N/ha)	Range in rate of denitrification loss (kg N/ha/day)	Estimated total N loss (kg N/ha)	Proportion lost as N_2O (%)	Technique used
Yolo loam (California)	Uncropped	−14.7	23	14.0	300	<1.0–2.2	8[b]	27.0	^{15}N
		−68.6	23	14.0	300	<1.0–1.2	4[b]	14.3	^{15}N
	Ryegrass	−14.7	23	n.a.	300	<1.0–4.9	34[b]	12.5	^{15}N
		−68.6	23	n.a.	300	<1.0–2.2	9[b]	20.4	^{15}N
	Uncropped with manure	−14.7	23	16.5	300	<1.0–75	208[b]	4.7	^{15}N
		−68.7	23	16.5	300	<1.0–13	47[b]	11.3	^{15}N
	Ryegrass	−7.8	8	n.a.	300	≤1.0	19.1[b]	3.8	^{15}N
	Ryegrass	0 to −600	30	n.a.	285	0–0.5	0.7–1.5[c]	16–27	^{15}N and C_2H_2
	Ryegrass with straw	0 to −600	23	n.a.	285	0–10	1.8–6.4[c]	12–22	^{15}N
Salinas loam (California)	Irrigated vegetable production	−50 to −800	~20	7.8	620/year	0.012–1.92	155/year	18.5	C_2H_2
		−50 to −100	~15			0.100–2.88		9.4	C_2H_2
Arlington sandy loam (California)	Irrigated citrus production	n.a.[d]	~25	2.0	100	0.005–0.050	1.5[e]	61.4	C_2H_2
Wickham loam (S. England)	Ryegrass	n.a.	15–20	34.7	250/year	0.005–0.500	10.8/year	11.8	C_2H_2
		n.a.	2–6	34.7		0.001–0.008		1.8	
		n.a.	15–20	34.7	500/year	0.010–0.750	28.3/year	28.1	C_2H_2
		n.a.	2–6	34.7		0.005–0.015		5.4	C_2H_2
		−20 to −100	10–12	34.7	125	0.29–1.15	16.3[f]	16.8	C_2H_2

[a] Data for Yolo loam from Rolston et al. (1978, 1982), for Salinas loam and Arlington sandy loam from Ryden and Lund (1980) and for Wickham loam from Ryden (1981)
[b] During 10 to 20 days
[c] During 50 days
[d] Not available
[e] During 60 days
[f] During 28 days in spring

C and soil water potential. This probably reflected the more vigourous reducing conditions prevailing in these treatments, resulting in more complete reduction of N_2O to N_2. Similarly, proportionate loss as N_2O was higher for the wetting and drying cycles during irrigation of the Salinas and Yolo loams than that associated with the continuously wet soil conditions associated with winter rain. A similar observation was made for the winter and summer portions of the study on the Wickham loam. However, the higher proportionate N_2O loss during the summer may also be attributable in part to the higher soil nitrate concentrations associated with this period during which fertilizer was applied. Ryden and Lund (1980) also observed that N_2O emission increased from approximately 14 to 30% of the total denitrification loss from plots on the same soil but with pH values of 7.6 and 6.5, respectively.

The direct estimates of field denitrification loss available at the time of writing reflect the numerous factors affecting denitrification in soils. It is reassuring that patterns of denitrification loss observed in field studies are consistent with concepts of denitrification established in the laboratory. The dynamic nature of denitrification loss (Fig. 4.2) demands detailed field measurements to obtain estimates of annual N loss from different agricultural systems. However, if relationships can be developed between more easily measured soil conditions (e.g. nitrate concentration or soil water regime) and rates of denitrification loss for broad categories of soil type and land use, it may be possible to reduce the number of measurements of gas production while maintaining monitoring of soil conditions determining denitrification loss.

4.5. References

Allison, F.E. 1955 The enigma of soil nitrogen balance sheets. Adv. Agron. 7, 213–250.

Allison, F.E. 1966 The fate of nitrogen applied to soils. Adv. Agron. 18, 219–258.

Arnold, P.W. 1954 Loss of nitrous oxide from soil. J. Soil Sci. 5, 116–128.

Bailey, L.D. and Beauchamp, E.G. 1973 Gas chromatography of gases emanating from a saturated soil system. Can. J. Soil Sci. 53, 122–124.

Balderston, W.L., Sherr, B. and Payne, W.J. 1976 Blockage by acetylene of nitrous oxide reduction in *Pseudomonas perfectomarinus*. Appl. Environ. Microbiol.31, 504–508.

Ball, R., Brougham, R.W., Brock, J.L., Crush, J.R., Hoglund, J.H. and Carran, R.A. 1979 Nitrogen fixation in pasture I. Introduction and general methods. N.Z. J. Exp. Agric. 7, 1–5.

Beard, W.E. and Guenzi, W.D. 1976 Separation of soil atmospheric gases by gas chromatography with parallel columns. Soil Sci. Soc. Am. J. 40, 319–321.

Beer, K., Lippold, H. and Ackerman, W. 1977 Emission spectrometric determination of nitrogen-15 in trace amounts of N_2, NO and N_2O after gas chromatographic separation: a new combined method for investigating denitrification in soil. In: Stable Isotopes in the Life Sciences, pp 189–196. IAEA, Vienna.

Birch, H.F. 1958 The effect of soil drying on humus decomposition and nitrogen availability. Plant Soil 10, 9–31.

Blackmer, A.M. and Bremner, J.M. 1977a Gas chromatographic analysis of soil atmospheres. Soil Sci. Soc. Am. J. 41, 908–912.

Blackmer, A.M. and Bremner, J.M. 1977b Denitrification of nitrate in soils under different atmospheres. Soil Biol. Biochem. 9, 141–142.

Blackmer, A.M. and Bremner, J.M. 1978 Determination of nitrous oxide in air. Agron. Abst. p. 137.

Blackmer, A.M., Baker, J.H. and Weeks, M.E. 1974 A simple gas chromatographic method for separation of gases in soil atmospheres. Soil Sci. Soc. Am. Proc. 38, 689–690.

Bowman, R.A. and Focht, D.D. 1974 The influence of glucose and nitrate concentrations on denitrification rates in sandy soils. Soil Biol. Biochem. 6, 297–301.

Breitenbeck, G.A., Blackmer, A.M. and Bremner, J.M. 1980 Effects of different nitrogen fertilizers on emission of nitrous oxide from soil. Geophys. Res. Lett. 7, 85–88.

Bremner, J.M. 1965a Inorganic forms of nitrogen. In: Black, C.A. (ed.), Methods of Soil Analysis, Part 2. Agronomy 9, pp. 1179–1237. American Society of Agronomy, Madison.

Bremner, J.M. 1965b Isotope ratio analysis of nitrogen in nitrogen-15 tracer investigations. In: Black, C.A. (ed.), Methods of Soil Analysis, Part 2. Agronomy 9, pp. 1256–1286. American Society of Agronomy, Madison.

Bremner, J.M. and Blackmer, A.M. 1978 Nitrous oxide: Emission from soils during nitrification of fertilizer nitrogen. Science (Wash. D.C.) 199, 295–297.

Bremner, J.M. and Blackmer, A.M. 1979 Effects of acetylene and soil water content on emission of nitrous oxide from soils. Nature (Lond.) 280, 380–381.

Bremner, J.M. and Shaw, K. 1958 Denitrification in soil. II. Factors affecting denitrification. J. Agric. Sci. 51, 40–52.

Bremner, J.M., Robbins, S.G. and Blackmer, A.M. 1980 Seasonal variability in emission of nitrous oxide from soil. Geophys. Res. Lett. 7, 641–644.

Brice, K.A., Eggleton, A.E.J. and Penkett, S.A. 1977 An important ground surface sink for atmospheric nitrous oxide. Nature (Lond.) 268, 127–129.

Burford, J.R. 1969 Single sample analysis of N_2-N_2O-CO_2-Ar-O_2 mixtures by gas chromatography. J. Chromatogr. Sci. 7, 760–762.

Burford, J.R. and Hall, K.C. 1977 Fluxes of nitrous oxide from a ryegrass sward. Agricultural Research Council, Letcombe Laboratory Annual Report, 1976. 85–88.

Burford, J.R. and Millington, R.J. 1968 Nitrous oxide in the atmosphere of a red-brown earth. Trans. 9th Int. Congr Soil Sci. Adelaide. 11, 505–511.

Burford, J.R. and Stefanson, R.C. 1973 Measurement of gaseous losses of nitrogen from soils. Soil Biol. Biochem. 5, 133–141.

Burford, J.R., Dowdell, R.J. and Crees, R. 1981 Emission of nitrous oxide to the atmosphere from direct-drilled and ploughed clay soils. J. Sci. Food Agric. 32, 219–223.

Cameron, D.R., Kowalenko, C.G. and Ivarson, K.C. 1978 Nitrogen and chloride distribution and balance in a clay loam soil. Can. J. Soil Sci. 58, 77–88.

Cawse, P.A. 1967 The determination of nitrate in soil solutions by ultraviolet spectrophotometry. Analyst (Lond.) 92, 311–315.

Chan, Y.K. and Knowles, R. 1979 Measurement of denitrification in two fresh water sediments by an in-situ acetylene inhibition method. Appl. Environ. Microbiol. 37, 1067–1072.

Cho, C.M., Sakdinan, L. and Chang, C. 1979 Denitrification intensity and capacity of three irrigated Alberta soils. Soil Sci. Soc. Am. J. 43, 945–950.

Cicerone, R.J., Shetter, J.D., Stedman, D.H., Kelley, T.J. and Liu, S.C. 1978 Atmospheric N_2O: Measurements to determine its sources, sinks and variations. J. Geophys. Res. 83, 3042–3050.

Conrad, R. and Seiler, W. 1980 Field measurements of the loss of fertilizer nitrogen into the atmosphere as nitrous oxide. Atmos. Environ. 14, 555–558.

De Camargo, O.A., Grohmann, F., Salati, E. and Matsui, E. 1974 A technique for sampling the soil atmosphere. Soil Sci. 117, 173–174.

Delwiche, C.C. and Rolston, D.E. 1976 Measurement of small nitrous oxide concentrations by gas chromatography. Soil Sci. Soc. Am. J. 40, 324–327.

Delwiche, C.C., Bissell, S. and Virginia, R. 1978 Soil and other sources of nitrous oxide. In: Nielsen, D.R. and MacDonald J.G. (eds.), Nitrogen in the Environment, Vol. 1, pp. 459–476. Academic Press, New York.

Denmead, O.T. 1979 Chamber systems for measuring nitrous oxide emission from soils in the field. Soil Sci. Soc. Am. J. 43, 89–95.

Dick, W.A. and Tabatabai, M.A. 1979 Ion chromatographic determination of sulfate and nitrate in soils. Soil Sci. Soc. Am. J. 43, 899–904.

Doner, H.E., Volz, M.G. and McLaren, A.D. 1974 Column studies of denitrification in soil. Soil Biol. Biochem 6, 341–346.

Dowdell, R.J. and Crees, R. 1974 Measurement of the nitrous oxide content of the atmosphere. Lab. Pract. 23, 488–489.

Dowdell, R.J. and Webster, C.P. 1980 A lysimeter study using nitrogen-15 on the uptake of fertilizer nitrogen by perennial ryegrass swards and losses by leaching. J. Soil Sci. 31, 65–75.

Dowdell, R.J., Smith, K.A., Crees, R. and Restall, S.W.F. 1972 Field studies of ethylene in the soil atmosphere – equipment and preliminary results. Soil Biol. Biochem. 4, 325–331.

Edwards, R.M., Sexstone, A.J. and Tiedje, J.M. 1980 Simultaneous determination of parts-per-billion amounts of nitrate- and nitrite-nitrogen by high pressure liquid chromatography. Agron. Abst. p. 152.

Erich, M.S. and Duxbury, J.M. 1979 Denitrification in Histosols. Agron. Abst. p. 156.

Espinoza, W., Gast, R.G. and Adams, R.S. 1975 Charge characteristics and nitrate retention by two Andepts from south-central Chile. Soil Sci. Soc. Am. Proc. 39, 842–846.

Federova, R.I., Milekhina, E.I. and Il'Yukina, N.I. 1973 Evaluation of the method of 'gas metabolism' for detecting extra-terrestrial life. Identification of nitrogen fixing organisms. Izv. Akad. Nauk. SSSR, Ser. Biol. 6, 797–806.

Fiedler, R. and Proksch, G. 1975 The determination of nitrogen-15 by emission and mass spectrometry in biochemical analysis: A review. Anal. Chim. Acta. 78, 1–62.

Findlay, W.I. and McKenney, D.J. 1979 Direct measurement of nitrous oxide flux from soil. Can. J. Soil Sci. 59, 413–421.

Firestone, M.K., Firestone, R.B. and Tiedje, J.M. 1979a Nitric oxide as an intermediate in denitrification: Evidence from nitrogen-13 isotope exchange. Biochem. Biophys. Res. Commun. 91, 10–16.

Firestone, M.K., Smith, M.S., Firestone, R.B. and Tiedje, J.M. 1979b The influence of nitrate, nitrite and oxygen on the composition of the gaseous products of denitrification in soil. Soil Sci. Soc. Am. J. 43, 1140–1144.

Firestone, M.K., Firestone, R.B. and Tiedje, J.M. 1980 Nitrous oxide from soil denitrification: Factors controlling its biological production. Science (Wash. D.C.) 208, 749–751.

Focht, D.D. 1978 Methods for analysis of denitrification in soils. In: Nielsen, D.R. and MacDonald, J.G. (eds.), Nitrogen in the Environment. Vol. 2, pp. 433–490. Academic Press, New York.

Focht, D.D. and Verstraete, W. 1977 Biochemical ecology of nitrification and denitrification. Adv. Microbial Ecol. 1, 135–214.

Focht, D.D., Valoras, N. and Letey, J. 1980 Use of interfaced gas chromatography–mass spectrometry for detection of concurrent mineralization and denitrification in soil. J. Environ. Qual. 9, 218–223.

Freney, J.R., Denmead, O.T. and Simpson, J.R. 1979 Nitrous oxide emission from soils at low moisture contents. Soil Biol. Biochem. 11, 167–173.

Fried, M., Tanji, K.K. and Van de Pol, R.M. 1976 Simplified long-term concept for evaluating leaching of nitrogen from agricultural land. J. Environ. Qual. 3, 391–396.

Galbally, I.E. and Roy, C.R. 1978 Loss of fixed nitrogen from soils by nitric oxide exhalation. Nature (Lond.) 275, 734–735.

Galsworthy, A.M. and Burford, J.R. 1978 A system for measuring the rates of evolution of nitrous oxide and nitrogen from incubated soils during denitrification. J. Soil Sci. 29, 537–550.

Gast, R.G., Nelson, W.W. and MacGregor, J.M. 1974 Nitrate and chloride accumulation and distribution in fertilized tile-drained soils. J. Environ. Qual. 3, 209–213.

Germon, J.C. 1980 Etude quantitative de la dénitrification biologique dans le sol à l'aide de l'acetylene II. Ann. Microbiol. (Paris) 131B, 81–90.

Gersberg, R., Krohn, K., Peek, N. and Goldman, C.R. 1976 Denitrification studies with ^{13}N-labelled nitrate. Science (Wash. D.C.) 192, 1229–1231.

Gilbert, R.G., Lance, J.C. and Miller , J.B. 1979 Denitrifying bacteria populations and nitrogen removal in soil columns intermittently flooded with secondary effluent. J. Environ. Qual. 8, 101–104.

Gilliam, J.W., Dasberg, S., Lund, L.J. and Focht, D.D. 1978 Denitrification in four California soils: Effect of soil profile characteristics. Soil Sci. Soc. Am. J. 42, 61–66.

Goh, K.M., Edmeades, D.C. and Robinson, B.W. 1978 Field measurement of symbiotic nitrogen fixation in an established pasture using acetylene reduction and a ^{15}N method. Soil Biol. Biochem. 10, 13–20.

Goldan, P.D., Bush, Y.A., Fehsenfield, F.C., Albritton, D.L., Crutzen, P.J., Schmeltekopf, A.L. and Ferguson, E.E. 1978 Tropospheric N_2O mixing ratio measurements. J. Geophys. Res. 83, 935–939.

Goleb, J.A. 1970 The use of optical spectroscopy for nitrogen-15 tracer studies in biological matter. Argonne Natl. Lab. Rev. N-101, 11–16.

Guthrie, T.F. and Duxbury, J.M. 1978 Nitrogen mineralization and denitrification in organic soils. Soil Sci. Soc. Am. J. 42, 908–912.

Hahn, J. 1972 Improved gas chromatographic method for field measurement of nitrous oxide in air and water using a 5Å molecular sieve trap. Anal. Chem. 44, 1889–1892.

Hansen, E.H., Ghose, A.K. and Ruzicka, J. 1977 Flow injection analysis of environmental samples for nitrate using an ion-selective electrode. Analyst (Lond.) 102, 705–713.

Hauck, R.D. 1971 Quantitative estimates of N-cycle processes: Concepts and review. In: [15]N in Soil-Plant Studies, pp. 65–80. IAEA, Vienna.

Hauck, R.D. and Bremner, J.M. 1976 Use of tracers for soil and fertilizer nitrogen research. Adv. Agron. 28, 219–266.

Hauck, R.D. and Melsted, S.W. 1956 Some aspects of the problem of evaluating denitrification in soils. Soil Sci. Soc. Am. Proc. 20, 361–364.

Hoather, R.C. and Rackham, R.F. 1959 Oxidized nitrogen in waters and sewage effluents observed by ultra-violet spectrophotometry. Analyst (Lond.) 84, 548–551.

Hoglund, J.H. and Brock, J.L. 1978 Regulation of nitrogen fixation in a grazed pasture. N. Z. J. Agric. Res. 21, 73–82.

Hollis, O.L. 1966 Separation of gaseous mixtures using porous polyaromatic polymer beads. Anal. Chem. 38, 309–316.

Hollocher, T.C., Garber, E., Cooper, A.J.L. and Reiman, R.E. 1980 ^{13}N, ^{15}N isotope and kinetic evidence against hyponitrite as an intermediate in denitrification. J. Biol. Chem. 11, 5027–5030.

Hutchinson, G.L. and Mosier, A.R. 1979 Nitrous oxide emissions from an irrigated corn field. Science (Wash. D.C.) 205, 1125–1127.

Hutchinson, G.L. and Mosier, A.R. 1981 Improved soil cover method for field measurement of nitrous oxide fluxes. Soil Sci. Soc. Am. J. 45, 311–316.

Hynes, R.K. and Knowles, R. 1978 Inhibition by acetylene of ammonium oxidation in Nitrosomonas europaea. FEMS (Fed. Eur. Microbiol Soc.) Microbiol. Lett. 4, 319–321.

Jansson, S.L. 1971 Use of ^{15}N in studies of soil nitrogen. In: McLaren, A.D. and Skujins, J. (eds.), Soil Biochemistry. Vol. 2, pp. 129–166. Marcel Dekker, New York.

Jolley, V.D. and Pierre, W.H. 1977 Profile accumulation of fertilizer derived nitrate and total nitrogen recovery in two long-term nitrogen-rate experiments on corn. Soil Sci. Soc. Am. J. 41, 373–378.

Jorgenson, J.R. 1974 A simple apparatus for obtaining multiple atmosphere samples from a single bore hole. Soil Sci. Soc. Am. Proc. 38, 540–541.

Jury, W.A., Letey, J. and Collins, T. 1982 Analysis of chamber methods used for measuring nitrous oxide production in the field. Soil Sci. Soc. Am. J. 46, 250–256.

Kanemasu, E.T., Powers, W.L. and Sij, J.W. 1974 Field chamber measurement of CO_2 flux from a soil surface. Soil Sci. 118, 233–237.

Kaplan, W.A., Teal, J.M. and Valiela, I. 1977 Denitrification in salt marsh sediments: Evidence for seasonal temperature selection among populations of denitrifiers. Microb. Ecol. 3, 193–204.

Keeney, D.R., Byrnes, B.H. and Genson, J.J. 1970 Determination of nitrate in waters with the nitrate-selective ion electrode. Analyst (Lond.) 95, 383–386.

Keeney, D.R., Fillery, I.R. and Marx, G.P. 1979 Effect of temperature on the gaseous nitrogen products of denitrification in a silt loam soil. Soil Sci. Soc. Am. J. 43, 1124–1128.

Kimball, B.A. 1978 Critique on: 'Application of gaseous diffusion theory to measurements of denitrification'. In: Nielsen, D.R. and MacDonald, J.G. (eds.), Nitrogen in the Environment. Vol. 1, pp. 351–361. Academic Press, New York.

Kimball, B.A. and Lemon, E.R. 1971 Air turbulence effects upon soil gas exchange. Soil Sci. Soc. Am. Proc. 35, 16–21.

Kimble, J.M., Bartlett, R.J., McIntosh, J.L. and Varney, K.E. 1972 Fate of nitrate from manure and inorganic nitrogen in a clay soil cropped to continuous corn. J. Environ. Qual. 1, 413–415.

Kissel, D.E., Smith, S.J., Hargrove, W.L. and Dillow, D.W. 1977 Immobilisation of fertilizer nitrate applied to a swelling clay soil in the field. Soil Sci. Soc. Am. J. 41, 346–349.

Klemedtsson, L., Svensson, B.H., Lindberg, T. and Rosswall, T. 1977 The use of acetylene inhibition of nitrous oxide reductase in quantifying denitrification in soils. Swed. J. Agric. Res. 7, 179–185.

Kowalenko, C.G. and Cameron, D.R. 1978 Nitrogen transformations in soil-plant systems in three years of field experiments using tracer and non-tracer methods on an ammonium fixing soil. Can. J. Soil Sci. 58, 195–208.

La Hue, M.D., Axelrod, M.D. and Lodge, J.P. 1971 Measurement of atmospheric nitrous oxide using a molecular sieve 5Å trap and gas chromatography. Anal. Chem. 43, 1113–1115.

Lance, J.C. and Whisler, F.D. 1972 Nitrogen balance in soil columns intermittently flooded with sewage water. J. Environ. Qual. 1, 180–186.

Lance, J.C., Whisler, F.D. and Bouwer, H. 1973 Oxygen utilization in soils flooded with sewage water. J. Environ. Qual. 2, 345–350.

Legg, B.J. 1975 Turbulent diffusion within a wheat canopy I. Measurement using nitrous oxide. Q. J. R. Meteorol. Soc. 101, 597–610.

Lemon, E. 1978 Nitrous oxide exchange at the land surface. In: Nielsen, D.R. and MacDonald, J.G. (eds.), Nitrogen in the Environment. Vol. 1, pp. 493–521. Academic Press, New York.

Letey, J., Jury, W.A., Hadas, A. and Valoras, N. 1980 Gas diffusion as a factor in laboratory incubation studies on denitrification. J. Environ. Qual. 9, 223–227.

Lippold, H. and Förster, I. 1980 Messung der Denitrifizierung in Bodenmonolithen von Grünland und Ackerland mit Hilfe der ^{15}N-Technik. Arch. Acker.-Pflanzenbau Bodenkd, Berlin 24, 85–90.

Lund, L.J., Ryden, J.C., Miller, R.J., Laag, A.E. and Bendixen, W.E. 1978 Nitrogen balances for the Santa Maria Valley. In: Pratt, P.F. (ed.), Management of Nitrogen in Irrigated Agriculture, pp. 395–414. University of California, Riverside.

Mahendrappa, M.K. 1969 Determination of nitrate nitrogen in soil extracts using a specific ion activity electrode. Soil Sci. 108, 132–136.

Matthias, A.D., Yarger, D.N. and Weinbeck, R.S. 1978 A numerical evaluation of chamber methods for determining gas fluxes. Geophys. Res. Lett. 5, 765–768.

Matthias, A.D., Blackmer, A.M. and Bremner, J.M. 1980 A simple chamber technique for field measurement of emissions of nitrous oxide from soils. J. Environ. Qual. 9, 251–256.

McGarity, J.W. 1962 Effect of freezing of soil on denitrification. Nature (Lond.) 196, 1342–1343.

McGarity, J.W. and Rajaratnam, J.N. 1973 Apparatus for the measurement of losses of nitrogen as gas from the field and simulated field environments. Soil. Biol. Biochem. 5, 121–131.

McKenney, D.J., Wade, D.L. and Findlay, W.I. 1978 Rates of N_2O evolution from N-fertilized soil. Geophys. Res. Lett. 5, 777–780.

McKenney, D.J., Shuttleworth, K.F. and Findlay, W.I. 1980 Nitrous oxide evolution rates from fertilized soil: Effects of applied nitrogen. Can. J. Soil Sci. 60, 429–438.

McNaughton, G.S. and More, R.D. 1979 The use of a 3 MV Van der Graff accelerator for the production of ^{13}N labelled ammonium and nitrate ions for biological experiments. Int. J. Appl. Radiat. Isot. 30, 489–492.

Methods for Chemical Analysis of Waters and Wastes 1976 U.S. Environmental Protection Agency, Cincinnati.

Miles, D.L. and Espejo, C. 1977 Comparison between an ultraviolet spectrophotometric procedure and the 2,4-xylenol method in the determination of nitrate in ground waters of low salinity. Analyst (Lond.) 102, 104–109.

Misra, C., Nielsen, D.R. and Biggar, J.W. 1974 Nitrogen transformations in soil during leaching: III. Nitrate reduction in soil columns. Soil Sci. Soc. Am. Proc. 38, 300–304.

Morgansen, V.O. 1977 Field measurements of dark respiration rates of roots and aerial parts of Italian ryegrass and barley. J. Appl. Ecol. 14, 243–252.

Mosier, A.R. 1980 Acetylene inhibition of ammonium oxidation in soil. Soil Biol. Biochem. 12, 443–444.

Mosier, A.R. and Mack, L. 1980 Gas chromatographic system for precise rapid analysis of nitrous oxide. Soil Sci. Soc. Am. J. 44, 1121–1123.

Myers, R.J.K. and McGarity, J.W. 1972 Denitrification in undisturbed cores from a solodized Solonetz B horizon. Plant Soil 37, 81–89.

Nielsen, D.R., Biggar, J.W. and Erh, K.R. 1973 Spatial variability of field measured soil water properties. Hilgardia 42, 215–259.

Nömmik, H. 1956 Investigations on denitrification in soil. Acta Agric. Scand. 2, 195–228.

Olsen, R.V., Murphy, L.S., Moser, H.C. and Swallow, C.W. 1979 Fate of tagged fertilizer nitrogen applied to winter wheat. Soil Sci. Soc. Am. J. 43, 973–975.

Patriquin, D.G., Mackinnon, J.C. and Wilkie, K.I. 1978 Seasonal patterns of denitrification and leaf nitrate reductase activity in a corn field. Can. J. Soil Sci. 58, 283–285.

Patten, D.K., Bremner, J.M. and Blackmer, A.M. 1980 Effects of drying and air-dry storage of soils on their capacity for denitrification of nitrate. Soil Sci. Soc. Am. J. 44, 67–70.

Payne, W.J. 1973 The use of gas chromatography for studies of denitrification in ecosystems. Bull. Ecol. Res. Comm. – NFR (Statens Naturvetensk Forskningsrad) 17, 263–268.

Pearson, J.E., Rimbey, D.H. and Jones, G.E. 1965. A soil-gas emanation measurement system used for radon-222. J. Appl. Meteorol. 4, 349–356.

Perschke, H. and Proksch, G. 1971 Analysis of ^{15}N abundance in biological samples by means of emission spectrometry. In: Nitrogen-15 in Soil-Plant Studies, pp. 223–225. IAEA, Vienna.

Pheiffer-Madsen, P. 1977 Analysis of gaseous nitrogen-15 by emission spectrometry in nitrate reduction experiments in soil. In: Stable Isotopes in the Life Sciences, pp. 171–177. IAEA, Vienna.

Pierotti, D., Rasmussen, R.A. and Chatfield, R. 1978 Continuous measurements of nitrous oxide in the troposphere. Nature (Lond.) 274, 574–576.

Pilot, L. and Patrick, W.H. 1972 Nitrate reduction in soils: effect of soil moisture tension. Soil Sci. 114, 312–316.

Porter, L.K. 1969 Gaseous products produced by anaerobic reaction of sodium nitrite with oxime compounds and oximes synthesized from organic matter. Soil Sci. Soc. Am. Proc. 24, 696–702.

Pratt, P.F., Warneke, J.E. and Nash, P.A. 1976 Sampling the unsaturated zone in irrigated field plots. Soil Sci. Soc. Am. J. 40, 277–279.

Pratt, P.F., Lund, L.J. and Rible, J.M. 1978 An approach to measuring the leaching of nitrate from freely drained irrigated fields. In: Nielsen, D.R. and MacDonald, J.G. (eds.), Nitrogen in the Environment. Vol. 1, pp. 223–256. Academic Press, New York.

Rasmussen, R.A., Krasnec, J. and Pierotti, D. 1976 N_2O analysis in the atmosphere via electron-capture gas chromatography. Geophys. Res. Lett. 3, 615–618.

Reddy, K.R., Sacco, P.D. and Graetz, D.A. 1980 Nitrate reduction in an organic soil-water system. J. Environ. Qual. 9, 283–288.

Rible, J.M., Nash, P.A., Pratt, P.F. and Lund, L.J. 1976 Sampling the unsaturated zone of irrigated lands for reliable estimates of nitrate concentrations. Soil Sci. Soc. Am. J. 40, 566–570.

Richards, L.A. 1954 Diagnosis and Improvement of Saline and Alkali Soils. Handbook No. 60. United States Department of Agriculture. 160 pp.

Riga, A., Fischer, V. and van Praag, H.J. 1980 Fate of fertilizer nitrogen applied to winter wheat as $Na^{15}NO_3$ and $(^{15}NH_4)_2SO_4$ studied in microplots through a four-course rotation: I. Influence of fertilizer splitting on soil and fertilizer nitrogen. Soil Sci. 130, 88–99.

Rittenberg, D. 1948 The preparation of gas samples for mass spectrographic isotope analysis. In: Wilson, E.W., Nier, A.O.C. and Reinmann, S.P. (eds.), Preparation and Measurement of Isotopic Tracers, pp. 31–42. J.W. Edwards, Ann Arbor.

Rolston, D.E. 1978 Application of gaseous-diffusion theory to measurement of denitrification. In: Nielsen, D.R. and MacDonald, J.G. (eds.), Nitrogen in the Environment. Vol. I, pp. 309–335. Academic Press, New York.

Rolston, D.E. 1982 Soil-gas flux. In: Klute, A. (ed.), Methods of Soil Analysis, Part 2. American Society of Agronomy, Madison (in press).

Rolston, D.E. and Broadbent, F.E. 1977 Field measurement of denitrification. U.S. Environmental Protection Agency, Ada, Oklahoma. EPA-600/2-77-233, 75 pp.

Rolston, D.E., Broadbent, F.E. and Goldhamer, D.A. 1979 Field measurement of denitrification: II. Mass balance and sampling uncertainty. Soil Sci. Soc. Am. J. 43, 703–708.

Rolston, D.E., Fried, M. and Goldhamer, D.A. 1976 Denitrification measured directly from nitrogen and nitrous oxide gas fluxes. Soil Sci. Soc. Am. J. 40, 259–266.

Rolston, D.E., Hoffman, D.L. and Toy, D.W. 1978 Field measurement of denitrification: I. Flux of N_2 and N_2O. Soil Sci. Soc. Am. J. 42, 863–869.

Rolston, D.E., Sharpley, A.N., Toy, D.W. and Broadbent, F.E. 1982 Field measurement of denitrification: III. Rates during irrigation cycles. Soil Sci Soc. Am. J. 46, 289–296.

Ross, P.J., Martin, A.E. and Henzell, E.F. 1964 A gas tight growth chamber for investigating gaseous nitrogen changes in the soil: plant: atmosphere system. Nature (Lond.) 444–447.

Ross, P.J., Martin, A.E. and Henzell, E.F. 1968 Gas lysimetry as a technique in nitrogen studies on the soil: plant: atmosphere system. Trans. 9th Intern. Congress Soil Sci., Adelaide. 2, 487–494.

Ryden, J.C. 1981 Denitrification loss. Annual Rep. Grassland Res. Inst. 1980 (in press).

Ryden, J.C. 1982 Effects of acetylene on nitrification and denitrification in two soils during incubation with ammonium nitrate. J. Soil Sci. 33 (in press).

Ryden, J.C. and Dawson, K.P. 1982 Evaluation of the acetylene inhibition technique for the measurement of denitrification in grassland soils. J. Sci. Food Agric. 33 (in press).

Ryden, J.C. and Lund, L.J. 1980 Nature and extent of directly measured denitrification losses from some irrigated vegetable crop production units. Soil Sci. Soc. Am. J. 44, 505–511.

Ryden, J.C., Lund, L.J. and Focht, D.D. 1978 Direct in-field measurement of nitrous oxide flux from soils. Soil Sci. Soc. Am. J. 42, 731–738.

Ryden, J.C., Lund, L.J. and Focht, D.D. 1979a Direct measurement of denitrification loss from soils: I. Laboratory evaluation of acetylene inhibition of nitrous oxide reduction. Soil Sci. Soc. Am. J. 43, 104–110.

Ryden, J.C., Lund, L.J., Letey, J. and Focht, D.D. 1979b Direct measurement of denitrification loss from soils: II. Development and application of field methods. Soil Sci. Soc. Am. J. 43, 110–118.

Schultz, I.J., Randall, G.W., Dowdy, R.H. and Gast, R.G. 1978 Conditions favouring denitrification in a Webster clay loam. Agron. Abst. p. 145.

Schutz, K., Junge, C., Beck, R. and Albrecht, B. 1970 Studies of atmospheric N_2O. J. Geophys. Res. 75, 2230–2241.

Siegel, R.S., Hauck, R.D. and Kurtz, L.T. 1982 Determination of $^{30}N_2$ and application to measurement of N_2 evolution during denitrification. Soil Sci. Soc. Am. J. 46, 68–74.

Smith, K.A. and Dowdell, R.J. 1973 Gas chromatographic analysis of the soil atmosphere: Automatic analysis of gas samples for O_2, N_2, Ar, CO_2, N_2O and C_1-C_4 hydrocarbons. J. Chromatogr. Sci. 11, 655–658.

Smith, M.S. and Tiedje, J.M. 1979a The effect of roots on soil denitrification. Soil Sci. Soc. Am. J. 43, 951–955.

Smith, M.S. and Tiedje, J.M. 1979b Phases of denitrification following oxygen depletion in soil. Soil Biol. Biochem. 11, 261–267.

Smith, M.S., Firestone, M.K. and Tiedje, J.M. 1978 The acetylene inhibition method for short term measurement of soil denitrification and its evaluation using ^{13}N. Soil Sci. Soc. Am. J. 42, 611–615.

Sørensen, J. 1978 Denitrification rates in a marine sediment as measured by the acetylene inhibition technique. Appl. Environ. Microbiol. 36, 139–143.

Starr, J.L., and Parlange, J.Y. 1975 Non-linear denitrification kinetics with continuous-flow soil columns. Soil Sci. Soc. Am. Proc. 39, 875–880.

Starr, J.L., Broadbent, F.E. and Nielsen, D.R. 1974 Nitrogen transformations during continuous leaching. Soil Sci. Soc. Am. Proc. 38, 283–289.

Stefanson, R.C. 1970 Sealed growth chambers for studies of the effect of plants on the soil atmosphere. J. Agric. Eng. Res. 15, 295–301

Stefanson, R.C. 1972a Soil denitrification in sealed soil-plant systems. I. Effects of plants, soil-water content and organic matter content. Plant Soil 37, 113–127.

Stefanson, R.C. 1972b Soil denitrification in sealed soil-plant systems. II. Effect of soil water content and form of applied nitrogen. Plant Soil 37, 129–140.

Stefanson, R.C. 1972c Effect of plant growth, soil-water content and form of nitrogen fertilizer on denitrification from four south Australian soils. Aust. J. Soil Res. 10, 183–195.

Stefanson, R.C. 1973 Evolution patterns of nitrous oxide and nitrogen in sealed soil-plant systems. Soil Biol. Biochem. 5, 167–169.

Stout, J.D. and More, R.D. 1980 The measurement of denitrification in soil using $^{13}NO_3$. In: Trudinger, P.A., Walter, M.R. and Ralph, B.J. (eds.), Biogeochemistry of Ancient and Modern Environments, pp. 293–298. Springer-Verlag, Berlin.

Taylor, G.S. and Abrahams, J.H. 1953 A diffusion-equilibrium method for obtaining soil gases under field conditions. Soil Sci. Soc. Am. Proc. 17, 201–206.

Tiedje, J.M., Firestone, R.B., Firestone, M.K., Betlach, M.R., Smith, M.S. and Caskey, W.H. 1979 Methods for the production and use of nitrogen-13 in studies of denitrification. Soil Sci. Soc. Am. J. 43, 709–715.

Tiedje, J.M., Firestone, R.B., Firestone, M.K., Betlach, M.R., Kaspar, H.F. and Sørensen, J. 1981 Use of nitrogen-13 in studies of denitrification. In: Krohn, K.A. and Root, J.W. (eds.), Recent

Developments in Biological and Chemical Research with Short-lived Isotopes. Advances in Chemistry, Series 197, 295–317, American Chemical Society, Washington D.C.

Volz, M.G., Belser, L.W., Ardakani, M.S. and McLaren, A.D. 1975 Nitrate reduction and associated microbial populations in a ponded Handford sandy loam. J. Environ. Qual. 4, 99–102.

Walter, H.M., Keeney, D.R. and Fillery, I.R. 1979 Inhibition of nitrification by acetylene. Soil Sci. Soc. Am. J. 43, 195–196.

Watanabe, I. and de Guzman, M.R. 1980 Effect of nitrate on acetylene disappearance from anaerobic soil. Soil Biol. Biochem. 12, 193–194.

Westerman, R.L. and Tucker, T.C. 1978 Factors affecting denitrification in a Sonoran Desert soil. Soil Sci. Soc. Am. J. 42, 596–599.

Wentworth, W.E. and Freeman, R.R. 1973 Measurement of atmospheric nitrous oxide using an electron capture detector in conjunction with gas chromatography. J. Chromatogr. 79, 322–324.

Wentworth, W.E., Chen, E. and Freeman, R. 1971 Thermal electron attachment to nitrous oxide. J. Chem. Phys. 55, 2075–2078.

Wickramasinghe, K.N., Talibudeen, O. and Witty, J.F. 1978 A gas flow-through system for measuring denitrification in soils. J. Soil Sci. 29, 527–536.

Wijler, J. and Delwiche, C.C. 1954 Investigations on the denitrifying process in soil. Plant Soil 5, 155–169.

Yamane, I. and Sato, K. 1968 Initial rapid drop of oxidation-reduction potential in submerged air-dried soils. Soil Sci. Plant Nutr. 14, 68–72.

Yeomans, J.C. and Beauchamp, E.G. 1978 Limited inhibition of nitrous oxide reduction in soil in the presence of acetylene. Soil Biol. Biochem. 10, 517–519.

Yeomans, J.C. and Beauchamp, E.G. 1982 Acetylene as a possible substrate in the denitrification process. Can. J. Soil Sci. 62, 139–144.

Yoshinari, T. and Knowles, R. 1976 Acetylene inhibition of nitrous oxide reduction by denitrifying bacteria. Biochem. Biophys. Res. Commun. 69, 705–710.

Yoshinari, T., Hynes, R. and Knowles, R. 1977 Acetylene inhibition of nitrous oxide reduction and measurement of denitrification and nitrogen fixation in soil. Soil Biol. Biochem. 9, 177–183.

5. Micrometeorological methods for measuring gaseous losses of nitrogen in the field

O.T. DENMEAD

5.1 Introduction

While it has been recognised for many years that gaseous transfer is an important pathway in the terrestrial N cycle, it is only in recent times that direct field measurements have been made of the exchanges of nitrogenous gases between soils, plants and the atmosphere. Methods of three general kinds have been employed: those using diffusion theory to calculate gas transport in the soil profile; enclosure methods in which the flux density of the gas at the soil or water surface is calculated from changes in gas concentration in an enclosure placed over the surface; and micrometeorological techniques in which the vertical flux density of the gas is measured in the free air above the surface.

The substantial difficulties involved in applying soil diffusion theory in the field have been discussed by Ryden and Rolston in Chapter 4 and need not be considered further here. Those authors also discuss enclosure methods which are the most popular of all the field approaches because of their relative simplicity, their suitability for small experimental plots and the lower sensitivity required in measuring gas concentrations. Indeed, for some applications such as the measurement of N_2O flux, enclosures are presently the only feasible routine method. As Ryden and Rolston point out though, there are difficulties in their use, the main ones being the need to maintain gas concentrations in the enclosure close to ambient, and the avoidance of pressure excesses or deficits in the enclosure.

Some special problems arise in the use of enclosures to measure NH_3 fluxes. These are associated with the chemistry of the gas and the strong influence of environmental factors on the volatilization process. Ammonia is highly reactive and readily soluble in water and is thus likely to be retained on the walls of the enclosure and air pipes and to be dissolved in water condensed anywhere in the system. As discussed in Chapter 1, evaporation rate, temperature, windspeed, ambient NH_3 concentration, and even dew formation can all have important influences on the volatilization of NH_3 in natural systems. Matching the enclosure to the outside world in all these respects is a difficult, if not impossible task. It must be conceded therefore that a good deal of uncertainty can attach to enclosure measurements of NH_3 flux.

Of the three possible approaches, there is no doubt that micrometeorological

techniques are to be preferred in principle. They do not disturb the environmental or soil processes which influence gas exchange; they allow continuous
rapid measurement, thus facilitating the investigation of environmental effects;
and they provide a measure of the average flux over a large area, thereby
minimising the sampling problem created by point to point variation. The
difficulties of the micrometeorological techniques in common use are that their
successful application requires large experimental areas and very accurate, and
sometimes very rapid measurements of small gas concentrations. The final
section of the chapter is devoted to a consideration of new micrometeorological
approaches which have less stringent requirements and promise considerable
simplification in experimental technique. Micrometeorological methods have
so far been employed for estimating fluxes of NH_3, volatile amines and the
oxides of N: N_2O, NO and NO_2. It is unfortunate though, that while
significant gains and losses of N at the earth's surface occur as molecular
nitrogen itself, no micrometeorological methods have yet been developed for
measuring N_2 fluxes. Alternative techniques for N_2 exist, and these have been
discussed by Ryden and Rolston in Chapter 4.

5.2. General principles

Three descriptions of the transport of gases and vapours in air, each leading to
a different methodological approach, will be discussed. The basic concept is
that transport is accomplished by the eddying motion of the atmosphere, which
displaces parcels of air from one level to another. This leads directly to the eddy
correlation approach in which the vertical transport of a gas past a point in the
atmosphere is obtained by correlating the instantaneous vertical wind speed at
the point with the instantaneous concentration. In the natural environment, the
eddies which are important in the process occur with frequencies extending up
to 5 or 10 Hz; hence a rapid response gas detector is required. One such detector
for NO and NO_2, based on chemiluminescence, already exists (Wesely *et al.*
1982), and extension of the chemiluminescence technique to other gases, like
NH_3, is possible (e.g. Hales and Drewes 1979, Braman *et al.* 1982).

   If an appropriate fast detector is available, eddy correlation methods offer
many attractions. In particular, they require a minimum number of assumptions about the nature of the transport process, and measurements at only one
level above the surface are needed. Disadvantages are that sophisticated
recording and computing facilities are required to cope with the rapid data
acquisition rate and considerable operational experience is necessary. For some
time, it will probably be that more conventional micrometeorological approaches based on gradient diffusion will be employed.

   This second approach is based on the concept of diffusion of the gas along its mean concentration gradient. The method requires measurement of mean gas concentrations at a number of heights above the surface and knowledge of the appropriate diffusion coefficient. Measurements of the exchanges of N gases between soils, plants and the atmosphere using gradient diffusion methods have been reported by Denmead *et al.* (1974, 1976, 1978), Lemon and van Houtte (1980), Freney *et al.* (1981) and Harper *et al.* (1983) for NH_3, Lemon (1978) and Mosier and Hutchinson (1981) for N_2O, and Hutchinson *et al.* (1982) for NH_3 and volatile amines.

Practical applications of both eddy correlation and gradient diffusion approaches are limited to situations in which the air has blown over a homogeneous exchange surface for a long distance so that profiles of gas concentration in the air are in equilibrium with the local rate of exchange, and horizontal concentration gradients are negligible. In such circumstances, the vertical flux density of the gas will be constant with height in the air layers close to the ground and a one-dimensional (vertical) analysis can be made.

It is conventional to define the upper boundary of the air layer whose gas concentration is modified by an emission at the ground as the height, Z, at which the concentration is reduced to some negligibly small fraction, usually $1/10$, of its value at the ground. According to Sutton (1953),

$$Z = x^{7/9},$$ (1)

where x is the fetch, i.e. the distance the wind has travelled over the treated area. His work suggests that Z is about 10 m for a fetch of 100 m. The depth of the air layer in which the flux is constant with height is much smaller, suggested fetch/height ratios varying from about 200:1 (Dyer 1963) to 100:1 (Weseley *et al.* 1982). For a fetch of 100 m, these ratios would give maximum working heights of only 0.5 to 1 m. However, many eddy correlation instruments can only be operated successfully at heights of 5 m or more above the surface because of limitations to their frequency response. Likewise, the small size of the gradients in the atmosphere often requires that concentration differences be measured over a height interval of 2 or 3 m. Even with the less stringent fetch/height ratio of 100:1, it is evident that experimental areas need to be hundreds of metres in extent.

Treated plots of several hectares are not easily established. Large uniform sites can often be found, but applying the treatment is likely to be both expensive and time-consuming. Indeed, the time course of gas exchange immediately after the treatment is imposed may be a desired experimental result, as for instance, in the studies of the dynamics of NH_3 emission following injection of anhydrous NH_3 into soil reported by Denmead *et al.* (1977, 1982a). There is thus scope for micrometeorological methods which can be used for measuring gas exchange from smaller land areas: areas of, say, a few metres extent. An appropriate method has been in existence for a considerable time,

136

but only recently has it been exploited for studies of this kind. The method is based on mass balance. The horizontal gas flux across a vertical plane of unit width on the downwind edge of a treated area is equated with the surface flux from a strip of similar width upwind. The horizontal flux at any height is the product of horizontal wind speed and gas concentration. The total flux is obtained by integrating that product over the depth of the modified layer, i.e. the Z of Equation (1). Unlike the methods described above, this technique deliberately creates a non-equilibrium situation, and it is most successful when the experimental area is small so that Z is also conveniently small.

The influence of fetch on profile development is illustrated in Fig. 5.1 which shows profiles of NH_3 concentration and wind speed measured close to the surface in the course of experiments in which urea fertilizer was spread on plots of different size.[1] The plots were located in a large field of short grass (0.2 m high). Because of the large fetch, the wind profiles are equilibrium profiles,

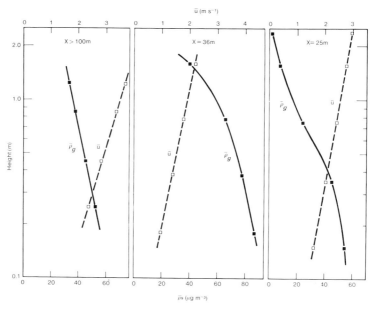

Fig. 5.1. Influence of fetch (x) on profile of windspeed ($\bar{u}$) and NH_3 density ($\bar{\rho}_g$) measured over plots of urea-treated pasture located in a large uniform field. Values of x refer to fetch over treated area.

[1] Unpublished experiments of V.R. Catchpoole, O.T. Denmead and D.J. Oxenham.

wind speed increasing with the logarithm of height.[2] When the treated plot (hence the region of NH_3 emission) extended for more than 100 m upwind, the NH_3 profiles were also equilibrium ones and like wind speed, the NH_3 concentration varied with log (height). As plot size decreased, the depths of the modified layer for NH_3 and the equilibrium layer also decreased until at a fetch of 25 m, the modified layer was about 2 m deep and the NH_3 profile appeared to be logarithmic for only about 0.2 m above the surface.

Mass balance approaches were employed in early studies of atmospheric dispersion from plane and line sources by Sutton (1947), Calder (1949) and Rider et al. (1963), for example. More recently the technique has been used to measure NH_3 losses from cross-wind line sources of limited upwind extent, and from small plane sources by Denmead et al. (1977, 1982a, 1982b), Beauchamp et al. (1978) and Wilson et al. (1982, 1983).

5.3. Eddy correlation methods

The instantaneous vertical flux density of a gas is simply the product of the vertical wind speed w and the gas density ρ_g. The mean flux density F is given by

$$F = \overline{w \rho_g} \tag{2}$$

where the bar denotes an average over an appropriate measuring period. Both w and ρ_g can be represented as sums of means, $\overline{w}$ and $\overline{\rho_g}$, and fluctuations about those means, w' and ρ'_g. Thus,

$$F = \overline{w}\,\overline{\rho_g} + \overline{w'\rho'_g}. \tag{3}$$

It has been common to assume that over a uniform, horizontal surface $\overline{w} = 0$, and hence $\overline{w}\,\overline{\rho_g} = 0$. The mean flux is then given by the second term on the right hand side of Equation (3). However, Webb et al. (1980) have recently pointed out that the fundamental premise should be not that $\overline{w} = 0$, but that the net flux of dry air at the surface, $\overline{w\rho_a}$, should be zero. The term ρ_a $(= \overline{\rho_a} + \rho'_a)$ is the density of dry air. Thus, $\overline{w\rho_a} = \overline{w}\,\overline{\rho_a} + \overline{w'\rho'_a} = 0$. This assumption leads to the conclusion that $\overline{w} = -\overline{w'\rho'_a}/\overline{\rho_a}$ and as Webb et al.

[2] Wind speed in the air layers above plant communities appears to approach zero not at the ground, but at some height above it. That height is called the zero-plane displacement d and is a substantial fraction of the height of the community. In the present case, $d = 0.15$ m. When dealing with plant communities, it is usual in micrometeorology to reckon height as height above the zero plane, i.e. as actual height above the ground, z, less the zero-plane displacement. The vertical axis in Fig. 5.1 is in fact z-d, and z-d should be substituted for z whenever height dependent relationships mentioned in this chapter are applied to plant communities. Thom (1975) discusses the physical significance of the zero-plane displacement and illustrates how it is determined in practice.

(1980) argue, this mean velocity will be non-zero whenever there is a flux of heat and/or water vapour between the surface and the atmosphere because the latter fluxes produce fluctuations in ρ_a which are correlated with w. Since $\overline{w}$ will generally be <1 mm/sec, it cannot be measured directly, but its magnitude can be calculated if the simultaneous flux densities of heat H and water vapour E are known. Webb *et al.* (1980) show that Equation (3) then becomes

$$F = \overline{w'\rho'_g} + (\overline{\rho_g}/\overline{\rho_a}) \{ \mu/1 + \mu\sigma \, (1 + \mu\sigma) \} E + (\overline{\rho_g}/\overline{\rho})H/c_p \, \overline{T} \qquad (4)$$

In Equation (4) $\mu = m_a/m_v$, m_a being the 'molecular weight' of dry air and m_v that of water vapour; $\sigma = \overline{\rho_v}/\overline{\rho_a}$, $\overline{\rho_v}$ being the density of water vapour; $\overline{\rho}$ is the mean density of air, i.e. $\overline{\rho} = \overline{\rho_a} + \overline{\rho_v}$; c_p is the specific heat of air at constant pressure; and $\overline{T}$ $(^\circ K)$ is air temperature.

The first term on the right hand side of Equation (4) is the eddy flux calculated on the assumption that $\overline{w} = 0$. The other two terms are the corrections for the density effects due to water vapour and heat transfer. The relative magnitudes of the correction terms in Equation (4) are examined in Section 5.4.1; see discussion to Table 5.1. However, it should be noted that attractive as the eddy correlation approach is in principle, these corrections become quite large for some N gases. In those cases, additional, simultaneous measurements of E and H will be required.

Table 5.1. Corrections to measurements of flux densities of N gases when gas concentrations are measured in situ. Environmental conditions are specified in Section 5.4.1.

Gas	Gas density $\overline{\rho_g}$ ($\mu g \, m^{-3}$)	Reference[a] for $\overline{\rho_g}$	Corrections For E ($\mu g m^{-2}s^{-1}$)	For H ($\mu g m^{-2}s^{-1}$)	Typical flux density F ($\mu g m^{-2}s^{-1}$)	Reference[a] for F	(Total corrections)/ $|F|$
NH$_3$	10	1	0.001	0.008	0.5	1	0.02
NH$_3$	600	2	0.06	0.5	10	2	0.06
NH$_3$	1200	3	0.1	0.8	103	3	0.01
NO$_2$	15	4	0.002	0.012	−0.076	4	0.18
NO	4	5	0.0004	0.003	0.007	5	0.53
N$_2$O	670	6	0.07	0.56	0.06	7	10.5
N$_2$O	670	6	0.07	0.56	0.75	8	0.84
N$_2$O	680	9	0.07	0.56	−2.5	9	0.25
N$_2$	10^9		10^5	8 × 10^5	70	10	10,000

[a] 1 Denmead *et al.* (1974); 2 Harper *et al.* (1982); 3 Hutchinson *et al.* (1982); 4 Wesely *et al.* (1982); 5 Galbally and Roy (1978); 6 Roy (1979); 7 Denmead *et al.* (1979); 8 Mosier and Hutchinson (1981); 9 Lemon (1978); 10 Rolston *et al.* (1978).

5.4 Gradient diffusion

5.4.1. General

Vertical transport of gases in the lower atmosphere may also be described by the relationship

$$F = - \overline{\rho_a} K_g \, \partial \overline{s} / \partial z \tag{5}$$

in which K_g is the eddy diffusivity for the gas, and $s = \rho_g / \rho_a$. For the evaporation of water, Equation 5 becomes

$$E = - \overline{\rho_a} K_g \, \partial \overline{r} / \partial z, \tag{5a}$$

where r is the mean mixing ratio of water vapour, defined formally as ρ_v / ρ_a and calculated as $0.622e / (p\text{-}e)$, with e being vapour pressure and p total pressure.

Unlike molecular diffusion which results from the random motion of molecules, eddy diffusion results from the movements of parcels of air from one level to another. Consequently, eddy diffusivities are usually several orders of magnitude greater than their molecular counterparts. Their actual magnitudes are determined by wind speed, height above the surface, the aerodynamic roughness of the surface, and the vertical temperature gradient. Formal relationships are examined in the next section.

Equation (5) follows Webb *et al.* (1980) in specifying the gas concentration in terms of its mixing ratio with respect to dry air, i.e. as ρ_g / ρ_a. The latter is essentially what is measured when the samples of air from different heights are pre-dried before measuring their gas concentrations at a common temperature and pressure. Webb *et al.* (1980) show that if gas concentrations are measured in this way, no corrections for the density effects associated with the simultaneous transfer of water vapour and heat are needed.

The more usual statement of the gradient diffusion equation is

$$F = - K_g \, \partial \overline{\rho_g} / \partial z \tag{6}$$

If $\overline{\rho_g}$ were measured in situ, i.e. on samples of moist air and in the presence of the mean temperature gradient, then the flux calculated from Equation (6) would correspond to the term $\overline{w' \rho'_g}$ in Equation (4) and the same corrections for E and H as in that equation would need to be added in order to obtain the true flux. This is the case, for instance, for the measurements of NH_3 flux described by Denmead *et al.* (1978), Lemon and van Houtte (1980), Hutchinson *et al.* (1982) and Harper *et al.* (1983).

If, as in the measurements of NH_3 flux described by Denmead *et al.* (1974, 1976), the air samples from different heights were brought to the same temperature before their gas concentrations were measured, the need to correct for the temperature gradient associated with the heat flux would be eliminated but the effects of the humidity gradient would still remain. From Webb *et al.* (1980) the corrected flux is then given by

$$F = (p/P_1)\,(T_1/\overline{T})\,\{\,-K_g\,\partial\overline{\rho}_{g1}/\partial z + (\mu\overline{\rho}_{g1}/\overline{\rho}_a)\,(1+\mu\sigma)^{-1}\,E\}, \qquad (7)$$

where p denotes the ambient pressure, P_1 and T_1 are the operating pressure and temperature of the measuring instrument, and $\overline{\rho}_{g1}$ is the gas density as measured in the instrument. If the air samples were pre-dried as well, as in the measurements of N_2O flux reported by Mosier and Hutchinson (1981), no corrections would be needed.

To provide an indication of the importance of these corrections in measurements of the flux densities of various N gases by either eddy correlation or gradient diffusion methods, Equation (4) has been evaluated for environmental conditions likely to be encountered about noon on a summer's day in a temperate climate, viz. $\overline{e} = 10$ mb, $\overline{T} = 298°$ K, $E = 1.2 \times 10^{-4}$ kg m^{-2} s^{-1}, and $H = 200$ W m^{-2}. For background gas concentrations, values of $\overline{\rho}_g$ reported by various workers have been used. The corrections are compared with published flux measurements for different gases in Table 5.1. The corrections for E and H are listed separately because of the different methods of gas sampling likely to be employed.

From Equation (4) it is evident that the relative importance of the corrections depends on the background concentration of the gas and the magnitude of the typical flux. For NH_3, for instance, it seems that the corrections could almost always be ignored, even in very NH_3-enriched situations such as the feedlot of Hutchinson *et al.* (1982). (This is fortunate because it is difficult to dry air without removing the NH_3 as well.) The corrections are more important for NO_2 and NO because of the smaller natural fluxes. For instance, they would be as much as 50% of the fluxes of NO measured by Galbally and Roy (1978) over grassland. For N_2O with a relatively high background concentration, the corrections appear to be always important, ranging from 25% to >1000% of published flux measurements. If it were possible to make micrometeorological measurements of the flux density of N_2 (present in air at some 78%) the corrections would be equivalent to an apparent flux of some 34,000 kg N ha^{-1} h^{-1}, or about ten thousand times the maximum flux to be expected.

To summarise, it appears that corrections for the density effects will seldom be necessary for micrometeorological measurements of NH_3 flux, but appear to be important for all the other N gases of interest. If eddy correlation methods are employed, the flux densities of water vapour and heat will need to be measured simultaneously, while if gradient diffusion approaches are used, air samples from the different heights should be dried and brought to a common temperature before gas concentrations are measured. Further, Table 5.1

indicates that the corrections for N_2O and N_2 would normally be so large that eddy correlation approaches would be inadvisable for these gases. The corrected flux would be a very small difference between very large numbers and the demands on instrumental accuracy would be prohibitive.

5.4.2. Flux-profile relationships

Equation (5) provides a means for calculating the flux density of a gas from measurements of its vertical concentration profile. The calculation requires knowledge of the eddy diffusivity, K_g, a function of wind speed, height, surface roughness and thermal stratification. The exact functional dependence of K on these factors has been the subject of much micrometeorological research over the last three decades. Here, it is only possible to consider some relevant results of that work, but more detailed discussion can be found in Dyer (1974) and Thom (1975).

In subsequent developments, it will be necessary to consider the simultaneous fluxes of momentum and heat, for which the counterparts of Equation (5) are respectively,

$$\tau = -\bar{\rho}\, K_m\, \partial \bar{u}/\partial z, \tag{8}$$

and

$$H = -c_p\, \bar{\rho}\, K_h\, \partial \bar{\theta}/\partial z. \tag{9}$$

In Equation (8), τ is the downward flux of momentum or shearing stress, K_m and K_h are eddy diffusivities for momentum and heat, $\bar{u}$ is mean horizontal wind speed, and $\bar{\theta}$ is the mean potential temperature ($= T + \Gamma z$, where Γ is the adiabatic lapse rate). In what follows, it will be more convenient to work with the friction velocity u_* than the shearing stress; $u_* = \sqrt{\tau/\bar{\rho}}$.

The relationships between K_m, K_h and K_g depend on the stability of the atmosphere, which can be specified by the Richardson number Ri; $\mathrm{Ri} = [(g/\bar{\theta})\,(\partial\bar{\theta}/\partial z)/(\partial\bar{u}/\partial z)^2]$, g being the acceleration due to gravity. Thom (1975), amongst many others, discusses the theoretical basis of the Richardson number, but in short, it provides a measure of the relative effects of buoyancy and wind shear on vertical diffusion. A negative Ri denotes an unstable atmosphere which usually occurs by day when buoyancy forces generated by heating at the ground augment the mechanical turbulence of the wind. A positive Ri, denoting a stable atmosphere, is characteristic of the nighttime when the temperature inversion suppresses mechanical turbulence. In a neutral atmosphere there is no sensible heat transfer so that $\partial\bar{T}/\partial z = -\Gamma$ and Ri is zero. The Richardson number is height dependent.

A second, height independent, stability parameter is the Monin-Obukhov stability length, L; $L = -\bar{\rho} c_p u_*{}^3 \bar{\theta}/k\, g\, H$, where k is the von Karman constant. The value of k in the free atmosphere is the subject of some speculation (Businger et al. 1971), but following Dyer and Hicks (1970), Dyer (1974) and Thom (1975), we have assumed it to be 0.41. As for Ri, L is negative in unstable conditions and positive in stable conditions. In a neutral atmosphere, however, L is infinite (because H is zero).

The influence of stability on vertical diffusion is expressed through stability functions φ_m, φ_h and φ_g which are dependent on Ri or L and are defined through the following expressions:

$$u_* = \frac{kz\,\partial\bar{u}/\partial z}{\varphi_m} \tag{10}$$

$$H = -\frac{\bar{\rho} c_p k\, u_* z\,\partial\bar{\theta}/\partial z}{\varphi_h} \tag{11}$$

and

$$F = -\frac{\bar{\rho}_a k u_* z\,\partial\bar{s}/\partial z}{\varphi_g} \tag{12}$$

When these flux-gradient relationships are compared with those employing eddy diffusivities, i.e. with Equations (5), (8) and (9), it is seen that

$$K_m = k\, u_*\, z/\varphi_m, \tag{13}$$

$$K_h = k\, u_* z/\varphi_h, \tag{14}$$

and

$$K_g = k\, u_* z/\varphi_g. \tag{15}$$

Many careful experiments have been performed in order to establish the functional dependence of φ_m, φ_h and φ_g on Ri and/or L. With one exception (Businger et al. 1971), these have given similar results and it now seems to be generally agreed (Dyer 1974, Thom 1975) that in neutral conditions,

$$\varphi_m = \varphi_h = \varphi_g = 1\,; \tag{16}$$

in stable conditions,

$$\varphi_m = \varphi_h = \varphi_g = (1-5\ \text{Ri})^{-1}; \tag{17}$$

and in unstable conditions,

$$\varphi_m = (1-16\ \text{Ri})^{-1/4}, \tag{18}$$

and

$$\varphi_h = \varphi_g = (1-16 \; \text{Ri})^{-1/2}. \tag{19}$$

In terms of L, the stability functions for stable conditions are

$$\phi_m = \phi_h = \phi_g = 1 + 5 \; z/L. \tag{17a}$$

For unstable conditions, Equation (19) and the definitions of L and R_i imply that $R_i = z/L$, so that

$$\varphi_m = (1-16 \; z/L)^{-1/4}, \tag{18a}$$

and

$$\varphi_h = \varphi_g = (1-16 \; z/L)^{-1/2}. \tag{19a}$$

These last relationships for unstable conditions were proposed by Dyer and Hicks (1970).

5.4.3. Aerodynamic methods

The foregoing developments lead to the so-called aerodynamic methods of flux measurement. In neutral conditions with $\varphi_m = 1$, Equation (10) becomes

$$u_* = k \; z \; \partial \bar{u}/\partial z, \tag{20}$$

which on integration leads to the familiar logarithmic wind profile

$$\bar{u} = \frac{u_*}{k} \; ln \; \frac{z}{z_0}, \tag{21}$$

z_0 being the roughness height determined as the intercept on a plot of $\bar{u}$ against $ln \; z$. Further, from Equations (10) and (12), we have that

$$F = - \frac{\bar{\rho}_a k^2 z^2 \left(\dfrac{\partial \bar{u}}{\partial z}\right)\left(\dfrac{\partial \bar{s}}{\partial z}\right)}{\varphi_m \; \varphi_g}$$

$$= - \frac{\bar{\rho}_a k^2 \left(\dfrac{\partial \bar{u}}{\partial \; ln \; z}\right)\left(\dfrac{\partial \bar{s}}{\partial \; ln \; z}\right)}{\varphi_m \; \varphi_g} \tag{22}$$

In neutral conditions, $(\varphi_m \; \varphi_g)^{-1} = 1$, and in finite difference form Equation (22) then becomes

$$F = \frac{\bar{\rho}_a k^2 \; (\bar{u}_2 - \bar{u}_1)(\bar{s}_1 - \bar{s}_2)}{\{ln \; (z_2/z_1)\}^2}. \tag{23}$$

The flux density of the gas can thus be evaluated from measurements of wind speed and gas concentration at two heights, z_1 and z_2. In stable conditions,

Equations (17) and (22) lead to the expression

$$F = \frac{\overline{\rho_a} k^2 (\overline{u}_2 - \overline{u}_1)(\overline{s}_1 - \overline{s}_2)}{\{ln\ (z_2/z_1)\}^2} \cdot (1\text{–}5\ \mathrm{Ri})^2, \tag{24}$$

while in unstable conditions Equations (18), (19) and (22) lead to

$$F = \frac{\overline{\rho_a} k^2 (\overline{u}_2 - \overline{u}_1)(\overline{s}_1 - \overline{s}_2)}{\{ln\ (z_2/z_1)\}^2} \cdot (1 - 16\ \mathrm{Ri})^{3/4}. \tag{25}$$

In practice, Ri would be evaluated from observations of $\overline{u}$ and $\overline{T}$ at z_1 and z_2. An appropriate procedure is described by Paulson (1970).

Accuracy will usually be improved by making measurements at more than two heights. Paulson (1970) outlines very useful procedures for evaluating Ri, L and the fluxes from more detailed profile measurements.

Equation (23) was first derived by Thornthwaite and Holzman (1939) for calculating evaporation rates. Although valid only for neutral conditions, it has frequently been employed to estimate gas fluxes regardless of stability. However, as Thom (1975) points out, it is only in a very narrow range of Richardson numbers, between -0.01 and $+0.01$, that corrections for stability effects are less than 10%. For most field situations they are much larger and cannot be neglected; see, for instance, Fig. 7 of Thom (1975). Measurement of the temperature gradient is thus an essential requirement of the aerodynamic method.

One other scheme for stability corrections should be mentioned. This is the KEYPS formulation for unstable conditions (Panofsky 1963). The assumption often made in its application is that

$$\varphi_m = \varphi_g = (1\text{–}18\ \mathrm{Ri})^{-1/4}, \tag{26}$$

e.g. Lemon (1978), Lemon and van Houtte (1980) and Harper $et\ al.$ (1983). This then leads to the finite difference relationship

$$F = \frac{\overline{\rho_a} k^2 (\overline{u}_2 - \overline{u}_1)(\overline{s}_1 - \overline{s}_2)}{\{ln(z_2/z_1)\}^2} \cdot (1 - 18\ \mathrm{Ri})^{1/2}. \tag{27}$$

The difference in flux estimates resulting from these various aerodynamic methods is illustrated for a particular case study in the next section.

A warning should be sounded about the use of aerodynamic methods over very rough surfaces, like forests. It has been found that when measurements are made close to such surfaces, say within one tree-height, the accepted stability functions fail in all stability conditions (Raupach and Thom 1981).

5.4.4. The energy balance method

An alternative approach for calculating gas fluxes, which obviates the need to apply stability corrections, can be made through the energy balance. The energy gained by natural surfaces from solar radiation is balanced by the loss of energy to the atmosphere through the transfer of sensible heat and water vapour and by a change in energy storage. Thus

$$R = H + \lambda E + G, \tag{28}$$

where R is the net radiation at the surface, i.e. the incoming short- and long-wave radiation less what is reflected and re-radiated, λ is the latent heat of evaporation of water, and G is the change in energy storage. This last term includes the flux of heat into the soil (or water) and if the surface is a plant community, the change in heat stored in the air and biomass of the community and the solar energy fixed in photosynthesis. Normally, the community energy storage terms are small and can be neglected.

From Equations (16), (17) and (19), it is apparent that $\varphi_h = \varphi_g$ in all stability conditions; hence $K_h = K_g = K_{h \cdot g}$. Rearranging Equation (28) and substituting for H and E from Equations (9) and (5a), we obtain

$$R - G = - K_{h \cdot g}(c_p \bar{\rho} \, \partial \bar{\theta} / \partial z + \lambda \bar{\rho}_a \, \partial \bar{r} / \partial z). \tag{29}$$

If we define an effective temperature, T_e, for the combined transport of sensible and latent heat as

$$T_e = \bar{\theta} + (\lambda / c_p) \bar{r} / (1 + \sigma), \tag{30}$$

then

$$K_{h \cdot g} = - (R - G) / (\bar{\rho} c_p \, \partial T_e / \partial z) \tag{31}$$

and

$$F = \frac{(R - G) \, \partial \bar{s} / \partial z}{c_p (1 + \sigma) \, \partial T_e / \partial z}, \tag{32}$$

or

$$F = \frac{(R - G)}{c_p (1 + \sigma)} \frac{\partial \bar{s}}{\partial T_e}. \tag{33}$$

The effective temperature defined in Equation (30) is very close to the equivalent temperature $(\bar{\theta} + (\lambda / c_p)q$, where q is the specific humidity, $\bar{\rho}_v / \bar{\rho})$ employed by Denmead et al. (1974, 1976) and Hutchinson et al. (1982).

In practice, simultaneous observations of gas concentration, temperature and humidity are made at a number of heights above the surface and $\partial \bar{s} / \partial T_e$ is evaluated as the slope of the plot of $\bar{s}$ against the corresponding T_e. Differences in $\bar{s}$ and T_e between two heights could also be used instead of $\partial\bar{s}/\partial T_e$ in Equation (33) with no loss of rigour. Various other equivalent formulations can be made in terms of vapour pressure and wet-bulb temperature, e.g. Denmead and McIlroy (1971) and Leuning *et al.* (1982).

As already noted, one advantage of the energy balance method is that no stability corrections are required. Another is that the measurements also permit calculation of evaporation rate, additional information which is often useful in interpreting the pattern of N loss as, for instance, in the studies of NH_3 loss reported by Denmead *et al.* (1976, 1978), Lauer *et al.* (1976) and Hutchinson *et al.* (1982).

A disadvantage of the method is that at night, both R and G become small and hence difficult to measure. From Equation (33), it can be seen that the accuracy of the flux estimate is directly proportional to the accuracy with which $R - G$ is measured. A further nighttime problem is that the condensation of dew on radiation instruments leads to erroneous measurements of R. The employment of both aerodynamic and energy balance methods would seem advisable at any time, but particularly so when the diurnal pattern of N loss is being investigated.

The efficacies of both aerodynamic and energy balance methods are illustrated by a practical example from investigations of the emission of NH_3 to the air following surface application of urea to pasture.[3] Simultaneous measurements were made of $\bar{u}$, $\bar{\theta}$, $\bar{p}_r$ and $\bar{p}_g$ at four heights above the surface,[4] and of R and G. From these, F has been calculated by four schemes: three aerodynamic methods described in the previous section, and the energy balance method. Results for two days when unstable conditions prevailed are shown in Fig. 5.2.

On both days, the agreement between the aerodynamic estimates and the energy balance estimate improves as the stability corrections become more elaborate, i.e. as we progress from the Thornthwaite-Holzman scheme with no corrections, through the KEYPS scheme with a single correction, to that of Dyer and Hicks with two different stability functions, φ_m and φ_g. The fluxes calculated with the Dyer-Hicks relationships agree very well with those calculated from the energy balance, but the other two schemes underestimate F on almost all occasions, the one exception being the neutral period on May 9 when no stability corrections are applied in any of the aerodynamic schemes. The need for stability corrections at other times is clearly evident even in the periods with strong winds and small Ri on May 15. The need is most obvious in the first observation periods on both days when highly unstable conditions

---

[3] See footnote 1, p. 136.
[4] See footnote 2, p. 137.

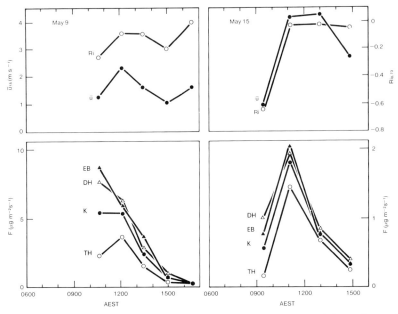

Fig. 5.2. Flux densities (*F*) of NH$_3$ over urea-treated pasture calculated by various micro-meteorological methods. Top: Richardson numbers (Ri) and wind speeds ($\bar{u}$) during the observation periods. Bottom: Flux densities (*F*) calculated by the energy balance method (EB) and three aerodynamic formulations: Dyer and Hicks (DH), KEYPS (K), and Thornthwaite and Holzman (TH).

prevailed. Then the Thornthwaite-Holzman scheme needs to be multiplied by a factor of 4 on one day and 6 on the other in order to equal the Dyer-Hicks or energy balance estimates.

5.5. Mass balance methods

The theoretical basis of these methods and their particular suitability for measuring gaseous emissions from small plots were discussed in Section 5.2. In brief, the flux of the gas into the atmosphere from a treated area of limited upwind extent is equated with the rate at which the gas is carried by the wind across a vertical plane at the downwind edge. If ρ_g is redefined as the density of the gas in excess of the background, then

$$F = (1/x) \int_0^Z \overline{u \rho_g}\, dz,\tag{34}$$

the term $\overline{u \rho_g}$ representing the time-averaged horizontal flux density at any particular level in the vertical plane.

Writing u and ρ_g as the sums of means, $\bar{u}$ and $\bar{\rho}_g$, and fluctuations about those means, u' and ρ'_g, Equation (34) becomes

$$F = (1/x) \int_0^Z (\bar{u}\,\bar{\rho}_g + \overline{u'\rho'_g})\,dz. \tag{35}$$

The first term within the integral in the flux due to horizontal convection while the second is that due to horizontal diffusion. The latter exists whenever there is a gas flux at the surface because then, both u' and ρ_g are correlated with w' and hence with each other. The diffusive flux is in a direction opposite to that of the convective flux.

It will be recognized that measurements of $\overline{u\rho_g}$ require the same fast-response instrumentation as would be required for eddy correlation measurements. Sufficient such instruments are unlikely to be available to measure $\overline{u\rho_g}$ at many heights, which evaluation of Equation (34) requires. More often, the quantities measured will be $\bar{u}$ and $\bar{\rho}_g$, and the diffusion term will be neglected. This is usually justified on the grounds that diffusion is expected to be small in comparison with convection, e.g. Wilson *et al.* (1982). Recent wind tunnel experiments (Raupach and Legg 1983) indicate that the diffusive flux may be about 10% of the convective flux, and so probably can be neglected for many purposes. For more exact work, it will usually suffice to apply an empirical correction of that amount. For convenience in subsequent discussion it will be assumed that

$$F \simeq (1/x) \int_0^Z \bar{u}\,\bar{\rho}_g\,dz. \tag{36}$$

Successful application of Equation (36) requires that measurements of $\bar{u}$ and $\bar{\rho}_g$ extend to the top of the air layer whose concentration is modified by the emission, i.e. to the Z of Equation (1). In practice, Z will depend on surface roughness and atmosphere stability as well as fetch. Experience suggests that for small fetches a good working approximation is that $Z \simeq 0.1x$, but stability does have a large influence. Figure 5.3 illustrates this point.[5] It shows profiles of NH_3 concentration at the downwind edge of a treated area for three stability regimes. In each case, the fetch was 30 m. Stability is specified in terms of L. As conditions varied from highly stable ($L = 4$ m) to highly unstable ($L = -3$ m), Z increased from 1.4 m to approximately 3.5 m. Unlike the eddy correlation method where F can be calculated from measurements of the primary variables at only one height, or gradient diffusion approaches where measurements at two heights suffice, the mass balance method requires that profiles be well

---

[5] From unpublished experiments of J.R. Simpson, J.R. Freney, R. Leuning and O.T. Denmead. The volatilization of NH_3 was measured after spraying a urea solution on a soil plot 25 m square, which had a very short, sparse, vegetation cover.

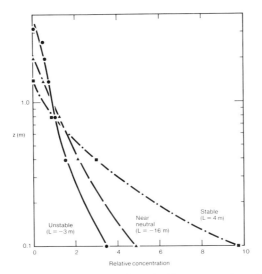

Fig. 5.3. Influence of stability on shapes of profiles of NH_3 concentrations developed over urea treated plots with fetch of 30 m. Abscissa is $\bar{\rho}_{gz} / \left\{ Z^{-1} \int_0^z \bar{\rho}_g(z)dz \right\}$. L is Monin-Obukhov length.

defined experimentally so that the integral in Equation (36) can be evaluated. Because of the changing profile shape, measurements of $\bar{u}$ and $\bar{\rho}_g$ will need to be made at five levels at least, and preferably more. If the experiment is within a crop canopy where the profiles have irregular shapes, many more points are required.

Figure 5.4 shows profiles of $\bar{u}$, $\bar{\rho}_g$ and their product obtained in two experiments concerned with emissions of NH_3 following fertilizer application. In one experiment, urea was spread onto short grass 0.2 m high.[6] In the other, anhydrous ammonia was dissolved in water applied by furrow-irrigation to a corn crop 2.1 m high (Denmead *et al.* 1982b). The profiles over the short grass had quite regular shapes and could be defined well by measurements at five levels, but those in the corn crop were variable in shape and measurements at twelve levels were required to define them.

It is reiterated that in Equation (36), $\bar{\rho}_g$ denotes the gas density in excess of the background. Except if the gas is normally present in trace quantities, that background must be measured although not, perhaps, at as many levels as in the treated area.

Another requirement of the method is that the fetch, x, be known precisely. If the experimental area is the usual agronomist's rectilinear plot, x will vary with wind direction. Its determination will require knowledge of plot geometry and frequent measurement of wind direction, e.g. Denmead *et al.* (1977, 1982b).

[6] See footnote 1, p. 136.

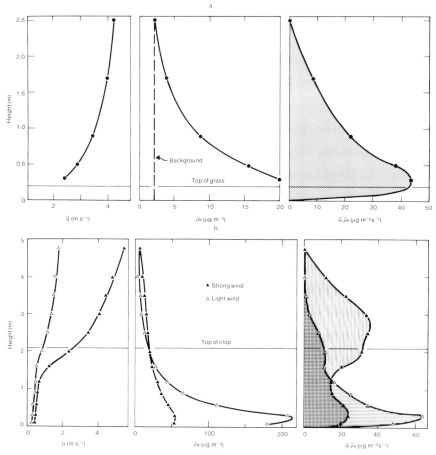

Fig. 5.4. Profiles of wind speed ($\overline{u}$), NH$_3$ density ($\overline{\rho}_g$) and horizontal flux density ($\overline{u}\,\overline{\rho}_g$) observed in experiments involving NH$_3$ release from small plots. (a) Short grass, 0.2 m high; fetch over plot, 25 m. (b) Corn crop, 2.1 m high; fetch over plot, 35 m.

However, as discussed by Wilson *et al.* (1982), this complication can be overcome by working with a circular plot, and measuring $\overline{u}$ and $\overline{\rho}_g$ at its centre. Regardless of compass direction, the wind will always blow towards the centre and *x* is always equal to the plot radius. Circular plots of 36 m radius have been used by Beauchamp *et al.* (1978) and 25 m radius by Wilson *et al.* (1983).

Apart from its eminent suitability for small plots, this method has some distinct advantages over the equilibrium methods discussed previously. First, it has a simple theoretical basis: it requires no special form for the wind profile and no corrections for thermal stratification. Second, if we are not concerned about the slight overestimation of *F* brought about by ignoring horizontal diffusion, the instrumentation is relatively simple. It needs neither to be of fast response nor to have the high precision required for gradient measurements.

A further advantage is that both theory and experiment indicate that in certain situations, it is possible to infer the surface flux from measurements of the horizontal flux at just one height within the modified layer. There are two requirements: although the treated plots should be small, the area in which they are located should be large and uniform so that the wind profiles are equilibrium ones. Second, the treated area should have no vegetative cover, or at least a very short one, so that virtually all of the horizontal flux occurs in the air layers above the surface. Then, the profile of horizontal flux density has a theoretically predictable shape which is determined by surface roughness, plot geometry and atmospheric stability (Philip 1959, Mulhearn 1977 and Wilson *et al.* 1982). From their analysis of the influence of stability on profile shape, Wilson *et al.* (1982) predict the existence of a particular height within the modified layer, at which the normalised horizontal flux, $\overline{u\,\rho_g}/F$, has almost the same value in all stability regimes. They call that height ZINST. If the appropriate value of $\overline{u\,\rho_g}/F$ at ZINST is known, measurements of $\overline{u}$ and $\overline{\rho_g}$ at only that height are sufficient to determine F.

Experimental evidence for the existence of a ZINST is provided by Fig. 5.3 which illustrates the effects of atmospheric stability on the shape of the concentration profile. Although greatly different in shape, the stable, unstable and neutral profiles all intersect within a narrow height interval whose mid-point is at about 0.8 m. Because the relative concentration at that height is insensitive to the stability regime, it can be anticipated that the same will be true of the normalised horizontal flux. That this is so can be seen by reference to Fig. 5.5 where $\overline{u\,\rho_g}$ measured at 0.8 m in the same experiments is plotted against the corresponding F evaluated from Equation (36). The observations are for one-hour sampling periods and come from two experiments at the same site, one in winter (June) and one in spring (September). The fetch was 30 ± 0.8 m and z_0 was 0.003 m. The measurements covered a wide range of stability conditions, as Fig. 5.3 indicates; wind speeds varied from <1 m s^{-1} to >13 m s^{-1}; and surface flux densities varied by more than two orders of magnitude. Yet all observations are fitted by the regression line shown in Fig. 5.5 with a correlation coefficient of 0.997. The intercept of that line is not significantly different from zero; that is, the ratio $\overline{u\,\rho_g}/F$ at 0.8 m was essentially constant.

More empirical evidence is provided by Fig. 5.6 which shows relationships between $\overline{u\,\rho_g}$ and F at three heights in experiments over short grass.[7,8] In that case, the treated plot was circular with a radius of 25 m, d was 0.15 m and z_0 was 0.005 m. Figure 5.6 distinguishes between measurements made at night when stable conditions prevailed and those made during the daytime when conditions were mostly unstable. Lines of best-fit which pass through the origin

---

[7] See footnote 1, p. 136.

[8] See footnote 2, p. 137.

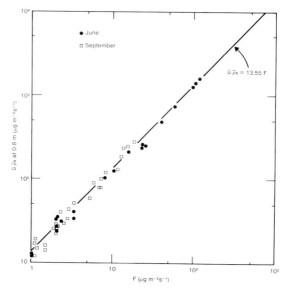

Fig. 5.5. Relationship between horizontal flux density of NH$_3$ at height of 0.8 m ($\bar{u}\,\bar{\rho}_g$) and surface flux density (F) in urea-treated plots with low, sparse vegetation. Points are 1-hour averages and come from two separate experiments at same site. Fetch over treated plots, 30 m.

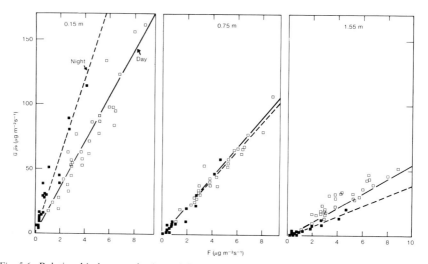

Fig. 5.6. Relationship between horizontal flux density of NH$_3$ at three heights ($\bar{u}\,\bar{\rho}_g$) and surface flux density (F) in urea-treated plots of short grass. Fetch over plot, 25 m. Points are 2-hour averages: circles and solid lines for daytime observations; dots and dashed lines for nighttime. Lines are lines of best-fit passing through origin.

have been calculated for the nighttime and daytime observations at each height and are shown in the figure. The influences of stability on profile shape manifest in Fig. 5.3 are again evident here: near the surface, at a z-d of 0.15 m the ratio of $\overline{u}\,\overline{p}_g$ to F is clearly higher in stable conditions than in unstable conditions, while in the top of the profile at 1.55 m, the reverse is true. At 0.75 m the scatter is much reduced and the ratio is virtually independent of stability.

For all the observations, the best-fit value for $\overline{u}\,\overline{p}_g/F$ at z-d of 0.75 m is 11.3. The generality of this particular value has been tested by using it to predict F in a second experiment conducted at the same site. Again the cover was short grass 0.2 m high, the treated area was a circular plot of 25 m radius, and d was 0.15 m. The surface was somewhat rougher with a z_0 of 0.02 m. The surface fluxes predicted from measurements at z-$d = 0.75$ m, i.e. from $(\overline{u}\,\overline{p}_g)_{0.75}/11.3$, are compared with F evaluated from the full profiles via Equation (36) in Fig. 5.7. For the 31 observations, which extended over 6 days, the regression of the predicted on the measured surface flux has an intercept of 0.24, which is not significantly different from zero, and a slope of 0.963, within 4% of unity. The correlation coefficient is 0.970. As in the first experiment, the agreement is equally good for the daytime and nighttime observations. There is thus strong experimental evidence for the existence of a ZINST. The simplification in measurement technique which this permits is obviously very great.

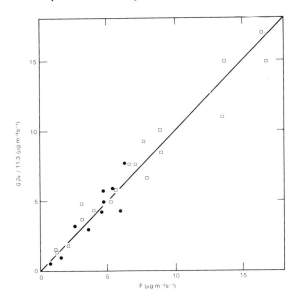

Fig. 5.7. Comparison of surface flux density of NH_3 calculated empirically from measurements of wind speed (u) and NH_3 density ($\overline{p}_g$) at 0.75 m with measured flux (F) in urea-treated plot of short grass. Fetch over plot, 25 m. Points are 2-hour averages: circles for daytime observations; dots for nighttime.

While the appropriate value of $\overline{u\,\rho_g}/F$ at ZINST can be determined empirically by 'calibrating' experiments like those just described, it can also be calculated *a priori* from models of atmospheric dispersion from plane sources such as those of Philip (1959), Mulhearn (1977) and Wilson *et al.* (1982). These analyses predict the development of the concentration profile downwind of a change in surface flux density. Without examining them in detail, all three confirm that at least in neutral conditions, the normalised horizontal flux profile has a fixed shape dependent only on the surface roughness and fetch. Hence $\overline{u\,\rho_g}/F$ at a given height can be calculated from knowledge of these two site parameters.

Only Wilson *et al.* (1982) deal specifically with the influence of atmospheric stability and their model is thus the only one in which ZINST itself can be determined theoretically. They define it as the height at which the profiles of the normalised flux density calculated for $L = +5$ m and $L = -5$ m intersect. In accordance with the experimental evidence of Figs. 5.5, 5.6 and 5.7, their model indicates that the value of $\overline{u\,\rho_g}/F$ at ZINST varies by only a few percent over the range of stability conditions likely to be encountered in the field. The model is a numerical one, requiring a computer to obtain solutions, but Wilson *et al.* (1982) do give values of ZINST and the corresponding values of $\overline{u\,\rho_g}/F$ for fetches of 20 and 50 m and for z_0 values from 0.0005 to 0.05 m. To these predictions can be added the empirical observations presented in this chapter, viz.,

$$x = 25\,\text{m}, \ 0.005\,\text{m} \leqslant z_0 \leqslant 0.02\,\text{m}, \ \text{ZINST} \simeq 0.75\,\text{m}, \ \overline{u\,\rho_g}/F = 11.3;$$

and

$$x = 30\,\text{m}, \ z_0 = 0.004\,\text{m}, \ \text{ZINST} \simeq 0.8\,\text{m}, \ \overline{u\,\rho_g}/F = 13.6$$

The predictive ability of all three models has been tested by using each to compute surface flux densities from the measurements of $\overline{u\,\rho_g}$ at 0.75 m shown in Fig. 5.6. Comparisons between the model predictions and the flux densities measured over three days and nights are shown in Fig. 5.8. The models of Philip (1959) and Mulhearn (1977) tend to overestimate the surface flux slightly by day, while that of Wilson *et al.* (1982) slightly underestimates it. However, all three models give predictions of quite acceptable accuracy, generally well within 10% of the measured flux.

In a separate study, Wilson *et al.* (1983) have calculated a theoretical ZINST of 1.08 m for the same observations. When applied at that height, their model overestimates the measured flux by about as much as it underestimates it when applied at the empirical ZINST of 0.75 m. There thus appears to be some latitude in the choice of ZINST.

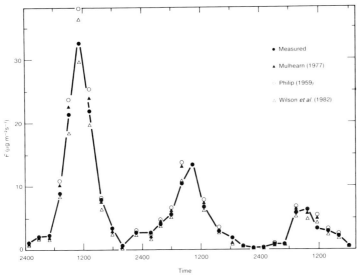

Fig. 5.8. Comparison of surface flux densities of NH₃ (*F*) calculated theoretically from measurements of wind speed and NH₃ density at 0.75 m with measured flux in urea-treated plot of short grass. Fetch over plot, 25 m. Points are for 2-hour observation periods.

Finally, if a trapping method is employed to measure gas concentration as was done, for example, by Denmead *et al.* (1976) and Ferm (1979) for NH₃ and by Ryden *et al.* (1978) for N₂O, one more simplification in experimental technique seems possible. The trap must be so designed that air flows through it at a rate which is linearly related to wind speed. If the cross-sectional area of the air intake is A and the mass of gas collected in the trap after time t is M, then $M/At \propto \overline{u\,\rho_g}$. This sampling scheme thus gives a direct and formally correct measurement of the horizontal flux density, i.e., $\overline{u\,\rho_g}$ instead of $\overline{u}\,\overline{\rho_g}$. Further, if the treated plot is circular and the intake of the trap is mounted at the centre of the plot at a height of ZINST, a single determination of M at the end of the experiment, which might conceivably last several days, permits calculation of the total gas loss. A special, isokinetic sampler is required to exploit this technique, but in view of the simplification that is possible, the development of such a device seems to be a worthwhile endeavour.

5.6. References

Beauchamp, E.G., Kidd, G.E. and Thurtell, G. 1978 Ammonia volatilization from sewage sludge applied in the field. J. Environ. Qual. 7, 141–146.

Braman, R.S., Shelley, T.J. and McClenny, W.A. 1982 Tungstic acid for preconcentration and determination of gaseous and particulate ammonia and nitric acid in ambient air. Anal. Chem. 54, 358–364.

Businger, J.A., Wyngaard, J.C., Izumi, Y. and Bradley, E.F. 1971 Flux-profile relationships in the atmospheric surface layer. J. Atmos. Sci. 28, 181–189.

Calder, K.L. 1949 Eddy diffusion and evaporation in flow over aerodynamically smooth and rough surfaces: a treatment based on laboratory laws of turbulent flow with special reference to conditions in the lower atmosphere. Q.J. Mech. Appl. Math. 2, 153–176.

Denmead, O.T., Freney, J.R. and Simpson, J.R. 1976 A closed ammonia cycle within a plant canopy. Soil Biol. Biochem. 8, 161–164.

Denmead, O.T., Freney, J.R. and Simpson, J.R. 1979 Studies of nitrous oxide emission from a grass sward. Soil Sci. Soc. Am. J. 43, 726–728.

Denmead, O.T., Freney, J.R. and Simpson, J.R. 1982a Atmospheric dispersion of ammonia during application of anhydrous ammonia fertilizer. J. Environ. Qual. 11, 568–572.

Denmead, O.T., Freney, J.R. and Simpson, J.R. 1982b Dynamics of ammonia volatilization during furrow irrigation of maize. Soil Sci. Soc. Am. J. 46, 149–155.

Denmead, O.T. and McIlroy, I.C. 1971. Measurement of carbon dioxide exchange in the field. In: Sestak, Z., Catsky, J. and Jarvis, P.G. (eds.), Plant Photosynthetic Production: Manual of Methods, pp. 467–516. W. Junk, The Hague.

Denmead, O.T., Nulsen, R. and Thurtell, G.W. 1978 Ammonia exchange over a corn crop. Soil Sci. Soc. Am. J. 42, 840–842.

Denmead, O.T., Simpson, J.R. and Freney, J.R. 1974 Ammonia flux into the atmosphere from a grazed pasture. Science 185, 609–610.

Denmead, O.T., Simpson, J.R. and Freney, J.R. 1977 A direct field measurement of ammonia emission after injection of anhydrous ammonia. Soil Sci. Soc. Am. J. 41, 1001–1004.

Dyer, A.J. 1963 The adjustment of profiles and eddy fluxes. Q.J.R. Meteorol. Soc. 89, 276–280.

Dyer, A.J. 1974 A review of flux-profile relationships. Boundary-Layer Meteorol. 7, 363–372.

Dyer, A.J. and Hicks, B.B. 1970. Flux-gradient relationships in the constant flux layer. Q.J.R. Meteorol Soc. 96, 715–721.

Ferm, M. 1979 Method for determination of atmospheric ammonia. Atmos. Environ. 13, 1385–1393.

Freney, J.R., Denmead, O.T., Watanabe, I. and Craswell, E.T. 1981 Ammonia and nitrous oxide losses following applications of ammonium sulfate to flooded rice. Aust. J. Agric. Res. 32, 37–45.

Galbally, I.E. and Roy, C.R. 1978 Loss of fixed nitrogen from soils by nitric oxide exhalation. Nature 275, 734–735.

Hales, J.M. and Drewes, D.R. 1979 Solubility of ammonia in water at low concentrations. Atmos. Environ. 13, 1133–1147.

Harper, L.A., Catchpoole, V.R., Davis R. and Weier, K.L. 1983 Ammonia volatilization: soil, plant, and microclimate effects on diurnal and seasonal fluctuations. Agron. J. (in press).

Hutchinson, G.L., Mosier, A.R. and Andre, C.E. 1982 Ammonia and amine emissions from a large cattle feedlot. J. Environ. Qual. 11, 288–293.

Lauer, D.A., Bouldin, D.R. and Klausner, S.D. 1976 Ammonia volatilization from dairy manure spread on the soil surface. J. Environ. Qual. 5, 134–141.

Lemon, E. 1978 Nitrous oxide (N_2O) exchange at the land surface. In: Nielsen, D.R. and MacDonald, J.G. (eds.), Nitrogen in the Environment. Vol. 1, pp. 493–521. Academic Press, New York.

Lemon, E. and van Houtte, R. 1980 Ammonia exchange at the land surface. Agron. J. 72, 876–883.

Leuning, R., Denmead, O.T., Lang, A.R.G. and Ohtaki, E. 1982 Effects of heat and water vapor transport on eddy covariance measurement of CO_2 fluxes. Boundary-Layer Meteorol. 23, 209–222.

Mosier, A.R. and Hutchinson, G.L. 1981 Nitrous oxide emissions from cropped fields. J. Environ. Qual. 10, 169–173.

Mulhearn, P.J. 1977 Relations between surface fluxes and mean profiles of velocity, temperature and concentration, downwind of a change in surface roughness. Q.J.R. Meteorol. Soc. 103, 785–802.

Panofsky, H.A. 1963 Determination of stress from wind and temperature measurements. Q.J.R. Meteorol. Soc. 89, 85–94.

Paulson, C.A. 1970 The mathematical representation of wind speed and temperature profiles in the unstable atmospheric surface layer. J. Appl. Meteorol. 9, 857–861.

Philip, J.R. 1959 The theory of local advection: I. J. Meteorol. 16, 535–547.

Raupach, M.R. and Legg, B.J. 1983 The uses and limitations of flux-gradient relationships in micrometeorology. J. Hydrol. (in press).

Raupach, M.R. and Thom, A.S. 1981 Turbulence in and above plant canopies. Ann. Rev. Fluid Mech. 13, 97–129.

Rider, N.E., Philip, J.R. and Bradley, E.F. 1963 The horizontal transport of heat and moisture – a micrometeorological study. Q.J.R. Meteorol. Soc. 89, 507–531.

Rolston, D.E., Hoffman, D.L. and Toy, D.W. 1978 Field measurement of denitrification: I. Flux of N_2 and N_2O. Soil Sci. Soc. Am. J. 42, 863–869.

Roy, C.R. 1979 Atmospheric nitrous oxide in the mid-latitudes of the southern hemisphere. J. Geophys. Res. 84, 3711–3718.

Ryden, J.C., Lund, L.T. and Focht, D.D. 1978 Direct in-field measurement of nitrous oxide flux from soils. Soil Sci. Soc. Am. J. 42, 731–737.

Sutton, O.G. 1953 Micrometeorology. McGraw-Hill Bachn Company Inc., New York. 333 pp.

Sutton, O.G. 1947. The problem of diffusion in the lower atmosphere. Q.J.R. Meteorol. Soc. 73, 257–281.

Thom, A.S. 1975. Momentum, mass and heat exchange of plant communities. In: Monteith, J.L. (ed.), Vegetation and the Atmosphere. Vol. 1. pp. 57–109. Academic Press, London.

Thornthwaite, C.W. and Holzman, B. 1939 The determination of evaporation from land and water surfaces. Mon. Weather Rev. 67, 4–11.

Webb, E.K., Pearman, G.I. and Leuning, R. 1980 Correction of flux measurements for density effects due to heat and water vapour transfer. Q.J.R. Meteorol. Soc. 106, 85–100.

Wesely, M.L., Eastman, J.A., Stedman, D.H. and Yalvac, E.D. 1982 An eddy-correlation measurement of NO_2 flux to vegetation and comparison to O_3 flux. Atmos. Environ. 16, 815–820.

Wilson, J.D., Catchpoole, V.R., Denmead, O.T. and Thurtell, G.W. 1983 Verification of a simple micrometeorological method for estimating ammonia loss after fertilizer application. Agric. Meteorol. (in press).

Wilson, J.D., Thurtell, G.W., Kidd, G.E. and Beauchamp, E.G. 1982 Estimation of the rate of gaseous mass transfer from a surface source plot to the atmosphere. Atmos. Environ. 16, 1861–1867.

6. Gaseous nitrogen losses from plants

G.D. FARQUHAR, R. WETSELAAR and B. WEIR

6.1. Introduction

The identification, measurement and understanding of the different loss pathways of N from the plant-soil system are prerequisites in efforts to increase the efficiency of N added to the system through biological and industrial fixation. Of the many pathways of loss, those from plants have in the main been ignored, or were thought to be extremely low (Allison 1955, 1966). Recently, however, Wetselaar and Farquhar (1980) have reviewed data that suggest that *net* losses from above-ground parts of plants can be substantial, up to 75 kg N/ha in 10 weeks (corresponding to 18 nmol $(m^2$ leaf surface$)^{-1}$ sec^{-1} from a crop with a leaf area index of 5), from a wide variety of species, in many geographical areas and under a variety of environmental conditions. They further pointed out that such losses are highest for plants with high N contents and take place mainly between anthesis and maturity. During this physiological period proteins break down in senescing leaves, liberating NH_3 (Thimann 1980). Gaseous losses of NH_3 are therefore possible, and have indeed been found. Gaseous losses of other reduced forms and of oxidized forms have also been found. In addition, losses as N_2 have been postulated. Stutte and coworkers (Stutte and Silva 1981, Stutte and Weiland 1978, Stutte *et al.* 1979, Silva and Stutte 1979a, b, 1981a, b, Weiland and Stutte 1978a, b, 1979a, b, 1980) have reported gaseous fluxes from leaves that would account for the above losses. However, Wetselaar and Farquhar (1980) have suggested that the magnitude may have been overestimated.

In this chapter, we present the history of research into gaseous N losses from plants, and then discuss the different forms and possible mechanisms of these losses, together with the problems associated with their measurement.

6.2. Historical background

During the 19th century contradictory results were reported from attempts to establish whether N was lost from plants in some gaseous form during germination and early growth. No losses could be found by Boussingault (1838), Hellriegel (1855) and Lawes *et al.* (1861), while Schulz (1862) and

Atwater and Rockwood (1886) reported such loss.

To resolve this controversy, Davidson (1923) germinated winter wheat seeds and cowpea seeds, sterilized or non-sterilized, in distilled water directly in the Kjeldahl flasks in which they were to be digested for assessment of N content. He selected this early physiological stage of plant growth because in this period intensive breakdown, transformation and transportation of N compounds takes place. After 10 to 14 days of growth no changes in total N content could be detected, and he concluded that N loss in the gaseous form during germination was unlikely.

The possibility of gaseous N losses from the above-ground parts of plants (hereafter referred to as 'tops') was suggested by Wicke (1862), who found trimethylamine in blossoms of *Crataegus oxyacantha* and in the shoots of *Chenopodium vulvaria*. In 1905 Wilfarth *et al.* observed a 20% net loss of N from the tops of spring wheat during the last three weeks before maturity; during the same period the N content (kg/ha) of the stubble plus roots decreased by 29%, while the dry matter in the tops increased. Similar N losses from tops were obtained with barley, peas and mustard. They concluded that the missing N was either lost in the gaseous form to the atmosphere or returned to the soil.

Klein and Steiner (1928) identified NH_3 qualitatively and quantitatively as a volatilization product from flowers of 70 species and from the leaves of 14 species. Their chemical techniques were sound, but their method for exposing plant parts to pretreated air was of doubtful quantitative value. Above-ground plant parts, cut off from their base, were put into a jar filled with distilled water and then put under a bell jar into which NH_3-free air entered by suction; the outlet passed through 0.1 N HCl to collect the volatilized NH_3. Several such arrangements were connected in series and each one was interspersed with a 10% NaOH scrubber to avoid transfer of acid droplets. This scrubber induced a CO_2-free atmosphere, while the HCl-containing collector ensured an NH_3-free atmosphere for the next arrangement; both features are unrepresentative of ambient air. Furthermore, the sole use of detached plant material may have affected the natural N metabolism of the material tested. Nevertheless, the qualitative differences in NH_3 volatilization rates could well have been correct. Flowers and leaves of all species produced NH_3, but the rates were higher for certain species, for younger leaves (senescing leaves were not tested), at high temperatures and at high light intensities. Amines were given off from 12 species, especially from their flowers.

Klein and Steiner (1928) gave the following biochemical possibilities for the formation of NH_3 and volatile amines:

1. Deamination of amino acids,

$$R \cdot CHNH_2 \cdot COOH + H_2O \rightleftharpoons R \cdot CHOH \cdot COOH + NH_3 \quad (1)$$

2. Oxidative deamination leading to the formation of unstable oxyamino-acids,

$$R \cdot CHNH_2 \cdot COOH + \tfrac{1}{2}O_2 \rightleftharpoons R \cdot COHNH_2 \cdot COOH \qquad (2)$$

which break down to give $RCOCOOH$ and NH_3

3. Reductive deamination,

$$R \cdot CHNH_2 \cdot COOH + H_2 \rightleftharpoons R \cdot CH_2 \cdot COOH + NH_3. \qquad (3)$$

4. Simple decarboxylation, leading to the formation of amines,

$$R \cdot CHNH_2 \cdot COOH \rightleftharpoons R \cdot CH_2 \cdot NH_2 + CO_2. \qquad (4)$$

5. The formation of methyl-, dimethyl-, or trimethylamine through the splitting off of substituted amines from secondary, tertiary or quaternary bases, for instance,

$$(CH_3)_3N \cdot OH \cdot CH_2CH_2OH \rightleftharpoons (CH_3)_3N + CH_2OH \cdot CH_2OH \, (5)$$

6. The formation of an amine through the methylation of amino acids, especially substituted aminopropionic acid,

$$\begin{aligned} &R \cdot CH_2CHNH_2 \cdot COOH + 3CH_3OH \rightleftharpoons \\ &R \cdot CH_2 \cdot CHN(CH_3)_3OH \cdot COOH + 2H_2O \end{aligned} \qquad (6)$$

$$\begin{aligned} &R \cdot CH_2 \cdot CHN(CH_3)_3OH \cdot COOH \rightleftharpoons \\ &R \cdot CH \cdot CH \cdot COOH + (CH_3)_3N + H_2O \, . \end{aligned} \qquad (7)$$

7. The saponification of phosphatides, forming phosphatide bases (such as choline), which in turn break down into an amine and a glycol.

The presence of NH_3 and some volatile amines, especially trimethylamine, *l*-amylamine and *l*-butylamine, in different plant parts was established in the same laboratory by Steiner and Löffler (1931), who thoroughly investigated 271 species for the presence of a wide variety of volatile N compounds.

Several years later, gaseous losses attributed to the van Slyke (1914) reaction were proposed (Eggleton 1935, Pearsall and Billimoria 1937, 1939). This reaction,

$$HNO_2 + R \cdot CHNH_2 \cdot COOH \rightarrow R \cdot COH \cdot COOH + N_2 + H_2O \qquad (8)$$

is generally very rapid, while according to Pearsall and Billimoria (1937) the pH in plant tissues is normally low enough for nitrous acid to be present. These authors floated leaves of *Narcissus pseudonarsiccus* on sterile media containing ammonium- or nitrate-N; they observed N losses from the plant tissues, which were greatest in older tissues where protein hydrolysis was in progress. On this basis, and because Irving and Hankinson (1908) had shown that the formation of suitable amounts of nitrous acid in plant tissues could occur, they proposed the following scheme to explain their results:

$$Amide\text{-}N \rightleftharpoons NH_3 \rightleftharpoons HNO_2 \rightleftharpoons HNO_3$$

$$protein\text{-}N \rightleftharpoons amino\text{-}N \rightarrow N_2$$

From similar experiments Pearsall and Billimoria (1939) concluded that in darkness most of the nitrite formed by reduction of nitrate was lost in amounts that were stoichiometrically consistent with the van Slyke reaction. Such losses occurred also from senescing plant material in the light.

Allison and Sterling (1948) repeated the experiments of Pearsal and Billimoria (1937, 1939), but changed their methods of plant drying to minimize possible N losses during this process, and of digestion to ensure that nitrate was included. The supplied N was readily taken up by the excised leaves, but there was no loss of N during the $2\frac{1}{2}$- or $3\frac{1}{2}$-day incubation period. They concluded that, in general, N losses due to the van Slyke reaction were unlikely; firstly because nitrite is rapidly reduced and thus seldom accumulates in appreciable concentrations, and secondly because the percentage of undissociated nitrous acid decreases rapidly from 69% at pH 3 to 2% at pH 5.

The possibility of plant-N losses through NH_3 volatilization was considered in 1968 by Martin and Ross, who used a gas lysimeter in which Rhodes grass (*Chloris gayana*) was grown in a soil to which ^{15}N-labelled potassium nitrate or ammonium sulphate was applied. Any NH_3 that was volatilized was trapped. After 27 days, 0.6% of the ammonium sulfate-N and 0.1% of the potassium nitrate-N had been volatilized as NH_3. In the case of ammonium sulfate, some of this may have escaped from the soil, in spite of a soil pH of 5.7. However, in the nitrate-treated system an NH_3 loss from the soil was less likely, and the authors speculated on possible NH_3 volatilization by the plants. In similar experiments (Craswell and Martin 1975), much larger amounts of NH_3 were trapped during leaf senescence, which strengthened speculation concerning NH_3 losses from leaves.

After Porter *et al.* (1972) had exposed maize plants to ^{15}N-labelled NH_3, they found trace amounts of label in condensates in the exposure chamber; the possibility of this N originating from NH_3 volatilization from the leaves could not be discounted.

Studies of gaseous losses from plants reported in more recent literature will be dealt with in later sections, but first the transfer of gases to and from leaves will be put into a quantitative framework.

6.3. Gas exchange between plant leaves and the atmosphere

The majority of gas exchange by terrestrial plants is *via* stomata in the leaves.

The molar flux density, J (mol m^{-2} sec^{-1}), of a gas entering the leaf is then determined (Cowan 1977) by

$$J = g(p_a - p_i)/P \tag{9}$$

where g (mol m^{-2} sec^{-1}) is the conductance to diffusion of the gas through stomata and the boundary layer surrounding the leaf, p_a is the ambient partial pressure of that gas, p_i is the partial pressure in the intercellular spaces of the leaf (strictly in the stomatal cavities) and P is the atmospheric pressure. Since the atmospheric pressure is often close to one bar, it is usually convenient to measure partial pressures as fractions of a bar, instead of the SI unit (Pascal).

We take, as an example, the flux of NH_3 into the leaf. Since the partial pressure of NH_3 in the atmosphere is typically 1 to 8 nbar in unpolluted areas (Junge 1956), it is convenient to scale p_a and p_i in terms of nbar with the result that the flux, J, becomes nmol m^{-2} sec^{-1}. Thus, with p_i typically being 2.5 nbar, taking p_a as 5 nbar, and given a typical conductance, g, (stomatal plus boundary layer) of 0.4 mol m^{-2} sec^{-1}, the resulting flux is

$$J = 0.4(5 - 2.5)/1 = 1 \text{ nmol m}^{-2} \text{ sec}^{-1}. \tag{10}$$

The easiest method for the determination of g is to first measure the conductance to diffusion of water vapour. Here p_i may be taken as the saturated vapour pressure at the temperature of the leaf (Farquhar and Raschke 1978). The ambient vapour pressure, p_a, is routinely measured by psychrometers, dew point hygrometers, variable capacitance humidity sensors, or infrared gas analysers. The efflux $(-J)$ is determined, for example, from the difference in the amount of water vapour between air entering and leaving a chamber containing the leaf, i.e.

$$-aJ = u(p_o - p_e)/P, \tag{11}$$

where a (m^2) is the leaf area, u (mol sec^{-1}) is the molar flux of air past the leaf, and p_o and p_e are the vapour pressures of air leaving and entering the chamber, respectively. In well-stirred chambers $p_a = p_o$, and in others p_a is usually taken as $(p_e + p_o)/2$. The conductance to water, $g(H_2O)$, is then determined from Equation (9). This conductance is dependent on the boundary layer conductance $g_b(H_2O)$, as well as on the stomatal conductance $g_s(H_2O)$, by

$$1/g(H_2O) = 1/g_s(H_2O) + 1/g_b(H_2O). \tag{12}$$

The boundary layer conductance is often much greater than the stomatal conductance, but may be determined, using Equations (9) and (11), from measurements made using a wet filter paper replica of the leaf. Strictly, $g_s(H_2O)$ consists of stomatal and cuticular components in parallel. The latter is usually small compared to the former.

The conductance to the gas of interest, and our example is again NH_3, is found from

$$1/g(NH_3) = 1/g_s(NH_3) + 1/g_b(NH_3), \tag{13}$$

where the two terms on the right side are determined from

$$g_s(NH_3) = g_s(H_2O) \frac{D(NH_3/air)}{D(H_2O/air)} \tag{14}$$

and

$$g_b(NH_3) = g_b(H_2O) \left[\frac{D(NH_3/air)}{D(H_2O/air)} \right]^{\frac{2}{3}}, \tag{15}$$

where $D(NH_3/air)$ is the diffusivity of NH_3 in air and $D(H_2O/air)$ is the diffusivity of water vapour in air. The $\frac{2}{3}$ power relationship in Equation (15) results from the Polhausen analysis of heat or mass transfer in laminar parallel flow (Kays 1966). In applications where extreme accuracy ($>90\%$) is required, ternary interactions need to be taken into account (Caemmerer and Farquhar 1981, Leuning private communication). These authors also describe the corrections to Equation (11) which take into account the dilution of the trace gas by transpired water vapour.

It is commonly believed that the diffusivity of a gas is inversely proportional to the square root of the molecular weight. This (Graham's law) is true for diffusion *in vacuo*. However the coefficients of interest here are binary diffusion coefficients and the reduced mass, $(m_1 m_2/(m_1 + m_2))^{\frac{1}{2}}$, is the appropriate parameter, where m_1 and m_2 are the molecular weights of the diffusing species. However, the dependence on reduced mass is not regular and the binary diffusion coefficients must be determined empirically. Fortunately, the diffusivities in air of a number of gases are readily available (Andrussow 1969). Since the coefficients are inversely proportional to pressure and approximately proportional to the 1.8 power of absolute temperature, their ratio as in Equation (14) should be determined using coefficients accordingly adjusted to a common temperature and pressure. The ratio of the diffusivities of water vapour and CO_2 in air is 1.6 which happens, coincidentally, to approximate the inverse ratio of the square root of the molecular weights. However NH_3, which is *lighter* than H_2O, has a diffusivity in air 0.92 times that of water vapour (Andrussow 1969). Many studies of the uptake and release of gases (other than CO_2) have ignored the factors outlined here.

From the above we see that measurements of the rate of uptake of a gas enable its intercellular partial pressure, p_i, to be determined by rearranging (9) as

$$p_i = p_a - JP/g. \tag{16}$$

This is useful for removing effects of gaseous diffusion in studies of gaseous uptake, since then uptake rate, J, may be plotted against p_i when the latter is varied by varying p_a.

The procedure outlined above is routinely carried out in studies of CO_2 assimilation by leaves. However, in the present context we are interested in predicting J. To do this we need to know its relationship with p_i. Many authors have found a linear relationship between ambient concentration and flux, implying a linear relationship between J and p_i. A resistance analog is often employed so the relationship becomes

$$J = \frac{p_i}{r_m P} \qquad (17)$$

where r_m is some sort of 'mesophyll resistance'. Since Equation (16) may be rewritten as

$$p_i = p_a - JP(r_b + r_s), \qquad (18)$$

where

$$r_b = 1/g_b \qquad (19)$$

and

$$r_s = 1/g_s \qquad (20)$$

Equation (17) becomes

$$J = \frac{p_a/P}{r_b + r_s + r_m}. \qquad (21)$$

However, it has been recognized for some time that Equation (17) is inadequate to describe the uptake of CO_2, since there is a finite partial pressure of CO_2, $p(CO_2)$, at which the net flux is zero. This partial pressure is called the CO_2 compensation point. Recently, the existence of an NH_3 compensation point was established (Farquhar et al. 1980). As a first approximation in such cases we can rewrite Equation (17) as

$$J = \frac{p_i - \gamma}{r_m P} \qquad (22)$$

where γ is the compensation point. Equation (21) is then replaced by

$$J = \frac{(p_a - \gamma)/P}{r_b + r_s + r_m}. \qquad (23)$$

In most studies of the interactions between gases and leaves, the imposed ambient partial pressures have been far greater than those occurring naturally on a global scale. Thus, the presence of compensation points may have been missed.

From Equation (23) it is apparent that there is a net efflux of NH_3 from the leaf when p_a is less than γ. Analogous equations by be written for other gases, and losses will occur when the compensation points are greater than the respective ambient partial pressures. In the next sections we look for possible cases where γ may exceed p_a.

6.4. Gaseous nitrogen losses as dinitrogen

Nitrogenous compounds that could be volatilized from leaves are in the oxidized form, the reduced form, or N_2. We discuss these separately, beginning with losses as N_2.

Chibnall (1939) has argued that while N_2 loss may be slow, it would likely be greatest in older leaves undergoing protein hydrolysis (i.e. senescing leaves), or in young leaves exposed to light. Also Vanecko and Varner (1955) have observed evolution of $^{15}N_2$ upon illumination of 10- to 12-day old wheat-seedling leaves infiltrated with $K^{15}NO_2$, and have attributed this loss to the van Slyke reaction (Equation (8)). Nevertheless, McKee (1962) has suggested that the reaction may not be important in physiological conditions because the nitrite content of plant tissue is usually very low, and the reaction requires rather low levels of pH as discussed in 6.2.

Since 78% of the earth's atmosphere is N_2 the detection of small amounts of N_2 lost from plants remains the major difficulty in assessing the importance of this loss pathway.

6.5. Losses in the oxidized form

Although plant species in acid and anaerobic environments may take up the bulk of their nitrogen in the ammoniacal form, most non-leguminous, cultivated plants take it up as nitrate (Pate 1980). This nitrate may be either reduced to NH_3 (or ammonium) and converted to amides in the roots, or may pass up the xylem and be reduced in the leaves (Pate 1980). Studying a range of species, Pate (1973) found considerable variation in this pattern, from 90% reduction in the roots of lupins to no reduction at all in *Xanthium* roots.

Temporary accumulation of nitrate can occur in plant tops, but this does not appear to be detrimental. We do not know whether some of this nitrate may be reduced to NO_2 in strongly reducing conditions. Normally, however, the nitrate is reduced to nitrite in the cytoplasm according to

$$NO_3^- + 2H^+ + 2e^- \xrightarrow[\text{reductase}]{\text{nitrate}} NO_2^- + H_2O, \tag{24}$$

after which the nitrite is reduced to NH_3 in the chloroplasts:

$$NO_2^- + 7H^+ + 6e^- \xrightarrow[\text{reductase}]{\text{nitrite}} NH_3 + 2H_2O. \tag{25}$$

Accumulation of nitrite is potentially detrimental, in that volatilization of N compounds may then occur.

A nonenzymatic reduction of nitrite by ascorbic acid and NADPH (reduced pyridine nucleotide) has been observed by Evans and McAuliffe (1956). Most of the nitrite-N (80%) appeared as NO and N_2O but N_2 was also evolved. At pH 6 this reduction was slow and the rate increased sharply with increasing acidity. Porter (1969) observed that when nitrite reacted with oximes, which are present in plants (McKee 1962), or organic matter, the major gaseous product was NO.

Nitrite reduction can be blocked by the application of certain photosynthetic inhibitors to soybean leaves, resulting in the evolution of NO, and to a lesser extent NO_2 (Klepper 1979a). In the dark, the application of 2,4-D is thought to enhance nitrate reduction, also causing nitrite assimilation and the emission of NO_x (NO and NO_2) from soybean (Klepper 1979a). The greatest rate of evolution observed was 2.4 μg NO and 0.07 μg NO_2 (g fresh weight)$^{-1}$ min^{-1}. If we assume a leaf fresh weight of 0.2 kg per m^2 of leaf surface, this evolution corresponds to a flux of 230 nmol m^{-2} sec^{-1} (14 kg N ha^{-1} day^{-1} at LAI 5). Some aspects of this work by Klepper (1979a) have been verified by Churchill and Klepper (1979) in wheat, but with less evolution of NO_x observed. NO_x emission from soybean can also be caused by hydration products of SO_2 (Klepper 1979b).

The evolution rates quoted above are extremely high, and Klepper's conclusion that untreated leaves evolve no NO_x needs examination in a different context; technique capable of measuring fluxes of 230 nmol m^{-2} sec^{-1} may not be sensitive enough to detect those of interest here, i.e. fluxes of about an order of magnitude less.

When NO_2 reacts with water it yields nitric and nitrous acids. Plants take up and metabolize NO_2, the reaction being apparently first order with respect to NO_2 concentration (Rogers et al. 1979a). The partial pressure of NO_2 in their experiments was never less than 80 nbar, and extrapolation of these data to normal ambient levels is difficult. In subsequent experiments, however, Rogers et al. (1979b) found no evolution into NO_2-free air.

In many studies of the uptake of trace gases the results are expressed as pseudo-'first order rate constants' or JP/p_a in the terminology of the previous section. It is apparent from Equation (23) that these constants should be less than $1/r_s$, the stomatal conductance to the diffusion of the particular gas. Many such studies include simultaneous measurements of the stomatal resistance to diffusion of water, but, unfortunately, few make the necessary quantitative comparisons. For example, we calculate from Rogers *et al.* (1979b) that JP/p_a was typically 0.08 mol m^{-2} sec^{-1} for *Zea mays*, while the total conductance to diffusion of water vapour was 0.084 mol m^{-2} sec^{-1}. The binary diffusivity of NO_2/air has not, as far as we know, been measured directly. However, using Equation 16.3–1 of Bird, Stewart and Lightfoot (1960) we calculate it to be 0.58 times that of H_2O/air. On this basis the total conductance to the diffusion of NO_2 through the stomata and boundary layer was apparently only 0.05 mol m^{-2}sec^{-1}, i.e. insufficient to sustain the observed flux. For soybean JP/p_a was typically 0.16 mol m^{-2} sec^{-1}. The 'rate constants' observed by Fuhrer and Erismann (1980) appear larger, by at least two orders of magnitude, than would be allowed by the (reasonable) values of conductance reported. Rogers *et al.* (1979b) noted a strong correlation between NO_2 uptake and stomatal conductance (or its inverse, the 'surface resistance'). In our notation this means that both γ and r_m were small. It is apparent from our previous discussion that such studies need not rely on correlations and that quantitative relationships between gaseous fluxes and stomatal conductances should be formalized. Thus, in the present example, if $r_m = \gamma = 0$, a surface conductance (to the diffusion of NO_2) of 0.1 mol m^{-2} sec^{-1} means that the uptake of NO_2 by a leaf is 0.1 nmol m^{-2} sec^{-1} per nbar $p(NO_2)$.

Plants also assimilate NO (Anderson and Mansfield 1979), although the extent to which the NO must first be converted to NO_2 appears to be unknown. The data of Galbally and Roy (1978) suggested that a compensation point also exists for NO. When a chamber was placed over grass, the soil-plant system gave off NO, and the partial pressure in the chamber rose, but levelled off at 15 nbar. However, Galbally (personal communication) has since observed little difference between bare and grass covered soil, implying that the metabolism of NO by the grass leaves is probably negligible.

Osretkar (1971) reported that *Chlorella* cultures were able to convert fixed N in the media to N_2O. However, we know of no reports of significant losses of N_2O from terrestrial plants. Hutchinson and Mosier (1979) measured an emission of less than 3 kg/ha during the growth of a well-fertilized crop of *Zea mays*, but even this amount could be accounted for by denitrification in the soil.

In summary, there is as yet no evidence for appreciable losses of oxides of N under normal physiological conditions. On the other hand, conditions where the capacity for nitrite reduction may be impaired deserve closer scrutiny. Such conditions might include leaf water deficits and senescence.

6.6. Losses in the reduced form

Franzke and Hume (1945) detected emission of trace amounts of HCN from sorghum plants, but the fluxes involved were insignificant – less than 2.5 pmol

m^{-2} sec^{-1} (0.15 g N ha^{-1} day^{-1} at LAI 5).

The most likely candidates for volatilization are some of the amines and NH_3. We deal with these separately, although current analytical techniques cannot completely distinguish between the two N forms.

As previously noted (Section 6.2), amines are widely distributed amongst plant species. The simple aliphatic amines are the most common, especially in flowers (Smith 1971), though diamines, polyamines and related compounds are also widely distributed (Smith 1980). Amines are known to be volatilized from both leaves and flowers (Klein and Steiner 1928). Such N losses may well be seasonal since several authors have reported plant amine contents to be closely correlated with the development stage of their flowers. Cromwell (1949), for example, found that methylamine content in leaves of *Mercurialis annua* was greatest during flowering and declined thereafter. Richardson (1966) found that the liberation of amines from *Sorbus aucuparia* and *Crataegus monogyna* was closely correlated with opening of the flower buds.

Many studies have been made of the uptake of NH_3 by leaves placed in NH_3-enriched atmospheres. These have been reviewed by Farquhar *et al.* (1980). Further studies using enriched atmospheres have since been made by Rogers and Aneja (1980), and Cowling and Lockyer (1981). The former loosely related uptake to stomatal conductance and the latter to a deposition velocity. Further analyses could usefully be made using the techniques outlined earlier in this chapter.

Our interest here is more with the processes taking place under normal ambient conditions and, particularly, those which potentially cause volatilization. We listed earlier many of the reactions leading to NH_3. An example of a reaction represented by Equation (1) is the following:

$$\text{asparagine} \xrightleftharpoons[\text{ADP} + P_i \qquad \text{ATP}]{} \text{aspartic acid} + NH_3. \qquad (26)$$

Other deaminations involve arginine, glycine, glutamine and cysteine (Mazelis 1980).

Quantitatively the most important of these reactions (at least as far as C_3 species are concerned) occurs in the photorespiratory C and N cycles (Woo *et al.* 1978, Keys *et al.* 1978) (see Fig. 6.1). Since the key steps (Equations (27)–(29)) are central to other aspects of N metabolism (in C_4 as well as C_3 species), we examine photorespiration in some detail.

In this process (reviewed by Lorimer and Andrews 1980) phosphoglycolate, one of the products of ribulose bisphosphate oxygenation, is dephosphorylated in the chloroplast, and then converted to glycine in a series of reactions in the peroxisome (see Fig. 6.1). This involves the transamination of glyoxylate from both serine and glutamate. Then, in the mitochondrion

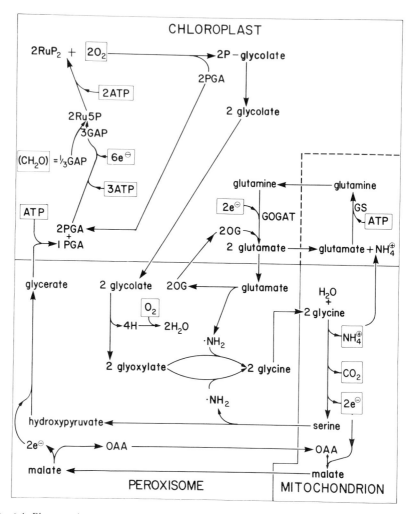

Fig. 6.1. Photorespiratory pathways of carbon and nitrogen. The net stoichiometry is:

$$3O_2 + ([CH_2O] = 1/3 \ GAP) + 8H^+ + 8e^- + 7ATP \rightarrow CO_2 + 5H_2O + 7ADP + 7P_i$$

The dashed lines indicate that glutamine synthetase (GS) may occur in the cytoplasm as well as in the chloroplast. There is also some uncertainty concerning the extent to which malate and OAA shuttle between the mitochondrion and the peroxisome (Woo and Osmond 1982). Abbreviations are as follows: ATP, adenosine triphosphate; GAP, glyceraldehyde-3-phosphate; GOGAT, glutamate synthase; GS, glutamine synthetase; OAA, oxaloacetic acid; 2OG, 2-oxoglutarate; PGA, 3-phosphoglycerate; RuP$_2$, ribulose-1,5-bisphosphate; Ru5P, ribulose-5-phosphate.

$$2 \ glycine + NAD^+ \xrightarrow[\text{decarboxylase}]{\text{glycine}} serine + CO_2 + NH_4^+ + NADH \qquad (27)$$

The stoichiometries of the reactions involved are such that one mole of oxygenation gives rise to the evolution of 0.5 mole CO_2 (photorespiration) and

0.5 mole NH_3. Since the ratio of oxygenation to carboxylation is given by $2\Gamma_*/p_c$, where Γ_* is the CO_2 photocompensation point and p_c is the $p(CO_2)$ in the chloroplast (Farquhar and Caemmerer 1982), the ratio of NH_3 release to primary carboxylation is given by Γ_*/p_c. At $25°$ C, Γ_* is thought to be 35 μbar and p_c typically about 190 μbar, so that about 0.18 mole of NH_3 is released for each primary carboxylation. However, because of photorespiratory and other evolution of CO_2, the ratio increases to 0.23 mole NH_3 released per mole net CO_2 fixation. Γ_* increases with temperature, which means that NH_3 release is much faster at high temperatures. The importance of this photorespiratory release can be assessed by comparing the ratio 0.23 with the value 0.05, which is the ratio of NH_3 incorporated to C fixed in a plant with N forming 2% of its dry weight.

The plant cannot afford to lose all the NH_3 released by glycine decarboxylation. The next step in the photorespiratory N cycle is an irreversible reaction catalysed by glutamine synthetase (GS):

$$\text{glutamate} + NH_4^+ + ATP \xrightarrow{Mg^{2+}} \text{glutamine} + ADP + P_i. \qquad (28)$$

This may occur either in the cytoplasm or in the chloroplast. Although there is variation amongst species recent evidence (Hirel et al. 1982) suggests that in C_3 species 80% or more is in the chloroplasts. Finally, the cycle is 'closed' in the chloroplast by the enzyme glutamate synthase (GOGAT), which irreversibly catalyses the reaction

$$\text{glutamine} + \text{2-oxoglutarate} + 2H^+ + 2e^- \rightarrow \text{2-glutamate}. \qquad (29)$$

However, for GS to function, there must be a finite concentration of NH_3 (or ammonium). The in vitro K_m of GS is 12 to 22 μM [NH^{+4}] (O'Neal and Joy 1974, Stewart and Rhodes 1977). Glutamate dehydrogenase (GDH) is another enzyme which can potentially be used to assimilate NH_3 by

$$\text{2-oxoglutarate} + NH_4^+ + 2e^- \xrightleftharpoons{GDH} \text{glutamate}. \qquad (30)$$

However, the K_m for this reaction is so high (10 to 50 mM [NH_4^+]) in higher plants (Davies and Teixeira 1975) that its physiological role is unclear. The cell appears to have another mechanism which may reduce loss of NH_3. Raven and Farquhar (1981) presented evidence suggesting that there is an ammonium transporter in the plasmalemma of Phaseolus vulgaris, probably a uniporter driven by an electrical potential gradient.

C_4 species have a photorespiritory N cycle, similar to that of C_3 species, but with reduced rates (Berger and Fock 1982) and this is shielded from the mesophyll cells which are adjacent to the intercellular gas spaces. Nevertheless, the mesophyll cells (like C_3 cells for that matter) have a number of other reactions involving NH_3 (or ammonium). Farquhar et al. (1979) found that at realistically low partial pressures of NH_3, no fluxes into or out of healthy leaves of *Zea mays* could be detected, and they inferred that a finite $p(NH_3)$ must exist in the substomatal cavities. The conclusion was consistent with observations by Meyer (1973) that when NH_3-free air was delivered to chambers containing *Z. mays* plants, the air leaving had a $p(NH_3)$ of 2 nbar. Other C_3 species had higher rates of evolution. Farquhar *et al.* (1980) varied the ambient $p(NH_3)$ and found positive compensation points for *Z. mays* and another C_4 species, *Amaranthus edulis*, as well as in the C_3 species *Phaseolus vulgaris* and *Eucalyptus pauciflora*.

The compensation point of *P. vulgaris* was 2.5 nbar at 26° C and increased to 5.5 n bar at 33° C. Coincidentally, perhaps, both compensation points corresponded to the same concentration of dissolved ammonium at the pH, 6.8, of the leaf apoplast (Raven and Farquhar, unpublished) once the temperature dependencies of the pK_a of ammonium and of the Henry constant for NH_3 were taken into account. This concentration was 46 μM, which is close to the estimated concentration of ammonium in the xylem of these nitrate-grown plants (Raven and Farquhar, unpublished). This is probably a coincidence as the relevant concentration is that of the leaf apoplast. The latter depends on that in the transpiration stream, and on the rates of transpiration, of ammonium uptake by the cells, and its possible transport in phloem, as well as on the rates of gaseous exchange of NH_3. If there were no abstraction of NH_3 by the leaf cells, NH_3 would evolve from the leaf at the rate at which it is supplied by the transpiration stream, a concentration of 50 μM corresponding to 0.9 μmol NH_3 volatilised per mol H_2O transpired. More data on the ammonium concentration in the xylem would be useful, particularly from plants which have access to ammonium in soil solution; it is important to determine whether ammonium ever forms a significant portion of the N exported from the roots.

At a pH of 7.4, approximating that of cytoplasm (Smith and Raven 1979), a $p(NH_3)$ of 2.5 nbar corresponds to 11 μM [NH_4^+]. At pH 8, as found in chloroplasts in the light (Heilmann *et al.* 1980) the same partial pressure corresponds to 3 μM [NH_4^+]. These values are comparable to or slightly lower than the K_m of GS, discussed earlier. However the $p(NH_3)$ may be higher inside the cell than in the apoplast, because of the ammonium transporter in the plasmalemma.

Farquhar *et al.* (1980) pointed out that the small fluxes of NH_3 through stomata under normal conditions (of the order of nmol m^{-2} sec^{-1}) are unlikely to affect the internal metabolism in the short term, since other fluxes through the internal NH_3 (or ammonium) pool are of the order of μmol m^{-2} sec^{-1}. Raven and Farquhar (unpublished) examined the fluxes of ammonium to and from leaf slices of *p. vulgaris* and found that the compensation point, γ, was 8 nbar. At concentrations deviating from this, uptake was described by Equation

(22) with $r_m = 0.06$ m^2 sec mol^{-1}. It is possible that both γ and r_m are overestimated by these *in vitro* techniques. Even so, the resistance is much less than those of the boundary layer around intact leaves, and of the stomata. Thus, under natural conditions the $p(NH_3)$ in the substomatal cavity, p_i, is likely to be close to γ and Equation (23) becomes

$$J = \frac{(p_a - \gamma)/P}{r_b + r_s} \tag{31}$$

or

$$J = g(p_a - \gamma)/P. \tag{32}$$

Similar approximations may be valid for the fluxes of other trace gases.

The laboratory results described above are consistent with the field micro-meteorological data of Denmead *et al.* (1976), who observed that a mixed pasture canopy absorbed NH$_3$ that had been released from the soil. Moving up the canopy the $p(NH_3)$ declined from 24 to 2 nbar; the temperature at the top of the canopy was 24° C. Lemon and van Houtte (1980) also concluded that a compensation point, γ, for NH$_3$ existed in the field. Their data suggest to us that γ was somewhat less than 3 to 4 nbar for *Agropyron repens* at 18° C and somewhat less than 5 nbar for *Glycine max* at 23° C. In some cases they observed evolution of NH$_3$ from the top and bottom leaves of the canopy, and absorption in midcanopy. The fluxes from the upper leaves were 5 to 6 nmol (m^2 ground area)$^{-1}$ sec^{-1}, into air with a $p(NH_3)$ or 5 to 8 nbar. Conditions for evolution included hot and sunny weather, ample soil moisture, ample N supply, and low ambient $p(NH_3)$. They appear to be consistent with conditions outlined earlier, i.e. high temperature causing high γ, low ambient $p(NH_3)$, and, probably, high stomatal conductance. Leaf water deficits were also implicated.

6.7. Ammonia evolution during senescence

The evolution of NH$_3$ observed by Lemon and van Houtte (1980) from the lower leaves of their canopies may have been from senescent leaves. Farquhar *et al.* (1979) observed fluxes of about 0.6 nmol m^{-2} sec^{-1} from senescing leaves of *Zea mays*. With senescence photorespiration declines in C$_3$ species, but proteolysis increases in both C$_3$ and C$_4$ species. Simpson and Dalling (1981) carried out a detailed study of senescence in attached wheat leaves (detached leaves show somewhat different characteristics) and found that, during grain-filling, activities of GS and GOGAT declined in parallel with other chloro-plastic, soluble protein. In a parallel study (Peoples *et al.* 1980) they found that the necessary ultrastructural integrity of the mitochondria was maintained beyond 25 days after anthesis. Simpson and Dalling (1981) found that the activity of GDH (glutamate dehydrogenase, a mitochondrial enzyme) was maintained during this period and reached a peak only late in the course of senescence. This should not, however, be taken as evidence for a rôle for GDH in reducing NH_3 loss during senescence; fluxes via GDH were not measured. This enzyme has a much higher K_m for NH_3 than GS and is unlikely to be involved in NH_3 refixation in non-senescent tissue (Lea and Miflin 1979); Simpson and Dalling (1981) only observed a 2.5 fold increase in ammonium levels over those before anthesis.

Hooker et al. (1980) followed the emission by winter wheat of NH_3 into NH_3-free air. After anthesis the rates increased to 0.2 nmol m^{-2} sec^{-1}. These, and the rates of volatilization reported by Farquhar et al. (1979), are very low considering the high K_m of GDH, and the compensation point into which it would translate, even allowing for the reduction of stomatal conductance during senescence. Thus the role of GDH, even in senescence, is unclear. The above losses of NH_3 are trivial in a crop context. Perhaps, such losses become more important in hot conditions, since $p(NH_3)$ in equilibrium with a particular ammonium ion concentration in solution has a Q_{10} of approximately 3 (see Farquhar et al. 1980).

6.8. Nitrogen loss measurements using pyro-chemiluminescence techniques

A new technique for measurements of total nitrogen (apart from N_2) has enabled measurement of total gaseous losses of nitrogenous compounds from leaves. Stutte and Weiland (1978) were the first to apply the technique, known as pyro-chemiluminescence, in this context. They enclosed leaves of various species in 'Saranex' plastic bags. Air was passed over the leaves, through a cold finger in dry ice for condensation, and then recycled over the leaves. The condensate was then analysed for N using this technique. Nitrogen losses of, typically, 9 nmol m^{-2} sec^{-1} were observed in a number of plant species. The losses increased with increasing temperature, the highest rate being 27 nmol m^{-2} sec^{-1} from Amaranthus palmeri at 35° C. A possible disadvantage of this procedure is that the partial pressures of the unknown nitrogenous compounds are close to zero in the air recirculating over the leaf. Overestimation of naturally occurring effluxes may then result unless the leaf is incapable of metabolizing the gases (Wetselaar and Farquhar, 1980). On the other hand, if the flow rate is small, a buildup of concentration may inhibit efflux, thus underestimating natural fluxes.

Many experiments followed with similar results. Stutte et al. (1979) noted that N loss was correlated with leaf temperature and transpiration rate. Analysis of all their data reveals that, under the majority of conditions, the ratio of the rates of N and water losses was between and 0.7 and 3 μmol N/mol

H_2O. Weiland and Stutte (1979b) estimated the amounts of N lost in oxidised and reduced forms by comparing the results obtained by substituting N_2 for O_2 in the pyro-chemiluminescent system. They concluded that typically 10 to 20% was in the 'oxidised form'. By this they meant combined with oxygen.

However, while this technique worked well with many standard compounds, it suggested that 10% of the N in ammonium chloride is oxidised. Nevertheless the tentative conclusion is that the large majority of N loss was in a reduced form. The most obvious candidate is NH_3, but the ratio of N loss to transpiration is considerably higher than that reported by others studying the gaseous fluxes of NH_3. The ratio is similar to that of ammonium and water in the xylem, and requires that net transport from the apoplast to the leaf cells be close to zero. This seems unlikely in healthy leaves and the discrepancy has yet to be resolved.

6.9. Isotopic exchange and fractionation

Because NH_3, and possibly other compounds, can be both absorbed and released, a plant grown with ^{15}N enriched fertilizer will tend to lose $^{15}NH_3$ and gain $^{14}NH_3$ even if the net flux of NH_3 is zero. Such isotopic fluxes may, in the future, be revealed in accurate studies of the N balance using gas lysimetry (see Craswell and Martin, 1975). Some useful information may be obtained from studies of isotopes at natural abundances, since the exchange of $^{14}NH_3$ is probably more rapid than that of $^{15}NH_3$. Depending on whether the plant is assimilating or volatilizing NH_3, the leaves could become depleted or enriched in ^{15}N compared to the xylem stream. We calculate the binary diffusivity of $^{15}NH_3$ in air to be 1.018 times less than that of $^{14}NH_3$, by comparing the square root of the reduced mass (discussed earlier) of $^{15}NH_3$ and air (MW 28.8) with that of $^{14}NH_3$ and air. Further the partial pressure of gaseous $^{15}NH_3$ in equilibrium with a certain aqueous concentration of $^{15}NH_3$ is 1.005 times less than that of $^{14}NH_3$ in equilibrium with the same concentration of aqueous $^{14}NH_3$ (Kirshenbaum et al., 1947). In turn the aqueous concentration of $^{15}NH_3$ in equilibrium with a certain concentration of $^{15}NH_4{}^+$ is 1.029 times less than that for the ^{14}N compounds (Kirshenbaum et al. 1947). These isotopic effects may explain the enrichment of ^{15}N sometimes observed at the top of the soil profile (Feigin et al. 1974). Mariotti et al. (1980) found the organic and total nitrogen of leaves of young Panicum americanum and P. mollisimum plants was enriched in ^{15}N compared to the roots. Volatilisation of ammonia from the leaves may explain, in part, these interesting results.

6.10. Deposition of particulate ammonium

The concentration of ammonium in particles in air is comparable to and often exceeds that of gaseous NH_3. Atmospheric particulate aerosol can be conveniently thought of as comprising two size fractions, partitioned at about 0.5 to 1 μM in radius. The sub-micrometre fraction usually comprises most of the aerosol by number, while the large particles comprise most of the aerosol by mass. Ammonium ions are predominantly in the sub-micrometre fraction, generally as ammonium sulfate (Galbally et al. 1982). Penkett et al. 1979 found a mean concentration of 1 nmol NH_4^+/mol air (a measure approximating 1 nbar $p(NH_3)$). Caiazza et al. (1978) estimated that 5.6 kg N ha^{-1} $year^{-1}$ (1.3 nmol $m^{-2}sec^{-1}$) was deposited in dry particulate form in central Alberta. Rodgers (1978) estimated 4 kg ha^{-1} $year^{-1}$ at Rothamsted and reviewed other estimates, which range up to 50 kg ha^{-1} $year^{-1}$. Tjepkema et al. (1981), claiming improved techniques, found mean concentrations of 2 nmol NH_4^+/mol air and estimated deposition of 2.5 kg N ha^{-1} $year^{-1}$. They observed that plastic leaves accumulated 2.8 μmol N m^{-2} in a one day period. The fate of ammonium deposited on real leaves is unknown. It is a potential source of error in interpretation of experiments described earlier.

6.11. Summary

Leaves can absorb and metabolise many nitrogenous gases, including NH_3. Ammonia is an important intermediate in the photorespiratory N cycle, in the conversion of nitrate to amino acids, and in the breakdown of proteins. The high activity of glutamine synthetase reduces ammonium levels inside leaf cells, and an ammonium transporter moves ammonium from the apoplast into the cells. However, continuous supply of low concentrations of ammonium in the transpiration stream, together, possibly, with some leakage from leaf cells result in a finite partial pressure of NH_3 in the substomatal cavities. When this partial pressure exceeds that in the atmosphere, net evolution of NH_3 occurs, this condition being more likely at higher temperatures and in senescing leaves. Studies by Stutte and colleagues using using pyro-chemiluminescent N-detection suggest losses averaging 9 nmol (m^2 leaf surface)$^{-1}sec^{-1}$ or 38 kg N ha^{-1} in 10 weeks from a crop with a leaf area index of 5, but other techniques indicate smaller losses. Assessment of gaseous NH_3 losses may be confounded by deposition of ammonium particles on leaf surfaces. Certain chemical inhibitors, and water stress combined with high irradiance, may cause the evolution of oxides of N. However the magnitude of gaseous N losses, and their importance in N yield, are still uncertain.

6.12. References

Anderson, L.S. and Mansfield, T.A. 1979 The effects of nitric oxide pollution on the growth of tomato. Environ. Pollut. 20, 113–121.

Andrussow, L. 1969 Dynamische Konstanten. In: Borchers, H., Hausen, H., Hellwege, K-H. and Schmidt, E. (eds.), Landolt-Bornstein Zahlenwerte und Funktionen aus Physik, Chemie, Astronomie, Geophysik und Technik. Vol. 2(5A), pp. 1–729 (6th ed.), Springer-Verlag, Berlin.

Allison, F.E. 1955 The enigma of soil nitrogen balance sheets. Adv. Agron. 7, 213–250.

Allison, F.E. 1966 The fate of nitrogen applied to soils. Adv. Agron. 18, 219–258.

Allison, F.E. and Sterling, de tar L. 1948 Gaseous losses of nitrogen from green plants. II. Studies with excised leaves in nutrient media. Plant Physiol. 23, 601–8.

Atwater, W.O. and Rockwood, E.W. 1886 On the loss of nitrogen by plants during germination and growth. Am. Chem. J. 8, 327–343.

Berger, M.G. and Fock, H.P. 1982 ^{15}N and inhibitor studies on the photorespiratory nitrogen cycle in maize leaves. Photosynthesis Res. (in prrss).

Bird, R.B., Stewart, W.E. and Lightfoot, E.N. 1960 Transport Phenomena. Wiley, New York.

Boussingault, J.B. 1838 Recherches chimiques sur la végétation enterprises dans le but d'examiner si les plantes prennent l'azote a l'atmosphere. Ann. Chim. Phys. 67, 5–54.

Caemmerer, S. von and Farquhar, G.D. 1981 Some relationships between the biochemistry of photosynthesis and the gas exchange of leaves. Planta (Berl.) 153, 376–387.

Caiazza, R., Hage, K.D. and Gallup, D. 1978 Wet and dry deposition of nutrients in Central Alberta. Water, Air Soil Pollut. 9, 309–314.

Chibnall, A.C. 1939 Protein Metabolism in the Plant. Yale University Press, New Haven.

Churchill, K. and Klepper, L. 1979 Effects of ametryn on nitrate reductase activity and nitrite content of wheat. Pestic. Biochem. Physiol. 12, 156–162.

Cowan, I.R. 1977 Stomatal behavior and environment Adv. Bot. Res. 4, 117–227.

Cowling, D.W. and Lockyer, D.R. 1981 Increased growth of ryegrass exposed to ammonia. Nature (Lond.) 292, 337–338.

Craswell, E.T. and Martin, A.E. 1975 Isotopic studies of the nitrogen balance in a cracking clay. 1. Recovery of added nitrogen from soil and wheat in the glasshouse and gas lysimeter. Aust. J. Soil Res. 13, 43–52.

Cromwell, B.T. 1949 The micro-estimation and origin of methylamine in *Mercurialis perrinis* L. Biochem. J. 45, 84–86.

Davidson, J. 1923 Is gaseous nitrogen a product of seedling metabolism? Bot. Gaz. 76, 95–101.

Davies, D.D. and Teixeira, A.N. 1975 The synthesis of glutamate and the control of glutamate dehydrogenase in pea mitochondria. Phytochemistry 14, 647–656.

Denmead, O.T., Freney, J.R. and Simpson, J.R. 1976 A closed ammonia cycle within a plant canopy. Soil Biol. Biochem. 8, 161–164.

Eggleton, W.G.E. 1935 The assimilation of inorganic nitrogenous salts, including sodium nitrate, by the grass plant. Biochem. J. 29, 1389–1397.

Evans, H.J. and McAuliffe, C. 1956 Identification of NO, N_2O and N_2 as products of the nonenzymatic reduction of nitrite by ascorbate or reduced diphosphopyridine nucleotide. In: McElroy, W.D. and Glass, B. (eds.), Inorganic Nitrogen Metabolism, pp. 189–197. Johns-Hopkins Press, Baltimore.

Farquhar, G.D. and Raschke, K. 1978 On the resistance to transpiration of the sites of evaporation within the leaf. Plant Physiol. 61, 1000–1005.

Farquhar, G.D. and Caemmerer, S. von 1982 Modelling of photosynthetic response to environmental conditions. In: Lange. O.L., Nobel, P.S., Osmond. C.B. and Ziegler, H. (eds.), Physiological Plant Ecology II, Encyclopedia Plant Physiology. New Series, Vol. 12B. Springer-Verlag, Berlin, pp. 549–587.

Farquhar, G.D., Wetselaar, R. and Firth, P.M. 1979 Ammonia volatilization from senescing leaves

178

of maize. Science (Wash. D.C.). 203, 1257–1258.

Farquhar, G.D., Firth, P.M., Wetselaar, R. and Weir, B. 1980 On the gaseous exchange of ammonia between leaves and the environment: determination of the ammonia compensation point. Plant Physiol. 66, 710–714.

Feigin, A., Shearer, G., Kohl, D.A. and Commoner, B. 1974 The amount and nitrogen-15 content of nitrate in soil profiles from two Central Illinois fields in a corn-soybean rotation. Soil Sci. Soc. Am. Proc. 38, 465–471.

Franzke, C.J. and Hume, A.N. 1945 Liberation of HCN in sorghum. J. Am. Soc. Agron 37, 848–851.

Fuhrer, J. and Erismann, K.H. 1980 Uptake of nitrogen dioxide by plants grown at different salinity levels. Experientia (Basel) 36, 409–410.

Galbally, I.E. and Roy, C.R. 1978 Loss of fixed N from soils by nitric oxide emission. Nature (Lond.) 275, 734–735.

Galbally, I.E., Farquhar, G.D. and Ayers, G.P. 1982 Interactions in the atmosphere of the biogeochemical cycles of carbon, nitrogen and sulphur. In: Galbally, I.E. and Freney, J.R. (eds.), The Cycling of Carbon, Nitrogen, Sulphur and Phosphorus in Terrestrial and Aquatic Ecosystems, pp. 1–9. Australian Academy of Science, Canberra.

Heilmann, B., Hartung, W. and Gimmler, H. 1980 The distribution of abscisic acid between chloroplasts and cytoplasm of leaf cells and the permeability of the chloroplast envelope for abscisic acid. Z. Pflanzenphysiol. 96, 67–78.

Hellriegel, M. 1855 Beitrag zur Keimungsgeschichte der ölgebenden Samen. J. Prakt. Chem. 64, 94–109.

Hirel, B., Perrot-Rechenmann, C., Suzuki, A., Vidal, J. and Gadal, P. 1982 Glutamine synthetase in spinach leaves. Immunological studies and immunocytochemical localization. Plant Physiol. 69, 983–987.

Hooker, M.L., Sander, D.H., Paterson, G.A. and Daigger, L.A. 1980 Gaseous N losses from winter wheat. Agron. J. 72, 789–792.

Hutchinson, G.L. and Mosier, A.R. 1979 Nitrous oxide emissions from an irrigated cornfield. Science (Wash. D.C.) 205, 1125–1127.

Irving, H.A. and Hankinson, R. 1908 The presence of a nitrate reducing enzyme in green plants. Biochem. J. 3, 87–96.

Junge, C.E. 1956 Recent investigations in air chemistry. Tellus 8, 127–139.

Kays, W.M. 1966 Convective Heat and Mass Transfer. McGraw-Hill, New York.

Keys, A.J., Bird, I.F., Cornelius, M.J., Lea, P.J., Wallsgrove, R.M. and Miflin, B.J. 1978 Photorespiratory nitrogen cycle. Nature (Lond.) 275, 741–743.

Kirshenbaum, I., Smith, J.S., Crowell, T., Graff, J. and McKee, R. 1947 Separation of the nitrogen isotopes by the exchange reaction between ammonia and solutions of ammonium nitrate. J. Chem. Phys. 15, 440–446.

Klein, G. and Steiner, M. 1928 Stickstoffbasen im Eiweissabbau höherer Pflanzen. I. Ammoniak und flüchtige Amine. Jahrb. Wiss. Bot. 68, 602–712.

Klepper, L. 1979a Nitric oxide (NO) and nitrogen dioxide (NO_2) emissions from herbicide treated soybean plants. Atmos. Environ. 13, 537–542.

Klepper, L. 1979b (Author's reply to letter to the editor.) Atmos. Environ. 13, 1475.

Lawes, J.B., Gilbert, J.H. and Pugh, E. 1861 On the sources of nitrogen of vegetation. Philos. Trans. R. Soc. Lond. B. Biol. Sci. 15, 431–577.

Lea, P.J. and Miflin, B.J. 1979 Photosynthetic ammonia assimilation. In: Gibbs, M. and Latzko, E. (eds.), Photosynthesis II, Encyclopedia Plant Physiology, New Series. Vol 6, pp. 445–456. Springer-Verlag, Berlin.

Lemon, E. and van Houtte, R. 1980 Ammonia exchange at the land surface. Agron. J. 72, 876–883.

Lorimer, G.H. and Andrews T.J. 1980 The C-2 photo- and chemorespiratory carbon oxidation cycle. In: Hatch, M.D. and Boardman, N.K. (eds.), The Biochemistry of Plants. Vol 8, pp.

239–274. Academic Press, New York.

Mariotti, A., Mariotti, F., Amarger, N., Pizelle, G., Ngambi, J.-M., Champigny, M.-L. and Moyse, A. 1980 Fractionnements isotopique de l'azote lors des processus d'absorption des nitrates et de fixation de l'azote atmospheric par les plantes. Physiol. Vég. 18, 163–181.

Martin, A.E. and Ross, P.J. 1968 A nitrogen-balance study using labelled fertilizer in a gas lysimeter. Plant Soil. 28, 182–186.

Mazelis, M. 1980 Amino acid catabolism. In: Miflin, B.J. (ed.), Amino Acids and Derivatives, The Biochemistry of Plants. Vol 5, pp. 541–567. Academic Press, New York.

McKee, H.S. 1962 Nitrogen Metabolism in Plants. Clarendon Press, Oxford. 728 pp.

Meyer, M.W. 1973 Absorption and release of ammonia from and to the atmosphere by plants. Ph.D. thesis, University of Maryland, College Park. 52 pp.

O'Neal, D. and Joy, K.W. 1974 Glutamine synthetase of pea leaves. Divalent cation effects, substrate specificity, and other properties. Plant Physiol. 54, 773–779.

Osretkar, A. 1971 Gaseous and other nitrogen losses from cultures of *Chlorella*. Ph.D. thesis, University of Maryland, College Park. 65 pp.

Pate, J.S. 1973 Uptake, assimilation and transport of nitrogen compounds by plants. Soil Biol. Biochem. 5, 109–119.

Pate, J.S. 1980 Transport and partitioning of nitrogenous solutes. Annu. Rev. Plant Physiol. 31, 313–340.

Pearsall, W.D. and Billimoria, M.C. 1937 Losses of nitrogen from green plants. Biochem. J. 31, 1743–1750.

Pearsall, W.H. and Billimoria, M.C. 1939 The influence of light upon nitrogen metabolism in detached leaves. Ann. Bot. (Lond.) 3, 601–618.

Penkett, S.A., Atkins, D.H.F. and Unsworth, M.H. 1979 Chemical composition of the ambient aerosol in the Sudan Gezira. Tellus 31, 295–307.

Peoples, M.B., Beilharz, V.C., Waters, S.P., Simpson, R.J. and Dalling, M.J. 1980 Nitrogen redistribution during grain growth in wheat (*Triticum aestivum* L.) II. Chloroplast senescence and the degradation of ribulose-1,5-bisphosphate carboxylase. Planta (Berl.) 149, 241–251.

Porter, L.K. 1969 Gaseous products produced by anaerobic reaction of sodium nitrite with oxime compounds and oximes synthesized from organic matter. Soil Sci. Soc. Am. Proc. 33, 696–702.

Porter, L.K., Viets, F.L. and Hutchinson, G.L. 1972 Air containing N-15 ammonia: foliar absorption by corn seedlings. Science (Wash. D.c.) 175, 759–761.

Raven, J.A. and Farquhar, G.D. 1981 Methylammonium transport in *Phaseolus vulgaris* leaf slices. Plant Physiol. 67, 859–863.

Richardson, M. 1966 Studies on the biogenesis of some simple amines and quaternary ammonium compounds in higher plants. Phytochemistry 5, 23–30.

Rodgers, G.A. 1978 Dry deposition of atmospheric ammonia at Rothamsted in 1976 and 1977. J. Agric. Sci. 90, 537–542.

Rogers, H.H. and Aneja, V.P. 1980 Uptake of atmospheric ammonia by selected plant species. Environ. Exp. Both. 20, 251–257.

Rogers, H.H., Campbell, J.C. and Volk R.J. 1979a Nitrogen-15 dioxide uptake and incorporation by *Phaseolus vulgaris*. L. Science (Wash. D.C.) 206, 333–335.

Rogers, H.H., Jeffries, H.E. and Witherspoon, A.M. 1979b Measuring air pollutant uptake by plants: nitrogen dioxide. J. Environ. Qual. 8, 551–557.

Schulz, E. 1862 Chemische Beiträge zur Kentniss des Keimprozesses bei jungen Phanerogamen. J. Prak. Chem. 817, 128–173.

Silva, P.R.F. da and Stutte, C.A. 1979a Loss of gaseous N from rice leaves with transpiration. Arkansas Farm Res. 28, 3.

Silva, P.R.F. da and Stutte C.A. 1979b Response of rice to foliar application of 'cytozyme crop + '. Proc. Plant Regulator Working Group 6, 35–38.

Silva, P.R.F. da and Stutte, C.A. 1981a Nitrogen loss in conjunction with transpiration from rice leaves as influenced by growth stage, leaf position and N supply. Agron. J. 73, 38–42.

Silva, P.R.F. da and Stutte, C.A. 1981b Nitrogen volatilization from rice leaves. II. Effects of source of applied nitrogen in nutrient culture solution. Crop Sci. 21, 913–916.

Simpson, R.J. and Dalling, M.J. 1981 Nitrogen redistribution during grain growth in wheat (*Triticum aestivum L.*) III. Enzymology and transport of amino acids from senescing flag leaves. Planta (Berl.) 151, 447–456.

Smith, F.A. and Raven, J.A. 1979 Intracellular pH and its regulation. Annu. Rev. Plant Physiol. 30, 289–311.

Smith, T.A. 1971 The occurrence, metabolism and functions of amines in plants. Biol. Rev. Camb. Philos. Soc. 46, 201–241.

Smith, T.A. 1980 Plant amines. In: Bell, E.A. and Charlwood, B.V. (eds.), Secondary Plant Products, Encycl. Pl. Physiol., New Series, Vol. 8, pp 433–460. Springer-Verlag, Berlin.

Steiner, M. and Löffler, H. 1931 Stickstoffbasen im Eiweisabbau höherer Pflanzen II. Histochemische Studien über Verbreitung, Verteilung und Wandel des Ammoniaks und der flüchtigen Amine. Jahrb. Wiss. Bot. 71, 463–532.

Stewart, G.R. and Rhodes, D. 1977 A comparison of the characteristics of glutamine synthetase and glutamate dehydrogenase from *Lemna minor* L. New Phytol. 79, 257–268.

Stutte, C.A. and Silva P.R.F. da 1981 Nitrogen volatilization from rice leaves. I. Effects of genotype and air temperature. Crop Sci. 21, 596–600.

Stutte, C.A. and Weiland, R.T. 1978 Gaseous N loss and transpiration of several crop and weed species. Crop Sci. 18, 887–889.

Stutte, L.A., Weiland, R.T. and Blem, A.R. 1979 Gaseous N loss from soybean foliage. Agron. J. 71, 95–97.

Thimann, K.V. 1980 The senescence of leaves. In: Thimann, K.V. (ed.), Senescence of Plants, pp. 85–115. CRC Press, Boca Raton.

Tjepkema, J.D., Cartica, R.J. and Hemond, H.F. 1981 Atmospheric concentration of ammonia in Massachusetts and deposition on vegetation. Nature (Lond.) 294, 445–446.

Vanecko, S. and Varner, J.E. 1955 Studies on nitrite metabolism in higher plants. Plant Physiol. 30, 388–390.

van Slyke, D.D. 1914 Gasometric determination of aliphatic amino nitrogen in minute quantities. J. Biol. Chem. 16, 121–124.

Weiland, R.T. and Stutte, C.A. 1978a N loss with transpiration in several crop and weed species. Arkansas Farm Res. 27 (2), 16.

Weiland, R.T. and Stutte, C.A. 1978b Chemically-bound N loss from soybean foliage as affected by several synthetic chemicals. Proc. Plant Growth Regulator Working Group 5, 78–85.

Weiland, R.T. and Stutte, C.A. 1979a Net CO_2 uptake and N loss responses in soybeans to growth regulators and temperature. Proc. Plant Growth Regulator Working Group 6, 17–23.

Weiland, R.T. and Stutte, C.A. 1979b Pyro-chemiluminescent differentiation of oxidized and reduced N form evolved from plant foliage. Crop Sci. 19, 545–547.

Weiland, R.T. and Stutte, C.A. 1980 Concomitant determination of foliar nitrogen loss, net carbon dioxide uptake, and transpiration. Plant Physiol. 65, 403–406.

Wetselaar, R. and Farquhar, G.D. 1980 Losses of nitrogen from the tops of plants. Adv. Agron. 33, 263–302.

Wicke, W. 1862 Beobachtungen an *Chenopodium vulvaria* über die Ausscheidung von Trimethylamin. Bot. Zeitung. 20, 393.

Wilfarth, M., Römer, M. and Wimmer, G. 1905 On the Assimilation of the Elements of Nutrition of Plants During Different Periods of Their Growth. Vinton, London. 72 pp.

Woo, K.C. and Osmond, C.B. 1982 Stimulation of ammonia, 2-oxoglutarate-dependent O_2 evolution in isolated chloroplasts by dicarboxylates and the role of the chloroplast in photorespiratory nitrogen recycling. Plant Physiol. 69, 591–596.

Woo, K.C., Berry, J.A. and Turner, G.L. 1978 Release and refixation of ammonia during photorespiration. Carnegie Inst. Wash. Year Book 77, 241–245.

7. Nitrogen loss from sewage sludges and manures applied to agricultural lands

E.G. BEAUCHAMP

7.1. Introduction

It has been known for some time that losses of ammonia (NH_3) from manures, sewage sludges or other organic substances occur after their application to land. A short review of this topic was included recently in an article published by Terman (1979). The purpose of the present review is to deal mainly with NH_3 volatilization losses from manures and sewage sludges as affected by climatic and soil factors. Information on NH_3 volatilization from other organic substances is meagre and is covered by Terman (1979).

Some research on NH_3 loss from manures had been completed before the widespread use of inorganic N fertilizers. Heck (1931a,b) and Slater and Schollenberger (1938) reported on crop response to applied manures with variable intervals following application. They concluded that, to conserve N as much as possible, it was necessary to incorporate the manure immediately into the soil. This permited sorption of volatile N compounds by the soil and largely prevented volatilization losses. Thus researchers 40 to 50 years ago were quite aware of N volatilization losses in practical terms. As fertilizer N prices began to escalate recently, a renewed interest in N volatilization has occurred.

7.2. Methods for measuring ammonia volatilization

Terman (1979) reviewed various methods used to measure volatilization losses of NH_3 from substances applied to the soil. These methods will not be reviewed here, except briefly to compare closed and open systems used in the field, since another review on methodology is included in this book. Closed systems essentially involve pulling air through an enclosure on the soil surface and then through a dilute acid solution. Open systems usually involve an aerodynamic approach using micrometeorological measurements and have been variously described by Denmead et al. (1974), Beauchamp et al. (1978), Lemon and Van Houtte (1980) and Wilson et al. (1982). Beauchamp et al. (1978), while comparing their data with those obtained by McGarity and Rajaratnam (1973), suggested that an open system for measurement may provide a truer estimate of NH_3 volatilization. Hoff et al. (1981) recently concluded that NH_3 volatilization in a field study involving a closed NH_3 collection system measured only 56% of the NH_3 volatilized under windy field conditions.

Denmead *et al.* (1974) used a micrometeorological method in the field based on the energy balance near the ground surface. This method involves the conservation of energy at the ground surface and the calculation of a transfer coefficient. This along with measurements of atmospheric NH_3 at several heights were used to determine flux densities. Beauchamp *et al.* (1978) also employed micrometeorological measurements along with a diffusion model to predict NH_3 flux densities. Measurements of windspeed and atmospheric NH_3 concentration profiles near the ground surface were required to determine flux density of NH_3 from applied sewage sludge. More recently, Wilson *et al.* (1982) have outlined a further development of the approach used by Beauchamp *et al.* where measurements of windspeed and atmospheric NH_3 at a single pre-determined height would provide satisfactory estimates of NH_3 flux density from a substance applied to the soil.

7.3. Other volatile nitrogen compounds

In addition to NH_3, N_2O and N_2, which are the forms of N generally lost to the atmosphere, other forms of N may also be volatilized. These other volatile N compounds may be particularly important in determining N losses from applied organic substances. Denmead *et al.* (1974) reckoned that, under some conditions in a sheep pasture, up to one-half the N collected in their acid solution traps was in the form of other volatile basic compounds, presumably amines. The proportion of NH_3 of the total N collected varied from day to day but on cool, humid days, virtually pure NH_3 was collected. In studies on NH_3 exchange between the atmosphere, the crop canopy and the soil, Lemon and Van Houtte (1980) suggested volatile amines might account for 10% of the N trapped in acid solution. They also suggested that 'particulate' NH_3 might also be collected. Miner *et al.* (1975) detected 3-amino pyridine, trimethyl amine and possibly isopropyl amine in atmospheres of swine buildings. They also found that non-NH_3 nitrogen evolution rates from various surfaces associated with dairy and swine operations varied from 0.25 to 0.75 of the NH_3 evolution rate. Non-NH_3 nitrogen values were consistently low from a swine manure pond surface.

Luebs *et al.* (1974) reported that substantial quantities of N trapped in 0.01 N H_2SO_4 solution were not NH_3 or 'distillable' N. They found aliphatic amines to be present although their method was not suitable for measuring secondary and tertiary amines which may have been present. They also found that 60 to 100% of the total N in air sampled near a dairy operation was distillable (NH_3) N.

Mosier *et al.* (1973) identified seven basic aliphatic organic N compounds, namely methyl, dimethyl, ethyl, n-propyl, isopropyl, n-butyl, and n-amyl amines, emanating from a high density cattle feedyard. They estimated that the identified amines constituted 2–6% of the basic materials volatilized from a cattle feedlot relative to NH_3.

It is evident from the rather sparse information available that research on the quantities and kinds of volatile non-NH_3 N compounds emanating from manures, sewage sludges and other organic substances is required. However, the following discussion generally deals with N losses in the NH_3 form.

7.4. Factors affecting NH_3 volatilization

Many factors have been cited to have an effect on the rate of NH_3 volatilization from manures or sewage sludges. These factors are discussed in the following discourse.

One obvious factors is the concentration of ammoniacal (NH_3 plus NH_4^+) N present in the organic substance. Koelliker and Miner (1973) suggested that the main 'driving force' for volatilization (desorption) of NH_3 from a liquid manure surface is the NH_3 partial pressure gradient (or difference) between the liquid phase and the ambient atmosphere. They expressed the NH_3 volatilization rate in the following mass flow relationship (using English units):

$$\frac{d(NH_3 - N)}{dt} = AK \ (P_l - P_g)$$

where

NH_3-N	=	weight of NH_3-N volatilized (lb)
t	=	time (days)
A	=	area of the manure-air interface (ft^2);
K	=	overall transfer coefficient ($lb/day/ft^2/atm$);
P_l	=	partial pressure of NH_3 in manure (atm);
P_g	=	partial pressure of NH_3 in atmosphere (atm).

The effect of temperature on the partial pressure of NH_3 in a liquid was estimated by Koelliker and Miner (1973) from Kowalke *et al.* (1925) as follows (assuming that Henry's law holds and the heat of solution is independent of temperature):

$$\ln \frac{P_L}{m} = \frac{-4425}{T} + 10.82$$

where

$$P_L \quad = \quad \text{partial pressure of } NH_3 \text{ in liquid (atm);}$$
$$m \quad = \quad \text{molality of } NH_3 \text{ in liquid (moles/1);}$$
$$T \quad = \quad \text{temperature } (^\circ K).$$

This equation predicts that the partial pressure of NH_3 in the liquid phase would increase as temperature increases for a given NH_3 concentration. Thus the volatilization of NH_3 predicted by the equation of Koelliker and Miner (1973) is dependent on temperature.

Koelliker and Miner (1973) noted that K decreased slightly as temperature increased from 10 to 30° C. The effect of pH is also dependent on temperature as shown in Fig. 7.1. The proportion of the ammoniacal N as NH_3 increases as pH and temperature of the liquid phase increase. The pH of manures and sewage sludges often ranges between 7.0 and 7.5 so that the potential for NH_3 volatilization is considerable.

Lauer *et al.* (1976) also found that the relative partial pressures of NH_3 in manure and the ambient atmosphere was a very important factor affecting NH_3 volatilization. They presented an isothermic equilibrium diagram showing the relationships between NH_3 partial pressure, pH and the concentration of total ammoniacal N (TAN) as in Fig. 7.2. Lauer *et al.* (1976) hypothesized that a sequence of events occurs as NH_3 volatilization proceeds with drying of non-fresh manure. Initially, the TAN concentration probably increases with slight increases in NH_3 partial pressure. As NH_3 volatilizes, the composition of the

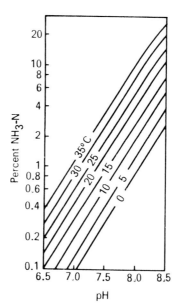

Fig. 7.1. Percent of ammoniacal nitrogen in the ammonia form in relation to pH and temperature (Koelliker and Miner 1973).

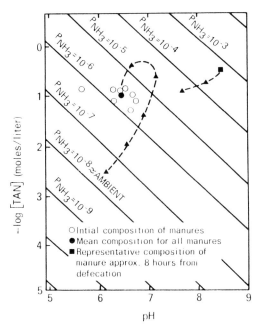

Fig. 7.2. Proposed changes in total ammoniacal nitrogen (TAN) and pH in manures with time following application. Triangular symbol with stippled curve represents pathway of changes (Lauer *et al.* 1976).

manure changes so that a lower pH occurs and the NH_3 partial pressure approaches that of the ambient atmosphere. Since the TAN concentration depends on the quantity of TAN and water in the manure, wetting and drying play significant roles in determining the rate of NH_3 volatilization. Lauer *et al.* (1976) also cited diffusion rates of NH_3 (and NH_4^+), temperature and possibly microbial transformations as affecting NH_3 volatilization rates.

The role of microorganisms is not clear insofar as the NH_3 volatilization rate is concerned. It may be surmised that microorganisms will degrade organic compounds to produce ammoniacal N.

Microbial activity may affect the CO_2 concentration in manure and thereby affect the equilibrium NH_3 concentration in the liquid phase (Marion and Dutt 1974). Vlek and Stumpe (1978) indicated that the NH_3 volatilization capacity of aqueous systems is dependent on the buffering capacity. Also, the loss of NH_3 was proportional to the loss of alkalinity which is defined as consisting of NH_3 (aqueous), hydroxyl, bicarbonate and carbonate ions. Thus the presence of a bicarbonate or carbonate source such as $CaCO_3$ enhances the NH_3 volatilization potential.

Fenn and Kissel (1973) suggested the following relationship involving ammonium fertilizers and a calcareous soil.

$$(NH_4)_2CO_3 + H_2O \rightleftarrows 2NH_3 \uparrow + H_2O + CO_2 \uparrow$$

This relationship suggests that the ratio of evolved NH_3 to CO_2 would be 2/1. On the other hand, Feagley and Hossner (1978) suggested the following relationship involving bicarbonate ions was more likely to predominate.

$$NH_4HCO_3 \rightleftarrows 2NH_3 + 2CO_2 + H_2O.$$

This relationship would provide for an NH_3/CO_2 ratio of 1/1. Research at the Univerity of Guelph on NH_3 and CO_2 evolution during oven drying of swine manure has resulted in an NH_3/CO_2 ratio of 0.96 (V. Pavlicik and J.B. Robinson, personal communication). Further research in this area of chemical composition may provide additional clues on the role of the carbonate system and hence the mechanisms involved in NH_3 volatilization from manures or sludges.

The rôle of the cation exchange capacity of manures or sludges on NH_3 volatilization remains to be elucidated. One may expect that the NH_3 volatilization rate would decrease as the cation exchange capacity increases in the same manner as for soils when N fertilizer is applied (Faurie and Bardin 1979a). Beauchamp et al. (1978) noted that the proportion of 'exchangeable' ammonium N extracted from two sewage sludges was 0.6 and 8.9% of the total ammoniacal N. The importance of this difference was not determined. Also, the influence of the cation exchange capacity of the underlying soil on NH_3 volatilization from manures or sludges is unknown. In a laboratory study, Curnoe (1975) found that NH_3 volatilization from sewage sludge increased as the silt-plus-clay content of several soils decreased. The rôle of the cation exchange capacity within this effect is unclear. However, one would expect that the effect of soil characteristics may not be as great as with N fertilizers because NH_3 volatilization would occur largely from the manure or sludge surface.

The effect of organic matter content on NH_3 volatilization from manures and sludges has been considered. Hashimoto and Ludington (1971) found that the dissociation constant (K_a) for ammoniacal N in concentrated chicken manure slurries to be about one-sixth the theoretical value for dilute NH_3 solutions. The K_a value is defined as follows:

$$NH_4^+ \rightleftarrows NH_3 + H^+$$

$$K_a = \frac{[NH_3][H^+]}{[NH_4^+]}$$

Hashimoto and Ludington also suggested that ionic strength and dielectric constant would influence the K_a value. On the other hand, it is noteworthy that English et al. (1980) found that the NH_3 concentration in sewage sludge was apparently not affected by the 'organic strength' of sewage sludge. This aspect requires further attention since K_a values may vary considerably not only in relation to organic matter content but also in the kinds of organic substances present in manures or sludges.

English *et al.* (1980) found that the pH of anaerobically digested sludge increases rapidly following exposure to the atmosphere. They suggested that this increase was due to a loss of dissolved CO_2 from the sludge. Sludge undergoes an approximate thousand-fold decrease in its partial pressure of CO_2 when transferred from the anaerobic digestor and exposed to the atmosphere. It has been observed in the author's laboratory that the pH of bottle-stored (at 2 C) liquid cattle manure increased from 6.8 to 7.3 within one-half hour following exposure to the atmosphere in a petri dish. There was a slight decrease to 7.2 during the next 24 hour period. These observations point to the need to carefully monitor pH during NH_3 volatilization studies since it will have considerable bearing on the NH_3 concentration in the liquid phase.

Although NH_3 volatilization and water evaporation rates have been studied together, an expected association has not always been evident. Chao and Kroontje (1964) did not find any relationship between NH_3 volatilization and water evaporation rates. Also, Ryan and Keeney (1975) obtained similar NH_3 volatilization losses when dry or moist air was passed over wastewater sludge applied to quartz sand. In a laboratory study, Curnoe (1975) found only a small increase in NH_3 volatilization as the relative humidity decreased. He controlled the relative humidity in the closed experimental units by controlling the concentration of H_2SO_4 in the sorption solution. In a study of NH_3 volatilization losses from sewage sludge in the field, Beauchamp *et al.* (1978) could find no obvious relationship with water vapour pressure deficits. It was suggested by Lauer *et al.* (1976) that, as water evaporation proceeds, the concentration of ammoniacal nitrogen in the manure or sludge will increase with a possible concomitant increase in the NH_3 concentration. This, in turn, could influence the NH_3 partial pressure gradient and hence the volatilization rate.

Hashimoto and Ludington (1971) and Hoff *et al.* (1981) found that air currents caused more rapid NH_3 volatilization. Vlek and Stumpe (1978) suggested that NH_3 volatilization from a solution increases with rising temperature, water turbulence and wind speed. They cautioned, however, that measurements of NH_3 volatilization losses with closed systems should be carefully considered because air movement is usually much lower than with open field conditions. Air movement probably influences NH_3 volatilization by changing the NH_3 vapour pressure gradient from the manure or sludge to the atmosphere. The effects of windspeed and atmospheric humidity were reviewed by Faurie and Bardin (1979b).

7.5. Ammonia volatilization from sludges

King and Morris (1974) were amongst the first to observe rather large losses of NH_3 from applied sewage sludge. In a greenhouse experiment, they found ammoniacal N losses of 36% from bare soil and 24% from grass stubble. They suggested that air movement over and through grass stubble was lower to account for the lower loss. King (1973) previously suggested that N losses from applied sludge occurred via denitrification. Ryan and Keeney (1975) determined that from 11 to 60% of the applied ammoniacal N was lost by NH_3 volatilization in a laboratory study with anaerobically digested sludge. They noted that NH_3 volatilization losses decreased as the clay content of the soil was increased from 5 to 30% generally agreeing with the results of Curnoe (1975). They also noted that, the greater the quantity of sludge applied, the greater the quantity and percentage of ammoniacal N lost through volatilization. Ryan and Keeney (1975) suggested that the degree of ammonium adsorption by the soil affects the availability of NH_3 for volatilization. A higher rate of application or repeated applications would result in a lower proportion of the ammoniacal N being adsorbed. However, Curnoe (1975) found that increasing the NH_4^+ concentration in sewage sludge three fold did not result in a higher volatilization loss. The associated change in alkalinity as discussed earlier may have had a rôle to play in this kind of observation.

In 1978, Beauchamp *et al.* reported NH_3 volatilization measurements obtained with anaerobically digested sewage sludge applied to bare soil in the field near Guelph, Ontario. Measuring windspeed and atmospheric NH_3 concentration profiles above the applied sludge surface in conjunction with an aerodynamic diffusion model (Wilson *et al.* 1982), a 60% loss of ammoniacal N from the sludge was estimated to occur over a 5-day period in early May. The ammoniacal N application rate in the sludge was 150 kg/ha and comprised 40% of the total N applied. In a subsequent experiment in October, 56% of an application of 89 kg ammoniacal N/ha was volatilized during a 7-day period; the ammoniacal N comprised 34% of the total N in the sludge in this case. The flux density of NH_3, while following a diurnal pattern, was greatest during the first and second days after application and decreased thereafter. Although there was some evidence that the occurrence of rainfall tended to depress NH_3 volatilization, ambient temperature measured at the one metre height appeared to be directly related to volatilization. The diurnal pattern showed that greatest volatilization (maximum rate) usually occurred shortly after midday whereas the least volatilization (minimum rate) occurred during the early morning hours. The diurnal atmospheric temperature pattern generally tended to correspond with the NH_3 flux pattern. There was no evidence that NH_3 flux was closely related to the atmospheric water vapour deficit. Assuming that the rate of NH_3 flux depended on the concentration of NH_3 present (first order reaction rate process) the 'half life' for the May experiment was 3.6 days whereas that for the October experiment was 5.0 days. The longer half-life for the October experiment was probably due to generally lower temperatures and the occurrence of more rainfall. An attempt was made to sample directly the sludge layer, and soil layers beneath the sludge layer. The ammonium and nitrate concentrations in these samples were found to be so variable that no conclusive interpretations could be made. Thus it appeared that direct sampling of sludge, and the soil beneath the sludge layer did not provide a satisfactory alternative to the atmospheric sampling approach used in this study.

English *et al.* (1980) developed a mass transfer model employing total alkalinity, ionic strength, NH_3 concentrations and partial pressures, pH and the NH_3/NH_4^+ equilibrium constant (K_a). They predicted a loss of 34% of the original NH_3 in the first four hours following application. This predicted estimate, while awaiting experimental verification by the authors, is much greater than the measured estimates by Beauchamp *et al.* (1978).

The volatilization of NH_3 from sludges produced at six sewage treatment plants in Ontario was studied in the laboratory by Curnoe (1975). Each sludge had been treated with either lime, $FeCl_3$ or $Al_2(SO_4)_3$ to remove phosphorus. No difference in NH_3 volatilization was observed between these sludges.

7.6. Ammonia volatilization losses from manures

Relatively few studies have been done on the losses of volatile N during storage and handling or from applied manures. Koelliker and Miner (1973) and Hashimoto and Ludington (1971) have reported on losses during storage but these are not well defined quantitatively. It is expected that the principles and factors involved with such losses are similar to those for applied manure. It may be suggested that NH_3 volatilization losses from a manure pile during storage will be proportional to the specific surface area (area/volume) exposed to the atmosphere (Koelliker and Miner 1973). Although not confirmed by research, it may be speculated that periodic addition of fresh solid manure to the top of the pile (top loading) would result in greater NH_3 losses than pumping manure from below (bottom loading). The latter procedure results in relatively little exposure of fresh manure to the atmosphere. Also, crusting on liquid manure storage ponds or tanks may significantly limit NH_3 volatilization losses but this too has not been confirmed experimentally. In any case, it would be expected that volatilization losses from a manure surface exposed to the atmosphere would decrease with time. Luebs *et al.* (1974) and Miner *et al.* (1975) have done studies on volatile N emanating from high animal concentrations. Their results, however, do not provide estimates of the magnitude of the losses which may occur. In view of the rather inconclusive quantitative information available on volatile N losses during handling and storage of manures, most attention will be focussed on losses following application in the field.

Slater and Schollenberger (1938) cited Danish research findings based on crop responses by oats or root crops to manure N in field experiments. They evaluated the results of 34 field experiments in which 'fermented' manure was spread and plowed under after various time intervals. An average nitrogen loss of 15% occurred with 6 hours exposure, increasing to 27% in 24 hours and 42% in 4 days. This research also showed that time of application was important, losses being much greater in spring than in winter.

Since crop response was used as an indicator, the data should be interpreted with caution because significant N losses from the soil via other avenues such as leaching or denitrification could occur during the long intervals between manure application and crop growth. Nevertheless, these data support the general principle that incorporation as soon as possible following application will result in greater NH_3 conservation. Heck (1931b) concluded that turning under 'complete' manure immediately following spreading resulted in 80% recovery of the urea or NH_3 it contained. However, he indicated that the addition of straw to manure decreased N recovery by the crop. Heck (1931a) stated that three-eights to one-half of the N in 'anaerobically stored' manure is easily lost by exposure to dry conditions.

Lauer et al. (1976) showed that NH_3 volatilization losses from solid dairy cattle manure may proceed for up to 25 days following application even under winter conditions in New York state. They suggested that the rate of volatilization followed a first order rate process; that is, the rate of NH_3 loss was proportional to the amount of ammoniacal N present. Expressing their results in terms of half lives, they found that the decrease in rate of loss over time was not constant but occurred in three stages in relation to the NH_3 partial pressure in the applied manure. They concluded that general evaporative conditions and precipitation were principal determinants of NH_3 volatilization losses and that these losses would occur over a wide range of weather conditions. In general, they observed that between 60 and 99% of the ammoniacal N in applied solid dairy cattle manure was volatilized during 5 to 25 days after application.

Recent research conducted at the University of Guelph has shown that NH_3 volatilization losses directly measured from applied liquid dairy cattle manure ranged from 24 to 33% of the ammoniacal N over periods of 6-7 days after application. Volatilization was measured in essentially the same manner as described by Beauchamp et al. (1978). Losses were somewhat less than those reported by Lauer et al. (1976) for solid dairy cattle manure. However, it must be kept in mind that the methods of measurement of NH_3 losses were different. Lauer et al. (1976) depended on periodic sampling and analysis of the applied manure.

A typical example of the volatilization loss pattern found for liquid dairy cattle manure applied in early May to bare cultivated soil is shown in Fig. 7.3. As found previously with sewage sludge (Beauchamp *et al.* 1978), NH_3 volatilization followed a diurnal pattern with the highest fluxes occurring at or shortly after midday and the lowest fluxes occurring in the early morning hours. This pattern of NH_3 flux has been observed by McGarity and Rajaratnam (1973), Denmead *et al.* (1974) and Hoff *et al.* (1981). As apparent in Fig. 7.3 and also suggested by Hoff *et al.* (1981), the observed diurnal pattern in NH_3 flux appears to be largely associated with ambient temperature. Although a general decrease in NH_3 flux occurred with time, the data in Fig. 7.3 show that rainfall depressed NH_3 volatilization.

Deposition of urine and faeces by animals in pastures represents a natural method of manure application. Denmead *et al.* (1974) studied NH_3 volatilization in a sheep pasture in Australia. They estimated that the sheep voided approximately 1 kg N/ha/day in their urine, and measured average NH_3

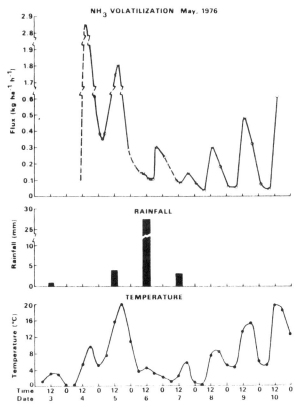

Fig. 7.3. Ammonia fluxes from liquid dairy cattle manure applied to bare, cultivated soil, daily rainfall and ambient temperature over a seven day period in May near Guelph, Ontario.

volatilization losses of 0.27 kg $NH_3 - N$/ha/day. Although NH_3 volatilization occurs readily from urine spots, little is known about the extent of subsequent foliar absorption of the volatilized NH_3 from such 'point' sources.

7.7. Absorption of volatilized nitrogen

Most of this chapter has dealt with the loss of NH_3 from applied sludges or manures. It must be considered, however, that such applications constitute 'point' sources or 'limited diffuse' sources of NH_3 for surrounding areas which did not receive applications. It is well known that NH_3 is absorbed by soils (Faurie and Bardin 1979a) and plant foliage (Lemon and Van Houtte 1980). It is expected that there would be considerable contact of soil and foliage with volatilized NH_3 molecules as they are transported horizontally downwind. Considerable dispersion of NH_3 molecules occurs in the vertical and horizontal directions normal to the direction of transport. Little is known about the degree of NH_3 absorption by soil or foliage which may result downwind from sites of application. Recent developments in mathematical modelling suggest that it may be possible to predict quantitatively NH_3 (or other substances) sorption by soil and plants downwind from an applied source (J. Wilson, personal communication).

7.8. Summary and conclusions

It is evident that volatilization losses of N especially NH_3, from applied manures or sludges can be large but are also quite variable. This variability depends on many factors reviewed in this chapter including temperature, rainfall, partial pressure gradients, air movement rate, application rate, chemical composition, and relative humidity. So far it is very difficult to predict with scientifically acceptable accuracy the N losses from applied sludges and manures. Rough estimates could be made, however, for practical purposes depending on the degree of accuracy required. A greater understanding of the phenomena involved with N volatilization losses would greatly enhance our ability to manage organic materials applied to soil with the objective of conserving and using N efficiently.

7.9. References

Beauchamp, E.G., Kidd, G.E. and Thurtell, G. 1978 Ammonia volatilization from sewage sludge applied in the field. J. Environ. Qual. 7, 141–146.

Chao, T-T. and Kroontje, W. 1964 Relationships between ammonia volatilization, ammonia concentration and water evaporation. Soil Sci. Soc. Am. Proc. 28, 393–395.

Curnoe, W.E. 1975 Ammonia volatilization from sewage sludge applied to the soil surface. Unpublished M.Sc. Thesis, University of Guelph.

Denmead, O.T., Simpson, J.R. and Freney, J.R. 1974 Ammonia flux into the atmosphere from a grazed pasture. Science (Wash. D.C.) 185, 609–610.

English, C.J. Jr., Miner, J.R. and Koelliker, J.K. 1980 Volatile ammonia losses from surface-applied sludge. J. Water Pollut. Control. Fed. 52, 2340–2350.

Faurie, G. and Bardin, R. 1979a La volatilisation de l'ammoniac. I. Influence de la nature du sol et des composés azotes. Ann. Agron. Paris 30, 363–385.

Faurie, G. and Bardin, R. 1979b La volatilisation de l'ammoniac. II. Influence des facteurs climatiques et du couvert végétal. Ann. Agron. Paris 30, 401–414.

Feagley, S.E. and Hossner, L.R. 1978 Ammonia volatilization reaction between ammonium sulfate and carbonate systems. Soil Sci. Soc. Am. J. 42, 364–367.

Fenn, L.B. and Kissel, D.E. 1973 Ammonia volatilization from surface applications of ammonium compounds on calcareous soils: I. General theory. Soil Sci. Soc. Am. Proc. 37, 855–859.

Hashimoto, A.G. and Ludington, D.C. 1971 Ammonia desorption from concentrated chicken manure slurries. In: Livestock Waste Management and Pollution Abatement. Proc. Int. Symp. Livestock Wastes. Amer. Soc. Agric. Eng., Columbus.

Heck, A.F. 1931a Conservation and availability of the nitrogen in farm manure. Soil Sci. 31, 335–363.

Heck, A.F. 1931b The availability of the nitrogen in farm manure under field conditions. Soil Sci. 31, 467–481.

Hoff, J.D., Nelson, D.W. and Sutton, A.L. 1981 Ammonia volatilization from liquid swine manure applied to cropland. J. Environ. Qual. 10, 90–95.

King, L.D. 1973 Mineralization and gaseous loss of nitrogen in soil applied liquid sewage sludge. J. Environ. Qual. 2, 356–358.

King, L.D. and Morris, H.D. 1974 Nitrogen movement resulting from surface application of liquid sewage sludge. J. Environ. Qual. 3, 238–243.

Koelliker, J.K. and Miner, J.R. 1973 Desorption of ammonia from anaerobic lagoons. Trans. ASAE (Am. Soc. Agric. Eng.) 16, 143–151.

Kowalke, O.L., Hoegen, O.A. and Watson, K.M. 1925 Transfer coefficients of ammonia in absorption towers. Wis. Agric. Exp. Stn. Bull. 68.

Lauer, D.A., Bouldin D.R. and Klausner, S.D. 1976 Ammonia volatilization from dairy manure spread on the soil surface. J. Environ. Qual. 5, 134–141.

Lemon, E. and VanHoutte, R. 1980 Ammonia exchange at the land surface. Agron. J. 72, 876–883.

Luebs, R.E., Davis, K.R. and Laag, A.E. 1974 Diurnal fluctuation and movement of atmospheric ammonia and related gases from dairies. J. Environ. Qual. 3, 265–269.

Marion, G.M. and Dutt, G.R. 1974 Ion association in the ammonia-carbon dioxide-water system. Soil Sci. Soc. Am. Proc. 38, 889–891.

McGarity, J.W. and Rajaratnam, J.A. 1973 Apparatus for the measurement of losses of nitrogen as gas from the field and simulated field environments. Soil Biol. Biochem. 5, 121–131.

Miner, J.R., Kelly, M.D. and Anderson, A.W. 1975 Identification and measurement of volatile compounds within a swine building and measurement of ammonia evolution rates from manure-covered surface. In: Managing Livestock Wastes. Proc. Third Int. Symp. Livestock Wastes. Am. Soc. Agric. Eng., Urbana.

Mosier, A.R., Andre, C.E. and Viets, F.G., Jr. 1973 Identification of aliphatic amines volatilized from cattle feedyard. Environ. Sci. Technol. 7, 642–644.

Ryan, J.A. and Keeney, D.R. 1975 Ammonia volatilization from surface applied wastewater sludge. J. Water Pollut. Control Fed. 47, 386–393.

Slater, R.M. and Schollenberger, C.J. 1938 Farm Manure. In: Yearbook of Agriculture. U.S.D.A., Washington, D.C.

Terman, G.L. 1979 Volatilization losses of nitrogen as ammonia from surface-applied fertilizers, organic amendments and crop residues. Adv. Agron. 31, 189–223.

Vlek, P.L.G. and Stumpe, J.M. 1978 Effects of solution chemistry and environmental conditions on ammonia volatilization losses from aqueous systems. Soil Sci. Soc. Am. J. 42, 416–421.

Wilson, J.D., Thurtell, G.W. Kidd, G.E. and Beauchamp, E.G. 1982 Estimation of the rate of gaseous mass transfer from a surface source plot to the atmosphere. Atmos. Environ. 16, 1861–1867.

8. Ammonia loss from fertilizer applied to tropical pastures

L.A. HARPER, V.R. CATCHPOOLE and I. VALLIS

8.1. Introduction

Large amounts of nitrogen (N) can be lost from agricultural systems (Allison 1973), and those from pastures have varied between 20 and 80% of the fertilizer N applied (Burton and Jackson 1962, Catchpoole and Henzell 1975, Lemon 1978, Terman 1979, Power 1980). Possible loss pathways are well known but their relative importance is less clear. Nitrogen can be lost in surface runoff water, in subsurface vertical and lateral water transport, and as the gases dinitrogen (N_2), ammonia (NH_3), nitric oxide (NO), nitrogen dioxide (NO_2), nitrous oxide (N_2O), and various amines.

Fertilizer losses vary with the type of compound applied. For subtropical areas of Australia and southeastern U.S.A. the general ranking for N loss is sodium nitrate <ammonium sulphate <ammonium nitrate <urea (Burton and Jackson 1962, Gartner and Everett 1970, Scarsbrook 1970, Henzell 1971). In spite of the high losses associated with its use, urea has advantages in production and transportation costs and it is currently the most widely used solid nitrogenous fertilizer in the world (Bridges 1980).

In this chapter we discuss gaseous N transport as NH_3 above tropical pasture systems. Some properties of tropical soils and pastures predispose them to N losses by NH_3 volatilization. For example, the tropical soils used for pastures generally have low cation exchange capacities and low organic matter contents. They therefore have a low buffering capacity against change in pH. Tropical grass swards produce tall but relatively sparse canopies that may be less effective absorbers of fertilizer or soil-derived NH_3 than temperate grass canopies. In the tropics, high temperature, high evaporation potential, and high soil water contents often coincide whereas this is less likely in temperate regions. The number of studies of NH_3 losses from tropical pastures is very small. Consequently, we have had to rely heavily on information from an integrated study in southeast Queensland where NH_3 losses to the atmosphere from a grazed Nandi Setaria (*Setaria sphacelata*) pasture were studied. We examined these losses in relation to total fertilizer N losses, hydrolysis of urea in the soil, and NH_3 volatilization from urine patches.

8.2. Nitrogen balance studies on fertilized tropical pastures

In this section, we review N balance under non-grazed and grazed swards, including ^{15}N and gas-lysimeter studies, and emphasize the need to trace the causes and pathways of losses of N from fertilized tropical pastures.

Nitrogen balance studies on non-grazed pastures (clipped plots) have consistently indicated N losses. Apparent losses varied between 14 and 59% (depending on fertilizer type) of the fertilizer N applied to Rhodes grass (*Chloris gayana*) in Queensland (Henzell 1971), and between 21 and 29% of that applied to Bahiagrass (*Paspalum notatum*) in Florida (Blue 1970). However, losses associated with grazing animals, notably volatilization of gaseous N compounds from dung and urine patches do not occur from clipped plots. Thus fertilizer N losses from grazed tropical pastures have generally been much larger than those from clipped plots. In southeast Queensland for example, a grazed Nandi Setaria pasture fertilized with urea at the rate of 374 kg N/ha/year lost 60% of the applied N over a four year period and 80% over the next four years, and a grazed Biloela buffelgrass (*Cenchrus ciliaris*) pasture receiving 168 kg N/ha/year as urea lost 40% (Henzell 1972, Catchpoole and Henzell 1975).

More precise direct measurements of fertilizer N losses from tropical pastures have been made by applying fertilizer labelled with ^{15}N. On Rhodes grass pastures in Queensland, Catchpoole (1975) found a 38% mean recovery of the ^{15}N applied as ^{15}NH$_4$^{15}NO$_3$ between 4 and 40 weeks after application. Henzell (1971), also using Rhodes grass, obtained losses of ^{15}N ranging from 5 to 43% depending upon fertilizer rate. Vallis *et al.* (1973) found a recovery of 77% on a Townsville stylo (*Stylosanthes humilis*) pasture. Impithuksa *et al.* (1979) found 68 to 74% recovery for three warm season grasses in Florida. A ^{15}N recovery of 98% was observed by Vallis *et al.* (1973), but this high recovery was associated with very low urea application rates which would not favour NH$_3$ volatilization.

Appropriate action to improve the efficiency of fertilizer N on tropical pastures depends initially on identification of the major loss pathways. Losses by leaching do not appear to be important in southeast Queensland where no correlation could be found between N recovery and total rainfall or the discharge term in a water budget (Henzell 1971). Furthermore, movement of ^{15}N labelled fertilizer into deep soil layers was very small even during a relatively wet season (Catchpoole 1975).

Losses of N gases into the atmosphere have been studied in southeast Queensland by placing soil cores containing grass into gas lysimeters. When the soil water potential was −0.1 atmosphere the large field losses were not reproduced in the gas lysimeters (Henzell *et al.* 1970) but where cores were waterlogged immediately after application of ^{15}NH$_4$, ^{15}NO$_3$ losses reached

43% (Catchpoole 1975). The latter loss was thought to be due to bacterial denitrification, but this process is probably not a major cause of N loss in the field in southeast Queensland where most soils are not waterlogged for long periods. The gas lysimeter work did show that small amounts of NH_3 were released into the atmosphere over a wide range of conditions. This prompted efforts to measure NH_3 losses in the field.

Estimates of NH_3 losses from urea applied to forest soils in the field have been made by applying ^{15}N-urea and measuring its recovery in the soil-plant system during the first few weeks after application (Nömmik 1973). Most of the ^{15}N loss in this period was assumed to be due to [15]NH_3 loss since nitrate accumulations were small and denitrification losses appeared unimportant. This method was used in southeast Queensland (Catchpoole et al. 1983b) to measure losses from urea during 14-day periods after application to a Nandi Setaria pasture. Measurements were made during the four seasons of the year and ^{15}N recovery in autumn (Table 8.1) was far lower than in the other seasons. The low recovery of ^{15}N in autumn suggested that NH_3 volatilization was largest at this time. This was confirmed by the flux density measurements which are discussed in following paragraphs.

8.2.1. Microclimatological determinations of aerial ammonia transport

In this section we examine net NH_3 transport determined in the field by microclimatological methods. A general overview is followed by a review of recent studies of diurnal, seasonal, and annual transport.

Laboratory and chamber studies have shown that plants can absorb NH_3 from the air (Hutchinson et al. 1972, Meyer 1973, Farquhar et al. 1980) as well as lose NH_3 to the air (Stutte et al. 1979). Microclimatological studies have shown evidence of diurnal cycling (Denmead et al. 1974, Harper et al. 1983) and differences in aerial NH_3 absorption and evolution in grazed and ungrazed grass-clover pastures (Denmead et al. 1976). It appears from the latter studies that mechanisms of NH_3 production and loss in grazed and ungrazed pastures are quite different because grazing produces dung and urine as ammonia sources and removal of foliage creates an open canopy from which NH_3 can escape more easily. Fig. 8.1 shows NH_3 concentration profiles above a grazed pasture demonstrating periods of net upward, net downward, and no net movement of NH_3. At 1000 h on April 13, NH_3 was produced at the soil surface, and perhaps by the leaf surface, at a greater rate than could be absorbed by the soil-plant system, thus NH_3 was lost to the atmosphere. At 6000 h on February 21, the profile was reversed indicating NH_3 was being absorbed by the soil-plant system from the atmosphere. At other times (e.g. at 1000 h on April 12), the soil-plant system was in equilibrium with the atmosphere. The profiles of Fig. 8.1 were measured during 'background'

Table 8.1. Recovery of ^{15}N-labelled urea nitrogen in plant and soil 14 days after application to a tropical grass pasture (after Catchpoole *et al.* 1983b)

Season	Recovery (% of application)			Soil				Total
	Grass tops	Border grass tops	Litter	0–3 cm	5–10 cm	10–20 cm	20–30 cm	
Summer	32.9±3.8	5.4±2.9	8.5±1.4	18.1±1.4	2.8±0.2	2.9±0.3	n.d.[a]	70.6±4.9
Autumn	26.6±4.1	2.5±0.7	8.5±1.4	13.6±1.9	1.3±0.2	2.6±0.5	n.d.[a]	55.0±2.4
Winter	28.7±2.7	1.1±0.2	3.3±0.5	22.3±3.5	10.6±1.7	7.8±3.2	3.2±0.3	77.4±3.4
Spring	45.4±2.0	2.5±0.5	7.1±1.5	18.7±1.3	1.8±0.3	2.1±0.1	2.2±0.1	79.8±2.3

[a] Not determined

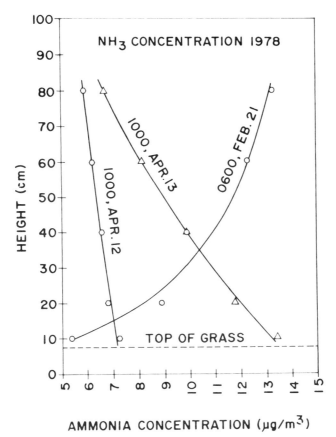

Fig. 8.1. Three types of ammonia concentration profiles obtained over a grass sward during background conditions (no fertilizer influence) at Samford, Queensland.

periods which were free from the influence of recent applications of urea fertilizer (Harper *et al.* 1983).

Aerial NH_3 concentrations above the sward increase sharply following applications of urea to the soil surface. In one of the above studies (Harper *et al.* 1983) the concentration 24 hours after the application was 100 times larger than the background levels.

8.2.2. *Diurnal transport*

Fluxes of NH_3 above pastures are commonly outward during the daytime and very small or inward at night. Denmead *et al.* (1974) demonstrated this cycling during a one-day study on a temperate, grazed alfalfa pasture. Harper *et al.* (1983) found that diurnal cycling occurred at all seasons during the year on a subtropical pasture except after recent applications of urea when NH_3 liberated from the urea masked the diurnal effects. The diurnal variations in NH_3 flux appear to offer an opportunity to study field factors that control emissions. However, it is difficult to identify these variables in the field because the flux density is the net effect of two opposed processes, namely absorption and emission of NH_3 by the soil-plant system.

Factors that influence losses of NH_3 have been examined in laboratory studies. Losses increased with temperature (Fenn and Kissel 1974, Feagley 1976) while soil water content and soil surface wetness have been both positively and negatively related to NH_3 losses (Fenn and Escarzaga 1976, Prasad 1976, Denmead *et al.* 1978). Other observations have suggested that NH_3 is not emitted in large amounts unless soil water is available for evaporation (Denmead *et al.* 1976, 1978). In southeast Queensland (Harper *et al.* 1983), multiple correlation analyses of NH_3 fluxes with respect to soil and microclimate factors gave a combined average r^2 of 0.92 ± 0.09 for the year. All parameters in the analyses were integrated over two-hour periods for seven days during each of the four seasons. The water content of the surface of the soil was not highly correlated with NH_3 flux density. During summer, NH_3 flux density was most highly correlated with soil temperature followed by windspeed and evapotranspiration. During the remainder of the year, evapotranspiration had the highest individual correlation followed by soil temperature and windspeed. The correlation with evapotranspiration does not suggest a causal relationship, since evapotranspiration and flux density both increase with soil temperature and windspeed.

Windspeed and thus turbulent transport were larger during the day than at night. This would increase daytime fluxes by moving NH_3 away from the source and by limiting the time for reabsorption of NH_3 by the soil-plant system. Likewise, the temperature of the surface of the soil increases during the day. This accelerates both the formation of ammonium from urea and diffusion of NH_3 to the surface. Increasing temperatures also shifts the equilibrium from ammonium in the soil towards gaseous NH_3, i.e.

$$NH_4^+ \rightleftarrows NH_3 \text{ (aqueous) } \rightleftarrows NH_3 \text{ (gaseous)}.$$

A net NH_3 influx into the soil-plant system was commonly observed (Fig. 8.2) near sunset and sunrise when the soil was dry. This influx did not correspond with any rapid changes in measured soil or microclimate factors. We speculate that the plant canopy influenced the absorption of NH_3 and discuss this later in the chapter.

Dew formed at night (by condensation and/or guttation) and subsequent evaporation has been suggested by Denmead *et al.* (1976) to explain diurnal cycling of NH_3 fluxes. However, Harper *et al.* (1983) found dew accumulation*

---

* Dew accumulation was a subjective evaluation of water on leaf surfaces using a scale of 0 to 6.

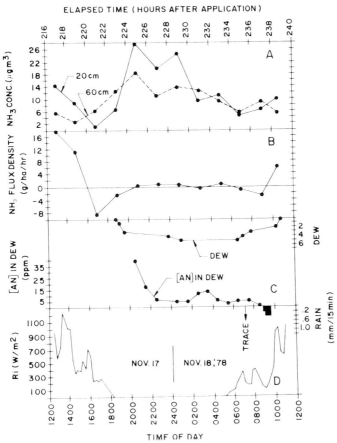

Fig. 8.2. Diurnal ammonia concentration and flux density with related data including incident radiation (Ri), rainfall intensity, dew, and ammoniacal nitrogen concentration ([AN]) at Samford, Queensland (after Harper *et al.* 1983).

did not explain the NH_3 influx observed at sunset and sunrise. Ammoniacal N concentration (ammoniacal N = ammonium N + ammonia N in the aqueous phase) in dew was generally highest in the first measurable sample (Fig. 8.2c). The initial high ammonium concentration in dew was probably due to ammonium in guttation (Goatley and Lewis 1966) and/or from ammonium salt residues left from previous dew evaporation (the maximum daily ammoniacal N concentration in dew tended to increase with time after each rainfall event). The high initial ammoniacal N concentration did not explain the rapid period of influx after sunset since the influx occurred before any dew was formed. Simultaneous measurements of dew pH were not made but subsequent measurements during summer in the same pasture (L.A. Harper, V.R. Catchpoole, and I. Vallis, unpublished data) have demonstrated pH values below 4.0 which are lower than those calculated for equilibrium between the aqueous phase and the measured partial pressures of NH_3 in the surrounding air. Thus some NH_3 absorption by dew could be expected which may help explain the long periods of small positive or negative flux density at night (Fig. 8.2b). The NH_3 efflux after sunrise was probably not influenced by ammonium in the dew since conversion of ammonium to NH_3 requires an alkaline pH.

8.2.3. Seasonal effects on ammonia transport

Losses of NH_3 from urea fertilizer broadcast onto pastures vary between seasonal applications. In southeast Queensland losses varied from 9% of the urea N applied in spring and summer through 13% in winter, to 42% in autumn (Table 8.2). This type of variation might help to explain the well-known variable growth-response of grass to applications of urea. It suggests that both the reliability of the response and the efficiency of use of urea N by grass could be improved if the conditions that caused the very heavy loss in autumn were

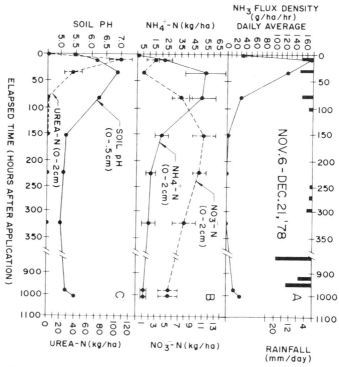

Fig. 8.3. Spring-summer daily ammonia flux density with rainfall, soil mineral nitrogen concentrations, and soil pH after a spring urea application to a tropical pasture at Samford, Queensland (after Harper et al. 1983).

avoided in practice. Progress towards describing these conditions is discussed below.

The variation in NH_3 flux densities was associated with seasonal changes in temperature and water content of the surface soil. In spring and autumn (Fig. 8.3 and 8.4) urea was applied to warm wet soil and flux densities increased immediately to 160 and 300 g/ha/h respectively. In winter (Fig. 8.5) the soil was cool and dry and fluxes reached only 70 g/ha/h. In the spring (Fig. 8.3) flux density decreased to background levels after 80 h whereas in the winter (Fig. 8.5) flux density remained at a high level for about 170 h. Total volatilization losses (Table 8.2) in spring were probably not significantly different from those in winter (errors of uncertainty in flux calculations of 10, 20 and 45% have been suggested (Lemon 1970, Harper *et al.* 1973, Denmead *et al.* 1977)). In autumn (Fig. 8.4), flux densities remained high for >100 h and the total volatilization losses were significantly larger than in the other seasons. The duration of high flux densities coincided with the number of rainless days after urea application. Similar rainfall effects on NH_3 volatilization after urea application have been noted by others (Carrier and Bernier 1971, Förster and Lippold 1975, Marshall

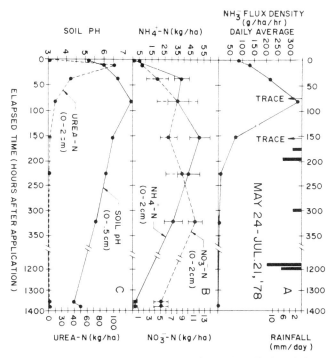

Fig. 8.4. Autumn-winter daily ammonia flux density with rainfall, soil mineral nitrogen concentrations, and soil pH after an autumn urea application to a tropical pasture at Samford, Queensland (after Harper *et al.* 1983).

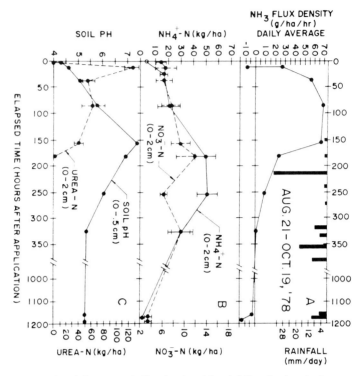

Fig. 8.5. Winter-spring daily ammonia flux density with rainfall, soil mineral nitrogen concentrations, and soil pH after a winter urea application to a tropical pasture at Samford, Queensland (after Harper *et al.* 1983).

and DeBell 1980). Rainfall may have reduced the NH_3 flux by moving the fertilizer N or initial hydrolysis products to below the soil surface where they could be more rapidly nitrified, immobilized, or utilized by plants. Thus, in practice, urea applications should be made prior to rain or irrigation in order to reduce volatilization losses.

8.2.4. *Annual ammonia transport*

The annual loss of NH_3 from grazed pastures fertilized with urea has two components, (1) volatilization directly from fertilizer input, and (2) volatilization from dung, urine, and decaying plant material. The first component is by far the largest. In southeast Queensland (Catchpoole *et al.* 1983a) the average NH_3 loss within 14 days of urea application for all four seasons was 19.1% of the urea N applied with individual seasonal losses ranging from 8.7 to 42.1%. During the remainder of the year, when the effects of recent application of urea were absent (i.e., from 14 days after application to the next application), the average loss was 5.3%. This latter figure varied among seasons between a 7.6% gain of NH_3 from the atmosphere and a 13.1% loss. The total loss as NH_3 was equivalent to 24% of the urea N applied.

Large amounts of N are apparently lost from urea-fertilized grazed pastures by pathways other than volatilization of NH_3. Between 1970 and 1974 the total N loss from the same pasture used in this NH_3 study was 80% (Catchpoole and Henzell 1975). Some variation could be expected between years but the NH_3 loss measured in 1978/79 never approached this level. Losses of nitrogenous gases other than NH_3 and of N compounds in water leaving the pasture system need to be measured.

8.3. Secondary losses of ammonia associated with grazing

About 20% of the annual loss of NH_3 from this pasture could not be attributed directly to NH_3 volatilization from the urea fertilizer, but came instead from dung, urine, and decaying plant material. Losses from the latter sources can be termed secondary losses, since the N contained in them has experienced at least one cycle of grazing. Urine patches probably represent the major source of these secondary losses, and the potential for loss by this pathway can be estimated from current knowledge of stock carrying capacities, pasture selection and intake by stock, and by partitioning N between animal products, faeces and urine. For example, assuming a pasture comsumption rate of 8000 kg dry matter ha/year (Anon 1980) with 2.5% N in the dry matter (Chacon *et al.* 1978), N intake by four beef steers/ha in a Nandi Setaria pasture would be about 200 kg/ha/year with urinary N excretion equalling approximately 60% of the N intake (Henzell and Ross 1973). Thus, about 120 kg/N/ha/year would be deposited in urine patches. This amount equals 32% of the applied fertilizer N on this pasture, so large losses from urinary N could be of some importance to the overall N balance of the pasture.

The major nitrogenous constituent of cattle and sheep urine is urea, so the basic processes of NH_3 formation and loss from urine patches in pastures are similar to those associated with urea fertilizer. However, the above processes are influenced by the other nitrogenous constituents of urine and by the characteristics of the urine patch. Doak (1952) found a range of 2.5–8.5 g N/l in cattle urine with the total N distributed as follows: urea 50–74%, allantoin 4.0–6.4%, hippuric acid 1.9–6.0%, creatine and creatinine 1.3–2.0%, NH_3 0.3–0.6%. Amino compounds comprise the remainder (Bathurst 1952). In our work urea accounted for 80–88% of the total N of cattle urine (Vallis *et al.* 1982). Laboratory tests by Doak (1952) showed that urea hydrolysis in a urine-soil mixture at 23° C was 82% complete after 24 hours compared with only 35% in a urea solution mixed with soil. It was shown that hippuric acid was mainly responsible for the more rapid urea hydrolysis in urine. An additional difference exists since solid urea fertilizer must be dissolved in soil water, dew or rain before hydrolysis can begin. We have observed 80% hydrolysis of urea within 2 hours in urine applied to a pasture with soil temperatures of 21–25° C (Vallis *et al.* 1982). Urine patches receive N at a high rate per unit area and this is another reason to expect greater NH_3 losses from urine patches than from urea fertilizer (Volk 1959, Kresge and Satchell 1960, Simpson and Melsted 1962).

The review of Whitehead (1970) showed that the characteristics of urine patches are quite variable; the concentrations of total N in cattle urine varies between 2.5 and 13 g N/l, volumes per urination being 1.6–3.5 l, and the areas covered per urination being 0.18–0.35 m^2. Using average figures, the calculated rate of total N addition was 575 kg N/ha. Consequently, the concentration of ammoniacal N in the surface soil following urea hydrolysis can approach 1000 ppm during the first day after urine application (Vallis *et al.* 1982). Penetration of urine into the soil is affected by the antecedent soil water content, being greater when the soil is moist than when dry (Doak 1952). For this reason an initially moist soil might be expected to lose less ammonia than a dry one. This contrasts with urea fertilizer, where greater losses occur from initially wet than dry soil (Catchpoole *et al.* 1983a). It is possible that under conditions of high evaporative demand NH_3 loss might be curtailed by the urine patch drying out (Ernst and Massey 1960). Cyclic wetting and drying of urine-treated soil in the laboratory was found to double the loss of NH_3 (Jewitt and Barlow 1949).

There is little published information on the rate and magnitude of NH_3 volatilization from urine patches in tropical pastures. Information for pastures in temperate climates has been gained in three ways: (1) Gaseous N losses have been calculated from total N balance measurements in lysimeters (Watson and Lapins 1969) but this method is not specific for NH_3. (2) Micrometeorological methods (described in Chapter 5) have been used by Denmead *et al.* (1974) to calculate NH_3 flux above a heavily grazed lucerne sward. The difference between this flux and the relatively small flux in the absence of the sheep was equivalent to 25% of the estimated output of urinary N by the sheep. (3) Most measurements of NH_3 volatilization from the urine of grazing animals have been made by drawing air through chambers placed over artificially applied urine and collecting the NH_3 in dilute acid. Ammonia losses measured in this way have ranged from 9 to 50% of the applied urine N (Doak 1952, Whitehead 1970, McGarity and Rajaratnam 1973, Ball *et al.* 1979).

Vallis *et al.* (1982) used chambers to study volatilization of NH_3 from urine patches in a pasture at Samford in south east Queensland. The effects of the chambers on ambient microclimatic conditions were minimised by leaving the chambers on the urine patches only 30 minutes each hour, and by adjusting the windspeed through the chambers to approximate that near the ground outside the chambers. They found that the onset of NH_3 volatilization following urine application to the tropical pasture at Samford was very rapid. Maximum loss rates occurred 4–8 hours after noon applications of urine in the summer. When urine was applied about one hour before sunset in the winter, the maximum rate of NH_3 loss occurred during the next day. As expected, losses were greatest during the day and smallest at night. Total daily losses declined exponentially with time, the rate of decline being more rapid in summer than in winter. This result is similar to that for temperate pastures (McGarity and Rajaratnam 1973, Ball et al. 1979). Integration of the exponential function of loss per day versus time over a 14-day period gave total losses of 18.8% in June, 14.4% in November–December, and 28.4% in February–March. The seasonal effects were probably the result of the combined effects of temperature and soil water status. The soil moisture was near field capacity for the June and February–March measurements and therefore the lower volatilization in June could be ascribed to the lower temperatures at that time. In contrast, the soil was near wilting point in November–December and the evaporative demand was high. Hence an early depletion of available water for evaporation could have restricted volatilization (Ernst and Massey 1960).

Based on our estimate of 120 kg N/ha for the annual return of urine N to the above pasture, a 14.4–28.4% loss as NH_3 represents 17–34 kg N/ha/year. This compares with the background NH_3 flux of 20 kg N/ha/year (i.e. 5.3% of the 374 kg N/ha annual application) that was determined by microclimatological methods. This suggests that most of the background NH_3 flux was from urine patches. However, the loss from urine equals only 4–8% of the annual application of 374 kg N/ha as fertilizer. Therefore for losses directly from urine to make an important contribution to the fertilizer N deficits in this pasture, substantial losses by routes other than NH_3 volatilization would be required. Other routes might include leaching or denitrification of the highly concentrated nitrate that ultimately occurs under urine patches, chemical decomposition of NO to N_2O and N_2 (Marshall and DeBell 1980), or transfer of N to unproductive areas where cattle congregate to drink or rest (Hilder and Mottershead 1963).

8.4. Concluding remarks

Annual NH_3 loss from a grazed tropical pasture fertilized with urea accounted for about one-third of the previously measured annual total N loss and about one-fourth of the annual fertilizer N input. This indicates that pathways of loss other than by NH_3 volatilization account for substantial losses of fertilizer N from grazed pastures. Further research is needed to trace these.

Net losses of NH_3 determined by microclimatological methods were usually less than N losses determined by ^{15}N techniques (Table 8.2). In three of four measurement series these net losses were about one half the concurrent ^{15}N losses. Since leaching losses were small (Catchpoole *et al.* 1982), the difference between ^{15}N and NH_3-N net losses was apparently gaseous N. We suggest two explanations for this difference between the two methods.

Firstly, the loss of ^{15}N would exceed NH_3-N losses if nitrogenous gases other than NH_3 were formed, for example by chemo-denitrification (see Chalk and Smith, this book) during the first weeks after urea application. Recent work by Marshall and DeBell (1980) on forest soils in Canada support this hypothesis. These other gases might include N_2, N_2O, NO and NO_2 but direct evidence is still needed.

Secondly, net NH_3 losses would be less than ^{15}N losses if a large proportion of NH_3 volatilized at the soil surface at one point was reabsorbed by the neighbouring pasture canopy. We have already seen that NH_3 fluxes measured by microclimate techniques are the net result of soil and plant evolution-absorption cycles. Although the microplots should have absorbed NH_3 at the same rate as the surrounding pasture, it is unlikely that all the reabsorption of $^{15}NH_3$ evolved from a m^2 microplot would occur within the same microplot. Some $^{15}NH_3$ would be absorbed by the adjacent pasture and thus would appear as a loss in a ^{15}N balance (Catchpoole *et al.* 1983b). This hypothesis could explain the seasonal variation shown in Table 8.2 between the ^{15}N and NH_3 losses. In summer and spring the apparent ^{15}N loss was much larger than the net NH_3 flux, and during these periods the canopy was in an actively growing state conducive to atmospheric NH_3 absorption. In autumn, when the ^{15}N loss and NH_3 flux were similar, the canopy was short and thin due to preceding drought conditions and heavy grazing and so was a poor absorber of NH_3. In winter the difference between the two losses was intermediate. The canopy was more active in winter than in autumn due to preceding rain, but less active than in spring and summer.

Research to test the above hypothesis could be rewarding. If reabsorption did substantially reduce the net loss of NH_3 a better understanding of the factors that control it in the field could give substantial savings of fertilizer N. For example, if, as our results suggested, it depends on the depth and density of the canopy, this condition could be arranged by suitable pasture management before urea was broadcast. Likewise, if losses of N_2 and N_2O caused a part of the discrepancy, factors that influence the formation of these gases in the field should be studied.

The net annual movement of NH_3 was from the soil-plant system to the atmosphere, but there were periods when the net movement was in the opposite direction. This occurred when urea had not been applied recently. Two distinct patterns were observed: one pattern was rapid net influx of NH_3 over relatively short periods near sunrise and sunset, and the other was slow net influx over

Table 8.2. Ammonia loss from a tropical pasture as determined by microclimatological and ^{15}N methods (after Catchpoole et al. 1983a.b)

Season	Application date	Measurement period (days after application)	Net NH$_3$-N loss	^{15}N loss	^{15}N loss/net NH$_3$-N loss
			(as % of urea N applied)		
Summer	Feb. 22	0–14	12	29	2.4
		15–91	9	[a]	[a]
Autumn	May 24	0–14	42	45	1.1
		15–91	6	[a]	[a]
Winter	Aug. 23	0–14	13	23	1.8
		15–91	–8	[a]	[a]
Spring	Nov. 8	0–14	9	21	2.3
		15–91	13	[a]	[a]

[a] No measurement made

relatively long periods, generally at night.

Knowledge of the factors that cause net NH_3 influx could lead to a better understanding of the N nutrition of pastures. The comparison between seasons given in Table 8.3 suggests that absorption of atmospheric NH_3 by grass is important because it occurred on 5 of the 8 days listed. The days in Table 8.3 were chosen here because they were free from the effects of recent applications of urea. Differences between the days in soil mineral N, aerial NH_3, soil surface pH and soil water content were generally not statistically significantly different. We therefore speculate that the NH_3 absorption and evolution characteristics were mainly due to changes in the characteristics of the grass, and not of the environment. To add to this speculation we note from Table 8.3 that NH_3 influx was common at the end of the seasons, while efflux was common in the mid seasons. This suggests that the grass was experiencing a deficiency of soil N by the end of the season, and this stimulated an affinity for aerial NH_3.

Even assuming completely random grazing patterns by animals, the proportion of area influenced by urine is very small in relation to the total area from which forage N is gathered and the concentration of urea in the urine patch is far too high to allow effective plant recovery before volatilization and/or leaching losses occur. Additionally, there is insufficient carbon substrate in the soil organic matter complex of developed soils to allow significant microbial immobilization in such a small area (Ball 1979). Thus animals cause substantial loss and much of the net NH_3 volatilization observed during near-background periods was due to urine patches.

Further research is needed on NH_3 transport in agricultural systems to define the interactions between NH_3 fluxes and plant, soil, and environmental factors. Plant factors would include the density and age of the canopy, and changes in the nitrogen status and affinity for NH_3 of the leaves. Soil factors would include the effects of fertilizer prills on the cation exchange capacity, urease activity, and concentration of ammoniacal N in microsites within the surface layer of soil. Environmental factors would include the effects of rainfall on the distribution of fertilizer in the soil.

8.5. Summary

Fertilizer N balance studies have shown unaccountable losses of N from tropical pastures ranging from 20 to 80% of the N applied. The purpose of this chapter was to review the importance of NH_3 transport in this loss. The chapter relies heavily on measurements of NH_3 fluxes from urea fertilizer and from urine, and on the recovery of ^{15}N in microplots made over four seasons on a grazed pasture at Samford in southeast Queensland.

Large increases in net efflux of NH_3 occurred during the two-week periods

Table 8.3. Seasonal atmospheric ammonia concentrations and fluxes, and extractable nitrogen content, water content and pH of surface soil at Samford, Queensland, 1978–1979 (Harper and Catchpoole, unpubl shed data)

Season	Day number	Atmospheric NH$_3$		Extractable soil N[a]			Water content (cm^3/cm^3)	pH
		Concentration at 20 cm (μg N/m^3)[b]	Integrated flux (g N/ha/day)[c]	NH$_4^+$ (kg/ha)	NO$_3^-$ (kg/ha)	Urea (kg/ha)		
Autumn	102	10.6±1.9	28.8	2.2±0.2	0.8±0.2	0	.41±.03	—
	142	16.6±6.2	−21.6	1.1±0.2	1.0±0.3	1.3±0.0	.27±.02	5.2±0.2
	201	17.8±5.0	235.2	1.9±0.4	5.1±1.3	0.4±0.1	.20±.04	4.7±.3
Winter	233	18.2±5.6	−206.4	2.6±0.6	4.9±1.1	0.5±0.2	.24±.02	4.3±.2
	291	23.7±9.4	−300.0	1.8±0.7	2.0±0.6	0	.30±.02	5.2±.2
Spring	310	32.7±10.5	−208.8	1.0±0.2	0.4±0.1	0.1±.01	.25±.01	5.4±.4
	353	74.2±31.4	384.0	6.2±1.6	5.3±1.6	0	.28±.02	5.3±.2
Summer	036	67.6±38.4	−64.8	1.7±0.4	1.1±0.6	1.1±0.4	.31±.02	5.1±.2

[a] Mean of 12 sampling sites; it is possible that some sites may have had contamination from urine-urea

[b] Mean of 3 reps by 12 periods per day

[c] Curvilinear integration of area under the curve determined by twelve 2-hour integrated NH$_3$ flux measurement periods. Negative sign denotes transport toward the soil-plant system

following urea applications. These losses ranged from 9% during spring when the grass was actively growing and rain fell soon after the fertilizer was applied, to 42% in autumn when growing conditions were relatively very poor and rain did not fall for 7 days after the application of fertilizer. Losses of NH_3 during background periods, when the effects of recent applications of urea were absent, came from urine patches. Annual losses of NH_3 into the atmosphere were calculated from curvilinear extrapolation of the data and were about 25% of the fertilizer N applied.

Total losses of N during the two-week periods after application of fertilizer calculated from the recovery of ^{15}N in microplots were usually about double the NH_3 losses. This discrepancy may be due to losses of nitrogenous gases other than NH_3, or to absorption of NH_3 liberated at the surface of the soil by the canopy of the pasture. The latter argument was supported by the observation that net NH_3 influx was common during short periods (2-4 hours) at sunset and sunrise, and for long periods (24 hours) prior to applications of urea.

8.6. References

Allison, F.E. 1973 Soil Organic Matter and its Role in Crop Production. Elsevier Scientific Publ. Co., Amsterdam. 673 pp.

Anon. 1980 The Nutrient Requirements of Ruminant Livestock. Commonwealth Agricultural Bureaux, Farnham Royal.

Ball, P.R. 1979 Nitrogen relationships in grazed and cut grass-clover systems. Ph.D. Thesis. Massey Univ., New Zealand. 217 pp.

Ball, P.R., Keeney, D.R. Theobald, P.W. and Nes, P. 1979 Nitrogen balance in urine-affected areas of New Zealand pasture. Agron. J. 71, 309–314.

Bathurst, N.D. 1952 The amino-acids of sheep and cow urine. J. Agric. Sci. 42, 476–478.

Blue, W.G. 1970 Fertilizer nitrogen uptake by Pensacola bahiagrass (*Paspalum notatum*) from Leon fine sand, a spodosol. Proc. 11th Intern. Grassl. Congr. Surfers Paradise 389–392.

Bridges, J.D. 1980 Fertilizer trends 1979. Nat. Fert. Devel. Cent. TVA Bull. Y-150.

Burton, G.W. and Jackson, J.E. 1962 Effect of rate and frequency of applying six nitrogen sources on Coastal Bermudagrass. Agron. J. 54, 40–43.

Carrier, D. and Bernier, B. 1971 Nitrogen losses through ammonia volatilization after fertilization of a jack pine forest. Can. J. For. Res. 1, 69–79.

Catchpoole, V.R. 1975 Pathways for losses of fertilizer nitrogen from a Rhodes grass pasture in south-eastern Queensland. Aust. J. Agric. Res. 26, 259–268.

Catchpoole, V.R. and Henzell, E.F. 1975 Losses of nitrogen from pastures. C.S.I.R.O., Trop. Agron. Div. Rept. 1974–75, pp. 82–83.

Catchpoole, V.R., Harper, L.A. and Myers, R.J.K. 1983a Annual losses of ammonia from a grazed pasture fertilized with urea. Proc. XIV Intern. Grassl. Congr. Lexington pp. 344–347.

Catchpoole, V.R., Oxenham, D.J. and Harper, L.A. 1983b Transformation and recovery of urea applied to a grass pasture in south-eastern Queensland. Aust. J. Exp. Agric. Anim. Husb. 23, 80–86.

Chacon, E.A., Stobbs, T.H. and Dale, M.B. 1978 Influence of sward characteristics on grazing behaviour and growth of Hereford steers grazing tropical grass pasture. Aust. J. Agric. Res. 29, 89–102.

Denmead, O.T., Freney, J.R. and Simpson, J.R. 1976 A closed ammonia cycle within a plant canopy. Soil Biol. Biochem. 8, 161–164.

Denmead, O.T., Nulsen, R. and Thurtell, G.W. 1978 Ammonia exchange over a corn crop. Soil Sci. Soc. Am. J. 42, 840–842.

Denmead, O.T., Simpson, J.R. and Freney, J.R. 1974 Ammonia flux into the atmosphere from a grazed pasture. Science (Wash. D.C.) 185, 609–610.

Denmead, O.T., Simpson, J.R. and Freney, J.R. 1977 A direct field measurement of ammonia emission after injection of anhydrous ammonia. Soil Sci. Soc. Am. J. 41, 1001–1004.

Doak, B.W. 1952 Some chemical changes in the nitrogen constituents of urine when voided on pasture. J. Agric. Sci. 42, 162–171.

Ernst, J.W. and Massey, H.F. 1960 The effects of several factors on volatilization of ammonia from urea in the soil. Soil Sci. Soc. Am. Proc. 24, 87–90.

Farquhar, G.D., Firth, P.M., Wetselaar, R. and Weir, B. 1980 On the gaseous exchange of ammonia between leaves and the environment: Determination of the ammonia compensation point. Plant Physiol. 66, 710–714.

Feagley, S.E. 1976 Ammonia volatilization from surface applications of ammonium sulfate to carbonate systems. M.S. Thesis. Texas A & M Univ., College Station.

Fenn, L.B. and Escarzaga, R. 1976 Ammonia volatilization from surface applications of ammonium compounds on calcareous soils: V. Soil water content and method of nitrogen application. Soil Sci. Soc. Am. J. 40, 537–541.

Fenn, L.B. and Kissel, D.E. 1974 Ammonia volatilization from surface applications of ammonium compounds on calcareous soils: II. Effects of temperature and rate of ammonium nitrogen applications. Soil Sci. Soc. Am. Proc. 38, 606–610.

Förster, I. and Lippold, H. 1975 Ammonia loss with urea fertilization; Communication 2. Determination of ammonia loss under field conditions as affected by the weather. Arch. Acker-Pflanzenbau Bodenk De 19, 631–639.

Gartner, J.A. and Everett, M.L. 1970 Effects of fertilizer nitrogen on a dense sward of kikuyu, paspalum, and carpet grass. 3. Nitrogen source. Queensl. J. Agric. Anim. Sci. 27, 73–87.

Goatley, J.L. and Lewis, R.W. 1966 Composition of guttation fluid from rye, wheat, and barley seedlings. Plant Physiol. (Bethesda) 41, 373–375.

Harper, L.A., Box, J.E. Jr., Baker, D.N. and Hesketh, J.E. 1973 Carbon dioxide and the photosynthesis of field crops. A tracer examination of turbulent transfer theory. Agron. J. 65, 574–578.

Harper, L.A., Catchpoole, V.R., Davis R. and Weier, K.L. 1983 Ammonia volatilization: Soil, plant and microclimate effects on diurnal and seasonal fluctuation. Agron. J. (in press).

Henzell, E.F. 1971 Recovery of nitrogen from four fertilizers applied to Rhodes grass in small plots. Aust. J. Exp. Agric. Anim. Husb. 11, 420–430.

Henzell, E.F. 1972 Loss of nitrogen from a nitrogen-fertilized pasture. J. Aust. Inst. Agric. Sci. 38, 309–310.

Henzell, E.F. and Ross, P.J. 1973 The nitrogen cycle of pasture ecosystems, In: Butler, G.W. and Bailey, R.W. (eds.), Chemistry and Biochemistry of Herbage. Vol. 2, pp. 227–245. Academic Press, London.

Henzell, E.F., Martin, A.E. and Ross, P.J. 1970 Recovery of fertilizer nitrogen by Rhodes grass. Proc. 11th Intern. Grassl. Congr., Surfers Paradise, pp. 411–414.

Hilder, E.J. and Mottershead, B.E. 1963 The redistribution of plant nutrients through free-grazing sheep. Aust. J. Sci. 26, 88–89.

Hutchinson, G.L., Millington, R.J. and Peters, D.B. 1972 Atmospheric ammonia: Absorption by plant leaves. Science (Wash. D.C.) 175, 771–772.

214

Impithuksa, V., Dantzman, C.L. and Blue, W.G. 1979 Fertilizer nitrogen utilization by three warm-season grasses on an alfic haplaquod as indicated by nitrogen-15. Soil Crop Sci. Soc. Fl. Proc. 38, 93–97.

Jewitt, T.N. and Barlow, H.W.B. 1949 Animal excreta in the Sudan Gezira. Emp. J. Exp. Agric. 17, 1–17.

Kresge, C.G. and Satchell, D.P. 1960 Gaseous loss of ammonia from nitrogen fertilizers applied to soils. Agron. J. 52, 104–107.

Lemon, E.R. 1970 Mass and energy exchange between plant stands and environment. Int. Biol. Program Symp. on Photosynthetic Productivity Proc., Trebon, pp. 199–205.

Lemon, E.R. 1978 Nitrous oxide exchange at the land surface. In: Neilsen, D.R. and MacDonald, J.G. (eds.), Nitrogen in the Environment, pp. 493–521. Academic Press, New York.

Marshall, V.G. and DeBell, D.S. 1980 Comparison of four methods of measuring volatilization losses of nitrogen following urea fertilization of forest soils. Can. J. Soil Sci. 60, 549–563.

McGarity, J.W. and Rajaratnam, J.A. 1973 Apparatus for the measurement of losses of nitrogen as gas from the field and simulated field environments. Soil Biol. Biochem. 5, 121–131.

Meyer, M.W. 1973 Absorption and release of ammonia from and to the atmosphere by plants. Ph.D. Thesis. Univ. of Maryland, College Park.

Nömmik, H. 1973 Assessment of volatilization loss of ammonia from surface-applied urea on forest soil by ^{15}N recovery. Plant Soil 38, 589–603.

Power, J.F. 1980 Response of semiarid grassland sites to nitrogen fertilization: II. Fertilizer recovery. Soil Sci. Soc. Am. J. 44, 550–555.

Prasad, M. 1976 Gaseous loss of ammonia from sulfur-coated urea, ammonium sulfate, and urea applied to calcareous soil (pH 7.3). Soil Sci. Soc. Am. J. 40, 130–134.

Scarsbrook, C.E. 1970 Regression of nitrogen uptake on nitrogen added from four sources applied to grass. Agron. J. 62, 618–620.

Simpson, D.M.H. and Melsted, S.W. 1962 Gaseous ammonia losses from urea solutions applied as a foliar spray to various grass sods. Soil Sci. Soc. Am. Proc. 26, 186–189.

Stutte, C.A., Weiland, R.T. and Blem, A.R. 1979 Gaseous nitrogen loss from soybean foliage. Agron. J. 71, 95–97.

Terman, G.L. 1979 Volatilization losses of nitrogen as ammonia from surface-applied fertilizers, organic amendments, and crop residues. Adv. Agron. 31, 189–223.

Vallis, I., Harper, L.A., Catchpoole, V.R. and Weier, K.L. 1982 Volatilization of ammonia from urine patches in a subtropical pasture. Aust. J. Agric. Res. 33, 97–107.

Vallis, I., Henzell, E.F., Martin, A.E. and Ross, P.J. 1973 Isotopic studies on the uptake of nitrogen by pasture plants. V. ^{15}N balance experiments in field microplots. Aust. J. Agric. Res. 24, 693–702.

Volk, G.M. 1959 Volatile loss of ammonia following surface application of urea to turf or bare soils. Agron. J. 51, 746–749.

Watson, E.R. and Lapins, P. 1969 Losses of nitrogen from urine on soils from south-western Australia. Aust. J. Exp. Agric. Anim. Husb. 9, 85–91.

Whitehead, D.C. 1970 The role of nitrogen in grassland productivity. A review of information from temperate regions. Commonw. Bur. Past. Fld Crops, Hurley, Berkshire, England. Commonw. Agric. Bureaux, Farnham Royal. Bull. 48, 202 pp.

9. Gaseous nitrogen exchanges in grazed pastures

J.R. SIMPSON and K.W. STEELE

9.1. Introduction

There is a well documented nitrogen (N) cycle in grasslands which becomes more complex under grazing (Whitehead 1970, Henzell and Ross 1973, Simpson and Stobbs 1981, Woodmansee *et al.* 1981, Steele 1982a). The cycle involves inputs of N from the atmosphere via biological or industrial fixation, and also mineralization of organic N in the soil, plant absorption of mineral N, consumption of plant material by grazers, return of excreta, decay of ungrazed material and losses by leaching of mineral N. It is becoming increasingly obvious, from work in the last 10 years, that the cycle also involves substantial returns of gaseous N compounds from the pasture to the atmosphere. These losses from the pasture system are of ecological, economic and environmental importance. It is therefore important to understand how they occur, their chemical nature and the magnitudes of the gaseous fluxes over various pasture ecosystems.

9.2. The circumstantial evidence for large losses of gaseous nitrogen in pastures

9.2.1. Balance studies

The N balance for any grazed pasture system during a defined time interval may be summarised as:

$$\text{N inputs} - \Delta \text{ total system N} - \text{N losses} = 0.$$

N losses can therefore be assessed by direct measurement, or indirectly from balance studies by measurement of N inputs and changes in total system N. The largest pool of N within a grazed pasture system is soil total N which may exceed 16 tonnes N/ha (Jackman 1964). Precise measurement of small changes in this large pool is not feasible due to the spatial variability of soil N (Biggar 1978), and interpretation of many balance studies in the literature is uncertain because changes in soil total N are not adequately recorded (Allison 1966). Few detailed balance studies have been reported for grazed pastures.

Balance experiments with commercial inorganic N fertilizers have shown low

($<50\%$) recoveries of the applied N (in plants and soil) over a few years in grazed pastures (Henzell *et al.* 1970, Henzell 1972, Catchpoole 1975, Simpson *et al.* 1974). This is due to factors in the field environment, since high recoveries of labelled ammonium nitrate have been observed in Rhodes grass and other species in controlled glasshouse environments (Henzell *et al.* 1970).

Low recoveries of ^{15}N labelled fertiliser N have also been reported for short term experiments (Simpson 1968a, Catchpoole 1975) and losses by NH_3 volatilization or denitrification have been implicated. Steele *et al.* (1980a) accounted for 70% of N applied as ammonium sulfate to an Ultic haplohumod after 84 days but could account for only 33% of N applied as calcium nitrate to the same soil.

Some New Zealand balance experiments on grass-clover pastures have shown very large losses. For example, Ball (1979) studied N balances in intensively grazed sheep pastures at Palmerston North which received various applications of fertilizer N during autumn, winter and spring (Table 9.1). He concluded that large unidentified losses of N from the soil-plant-animal systems occurred over a 3-year period.

Let us examine the probable extent of the major N fluxes in pasture systems.

9.2.2. Nitrogen input

In most pasture systems of temperate regions, biological fixation is a major input of N, which may be complemented with fertilizer N at rates depending on current economics for particular systems. Other inputs (e.g. in rainfall, snow or dry deposition) are generally less than 10 kg N/ha/year (Steele 1982a). In high producing pastures, biological fixation occurs mainly through the legume-rhizobium symbiosis, the annual input through free living soil microorganisms being only about 10–16 kg N/ha/year (Sears *et al.* 1965, Ball 1979). In pastures

Table 9.1. Annual nitrogen balances for grazed sheep pastures at Palmerston North, New Zealand

Treatment	Inputs				Outputs	
	Back-ground[a]	Legume[b]	Fertilizer (kg N/ha)	Δ Soil N	Animal	Losses[c]
0N	15	319	0	−69	20	383
112N	15	218	112	−55	20	380
448N	15	90	448	−36	25	564

[a] Principally non-symbiotic fixation
[b] Estimated from clover yields and a fixation index
[c] Estimated by difference.

of sub-tropical regions N_2 fixation associated with grasses may assume importance, since *Azospirillum* has been reported to fix 0.11–0.23 kg N/ha/day in association with *Bothriochloa insculpta*, *Urochloa bolboides* and *Sorghum plumosum* (Weier and MacRae 1981).

In soils of low N status annual inputs through symbiotic fixation may be in excess of 500 kg N/ha (Sears 1953, Sears *et al.* 1965, Weeda 1970). However, as soil N accumulates and mineral N concentrations increase, symbiotic fixation decreases and estimated inputs through symbiotic fixation generally lie in the range of 100–300 kg N/ha/year (Henzell and Norris 1962, Simpson 1976, Vallis *et al.* 1977, Hoglund *et al.* 1979). In addition to the effects of soil N status, symbiotic N_2 fixation is influenced by factors such as limiting soil moisture, insect pests, return of animal excreta, cyclic changes in the legume content of pastures (Steele 1982b) and botanical composition changes induced by management (Scott 1977).

There are relatively few reports of field measurements of symbiotic N_2 fixation, largely because of the difficulties in making such measurements (Knowles 1981). Prior to the 1970s estimates of biological fixation were based either on measurements of N uptake in pasture herbage and changes in soil N (Sears 1953) or on the amount of fertilizer N required to sustain equal production of herbage. These techniques, at best, provide only qualitative estimates over long periods.

During the last decade increasing interest has been shown in the use of acetylene (C_2H_2) reduction and ^{15}N techniques to measure symbiotic N_2 fixation. The C_2H_2 reduction assay is based on the observation that nitrogenase reduces C_2H_2 to ethylene (C_2H_4) at a rate proportional to its ability to fix N_2. Although the assay is sensitive and easily done, it has several limitations. Most reports use the theoretical value of 3 for the ratio of moles C_2H_2 reduced per mole N_2 fixed and few include experimental determination of this ratio. The wide variation in experimentally determined ratios makes adoption of the theoretical value of 3, particularly in field studies, doubtful (Bergersen 1970, Burris 1974, Sinclair 1975). The wide range of ratios may be at least partially explained by the failure to evaluate the ATP-dependent H_2 evolution, catalysed by nitrogenase (Schubert and Evans 1976), which can result in significant energy loss during N_2 fixation (Dixon 1972, Bergersen *et al.* 1973). This does not occur in the presence of C_2H_2. Many studies have used commercial grade C_2H_2 or have not reported the grade used, despite the well established fact that contaminants in commercial grade C_2H_2 significantly reduce C_2H_4 production (Tough and Crush 1979).

Choosing a truly representative pasture sample for C_2H_2 reduction assay is difficult because of the non-uniform distribution of legumes in mixed pastures. Diurnal and seasonal variation in N_2 fixation cause difficulties in extrapolating the short term kinetic measurements of C_2H_2 reduction to a seasonal value.

The considerably higher solubility of C_2H_2 in water compared with N_2 (Bergersen 1970) may produce erroneously high reaction rates under wetter conditions (Rennie *et al.* 1978) and adsorption of C_2H_2 by soil colloids may introduce an error in the reverse direction (Rinaudo *et al.* 1977).

The use of ^{15}N to measure biological fixation was first developed by McAuliffe *et al.* (1958), but it was not until the 1970s that ^{15}N techniques were used under field conditions. This was due in part to the high cost of ^{15}N, and also to the lack of suitable mass or photoemission spectrometers in many laboratories.

Techniques using ^{15}N for measuring biological fixation in the field are based either on the A-value concept (Fried and Broeshart 1975) or on isotope dilution principles (Fried and Middleboe 1977, Vallis *et al.* 1977), the latter being the preferred method for studies with pastures.

The soil mineral N pool is enriched by adding ^{15}N in readily mineralizable organic material or in small amounts of highly enriched inorganic materials. If all of the legume N is derived through biological N_2 fixation its isotopic composition will approach that of atmospheric N. If however, the legume derives its N entirely from soil solution, its isotopic composition will be similar to that of the soil mineral N. The proportion of N in the legume derived from symbiotic fixation ($\%N_f$) is calculated from:

$$\%N_f = \left[1 - \frac{\text{atom }\%\ ^{15}N \text{ excess (legume)}}{\text{atom }\%\ ^{15}N \text{ excess (soil solution)}} \right] \times 100 \ldots (1)$$

Because it is difficult to measure the N isotopic composition of the soil solution over time it is usually estimated from the isotopic composition of a control plant (ie. a non-fixing plant), usually a grass growing in the pasture with the legume. The procedure depends on several assumptions:

(1) No N_2 fixation is associated with the control plant.
(2) A constant and known amount of isotopic discrimination occurs during N_2 fixation.
(3) The proportional uptake of soil N with time by the legume and control plant is the same.
(4) The legume and control plants assimilate soil N of the same isotopic composition.
(5) N stored in the legume and control plants is translocated to the harvested portion in the same ratio in both species.
(6) No direct transfer of N occurs from the fixing to non-fixing species during the experimental period.
(7) Net absorption of atmospheric N compounds (e.g. NH_3), relative to total N assimilation is equal for the two species.

The ^{15}N methods offer an indirect, but quantitative, integrated measurement of the contribution of N_2 fixation to the total N harvested in the legume plant, accommodating diurnal and seasonal variations in fixation rate.

9.2.3. Nitrogen accumulation

Where the original nitrogen status is low, N generally accumulates in the surface soil when pastures are developed from virgin or cropped land. The extent of this accumulation has been studied by a number of researchers (Sears and Evans 1953, Donald and Williams 1954, Russell and Harvey 1959, Jackman 1960, Sears et al. 1965, Vallis 1972, Simpson et al. 1974, Kohn et al. 1977). Accumulations in the range of 40–80 kg N/ha/year are frequently observed in Australian environments, but higher values have been reported from wetter or irrigated temperate areas.

The rate of N accumulation and the time taken to approach an equilibrium, with equal rates of N input and output, depend on many factors. Phosphorus input (Donald and Williams 1954, Henzell et al. 1966), deficiency of other nutrients (Walker and Adams 1958, 1959, Walker et al. 1959), management (Sears and Evans 1953) and return or removal of pasture herbage (Sears et al. 1965) have all been implicated. A change in any of these factors may initiate a shift towards a new equilibrium level of soil N.

Intensive livestock grazing may reduce soil total N from a previously high level to a lower equilibrium level. Russell and Harvey (1959) resampled 40 reference sites in an area of South Australia which had been used for intensive dairy farming with irrigated grass/clover pastures. Soil of high initial total N content exhibited a decline over the 30-year period, while the initially low N sites had increased towards the same equilibrium level. Under intensive sheep grazing in New Zealand, Ball (1979) reported a decline of 182 kg N/ha over a 3-year period.

Henzell (1970) suggested that even though substantial inputs of symbiotically fixed N occurred in south-eastern Queensland, accumulation of N below some pastures may be nil or negative. A decrease in the total soil N of a pasture ecosystem may be interpreted as excess mineralization over immobilization, and a subsequent loss by conversion to gaseous N compounds, removal in animal products, or by leaching.

9.2.4. Leaching

Loss of N by leaching from pasture primarily results from the aggregation of N at high concentrations in small areas affected by animal urine return (O'Connor 1974). Concentrations of up to 58 mg N/l (mean 26.4 mg/l) have been reported in shallow aquifers under intensive dairy areas in New Zealand (Baber and Wilson 1972). Steele (1982b) suggested that about 100 kg N/ha/year must be leached to maintain such concentrations and successfully measured losses of this magnitude in leachates from grazed pastures in Northland, New Zealand (Steele 1982b).

The amount of N lost through leaching depends on rainfall (Owens 1960), age of pasture (Harmsen and Kolenbrander 1965), soil texture (Kolenbrander 1960) and other factors which control the concentration of nitrate in soil solution. Low nitrification activity in some soils may be an important mechanism for conserving N (Steele *et al.* 1980b).

The partitioning of N losses between the major pathways (leaching, NH_3 volatilization, denitrification) will depend on the prevailing climatic conditions. In most temperate pastures it is unlikely that leaching will remove more than 100 kg N/ha/year, but leaching losses may vary substantially between seasons, districts and soil types.

9.2.5. *Removal of nitrogen in livestock production*

The amount of N retained in grazing stock, removed in animal products or transferred to non-productive areas, depends on the type of stock and intensity of grazing. Milk, on average, contains 6.1 gN/kg (Agricultural Research Council 1965) and 10–15% of dairy cow excreta is deposited on non-productive areas such as milking sheds or races (During 1972). Therefore on an intensive dairy farm where 4.1 cows/ha consume 80% of the annual pasture dry matter production of 16,500 kg/ha and produce 10,800 kg milk/ha, 66 kg N will be removed in milk, 46 kg N transferred to unproductive areas and 8 kg N/ha will be retained in replacement stock, a total removal of 120 kg N/ha (Steele 1982a).

The removal of N from pastures grazed by beef cattle or sheep is much less. Cattle and sheep retain about 2.4% of their live weight gain as N (Agricultural Research Council 1965). A liveweight gain of 1000 kg/ha therefore represents a retention of 24 kg N/ha in stock. Greasy wool contains roughly 70% protein, containing 16.4% N, so that each kg of wool removes 0.11 kg N. Thus on intensively stocked pastures, carrying 20 sheep/ha, each producing 4 kg of wool, the wool represents a removal of about 9 kg N/ha. However, many sheep pastures are only extensively grazed due to climatic limitations and the wool produced would contain less than 1 kg N/ha.

9.3. Direct field measurements of gaseous emissions from pastures

The data reviewed in the previous section suggest that there are substantial unidentified gaseous losses of N from grazed pastures to the atmosphere, at least in moderately dry environments. These could be greater than 50 kg N/ha/year, but because of the differences in techniques used, in the pasture systems studied, and in the environments observed, it is not possible to obtain a clear assessment of either gross or net loss from the pasture. To do this, direct measurements of gaseous fluxes are required on typical grazed pasture eco-systems.

The spatially heterogeneous nature of grazed pastures and the complexity of the soil-plant-animal-atmosphere ecosystem involved create great difficulties in measuring the flux rates of emissions during grazing. One difficulty is in selecting an appropriately sized, representative area for the measurements. A second difficulty is in completing the measurements without disturbing the ecosystem and thereby changing the rates of emission. Because of these problems very few reliable direct measurements of gaseous N fluxes have been made over pastures.

9.3.1. Ammonia losses

The NH_3 component of the atmosphere (usually equivalent to $1-10$ μg N/m^3) has been calculated to originate primarily as NH_3 emissions formed by the hydrolysis of animal urine (Healy et al. 1970, Söderlund and Svensson 1976, Galbally et al. 1980). Much of this NH_3 emission would be associated with grazed pastures. For their United Kingdom and global estimates, Healy et al. (1970) and Söderlund and Svensson (1976) have assumed a 10% loss of urine N as NH_3. For Australia, Galbally et al. assumed a 26% loss.

The experimental basis for such estimates is rather limited. It depends on NH_3 flux measurements on (a) unfertilized, grazed pasture by Denmead et al. (1974, 1976); (b) NH_3 flux from solid urea fertiliser applied to tropical pastures by Harper et al. 1982; and (c) a number of small scale experiments in which NH_3 was trapped over individual urine patches covered by canopies of various kinds (Doak 1952, McGarity and Rajaratnam 1973, Ball and Keeney 1981, Vallis et al. 1982).

The measurements by Denmead et al. (1974, 1976) were made on un-disturbed lucerne and ryegrass/subterranean clover pastures grazed by sheep at stocking rates of 50/ha and 22.5/ha respectively. The average daily emission rates were 260 g N/ha at the high stocking rate in autumn, and 130 kg N/ha at the low stocking rate in spring. This was equivalent to about a 26% loss of the urine voided. By comparison, an adjacent, ungrazed ryegrass/subterranean clover pasture showed a daily emission of only 20 kg N/ha. These emission rates represent the whole pasture, in each case an area of about 4 ha. The technique used involved determining the equilibrium profile of NH_3 concen-tration up to 2 m above the surface, and the use of an energy balance to calculate a transfer coefficient and thus to obtain the vertical flux rate of NH_3. Consistent diurnal variations in rate of NH_3 emission were obtained with large fluxes on days of high evaporation and small fluxes on wet or humid days.

The approach of Harper *et al.* (this book) was complementary to that of Denmead *et al.* (1974, 1976) in that the emphasis was not on NH_3 flux from the whole, unfertilized pasture but from a defined small area (of ~ 0.2 ha) to which urea fertilizer had been applied. The horizontal flux of NH_3 from upwind to downwind of the treated area was determined and compared, under a range of experimental conditions, against the natural background. The results are reported in Chapter 8, and give valuable relationships with environmental factors.

The third approach to the problem of NH_3 loss from grazed pastures is the measurement of fluxes by canopy techniques over individual applications of urine, simulating natural urine patches (e.g. Vallis *et al.* 1982). Canopy experiments can give us some comparison of rates of NH_3 loss under different conditions, such as varying evaporation rates, types of vegetative cover or different soil cation exchange capacities. However, absolute rates of loss are better determined by techniques which do not affect windspeeds, temperatures or evaporation on the study area, since these factors are important determinants of the flux rate through their effects on gaseous transport and vapour pressure of NH_3 at the surface (Denmead *et al.* 1981; Beauchamp, this book).

Vallis *et al.* (1982), using a specially designed plastic hood as an incomplete canopy, with suction rates adjusted to approximate the natural windspeeds, found the greatest losses from simulated urine patches to be 28.4% in late summer on a *Setaria* pasture in a sub-tropical region of southern Queensland. Earlier experiments on temperate pastures using somewhat different types of apparatus had given losses in the range of 9–18% of the applied urine N.

Another recent study on New Zealand pastures by Ball and Keeney (1981) showed large variations in the NH_3 loss from urine patches in different seasons. Using a simple canopy system, and two rates of urine application, namely 30 and 60 g N/m^2, they found that the physical environment had a much bigger effect on losses than had the urine application rate. Taking the means for the two application rates, 5, 16 and 66% of the added urine N was volatilized as NH_3 under cool-moist (winter), warm-moist (spring) and warm-dry (summer) conditions respectively. The average of these figures, 29% loss, can be compared with the estimate of 26% given as an average under spring and autumn conditions by Denmead *et al.* (1974, 1976). Based on these data and a carrying capacity of 20 sheep/ha voiding an average of 400–500 g of urine N/ha/day, the annual loss of NH_3 from productive pastures grazed by sheep is likely to be 40–60 kg N/ha. Higher losses are possible from some intensively grazed New Zealand pastures (Ball and Keeney 1981).

9.3.2. Nitrous oxide emissions

Continuous emissions of N_2O are detectable in mown, ungrazed pasture swards under a wide range of temperatures and moisture conditions (Denmead *et al.* 1979). Daily losses varied from 0.05 mg N/m^2 (winter, dry soil) to 21.7 mg N/m^2 (spring, wet soil). The diurnal variations in flux rate closely reflected changes in soil surface temperatures. Emissions continued even when the soil surface was near air-dry, but flux rates increased dramatically on moistening a dry surface. It is therefore difficult to estimate the total loss of N_2O from a pasture soil for a whole year. In old grazed pastures the N_2O flux rates varied by $>100\%$ between sites a few metres apart (Roy, Denmead, Freney, unpublished). Daniel *et al.* 1980 have related the production of N_2O and N_2 in pastures to a denitrifying capacity of *Rhizobium* in its free-living state.

9.3.3. Losses of molecular nitrogen

Emissions of N_2 are extremely difficult to determine *in situ* for pasture systems. All the methods conceived so far involve some dissection or disturbance. In some cases the emission of labelled N_2 has been monitored after the addition of labelled fertilizer under a canopy. In other cases an N_2O reductase inhibitor such as acetylene, has been injected into soil to convert the N_2 emission into a more easily measurable one of N_2O (Ryden and Rolston, this book).

The natural emissions of N_2O which have been detected in undisturbed pastures are almost certainly associated with some liberation of N_2. The importance of the N_2 loss probably increases markedly in wet periods or as the soil nears moisture saturation. The processes producing N_2O have rather different optimum conditions from those producing N_2 (see Fillery, this book), and extrapolation from controlled laboratory experiments indicates that the $N_2O:N_2$ ratio of field emissions can be expected to vary widely (Gilliam *et al.* 1978). Thus pastures, which normally show great variations in soil conditions, within short distances and seasons, would be expected to show wide variations in the relative amounts of N_2O and N_2 emitted. Denitrification in urine patches, after the accumulation of nitrate to high concentrations and the onset of wet conditions, is one probable source of N_2 emission. The partitioning of losses from urine patches between NH_3 emissions, denitrification and leaching is obviously very sensitive to changes in moisture and temperature regimes. Below the organic surface layers, e.g. at 15–30 cm depth, the denitrification potential of many soils decreases to near zero (Ball and Keeney 1981).

9.3.4. Losses of amines

Some loss of N from pastures almost certainly occurs as amines although no satisfactory measurements have been recorded. The measurements of NH_3 fluxes already reported probably include inaccuracies due to the presence of, and interference by, amines but this point has not been properly checked.

Sources of volatile amines are senescent plants (Stutte and Wieland 1978, Wetselaar and Farquhar 1980), decomposing plant material and faecal matter (Floate and Torrance 1970, Beauchamp, this book). The presence of aliphatic amines, of varying molecular weight up to n-amylamine, in substantial quantities (2–6% relative to NH_3) has been recorded in air sampled above intensive cattle feed lots in Colorado (Mosier et al. 1973), but flux rates for pastures are unknown.

9.3.5. Losses of nitric oxide and nitrogen dioxide

Continual emissions of nitric oxide, NO, are known to occur from pasture soils at low rates, equivalent to about 2 kg N/ha/year (Galbally and Roy 1978). Burning of vegetation also produces NO_x (NO and NO_2) and N_2O in large quantities. Crutzen et al. (1979) estimated that 9 Tg N_2O-N and 14 Tg NO_x-N were emitted into the Earth's atmosphere annually from burning vegetation. Nitric oxide emissions in the normal cycle of pasture growth have been associated with nitrite decomposition (see Fillery, and Chalk and Smith, this book), and so should occur in urine patches, dunged areas or heavily fertilized zones where interrupted nitrification or nitrate reduction allows a transient accumulation of nitrite. Vallis et al. (1981) found nitrite concentrations up to ~ 10 kg/ha during the 7 days after application of urine to pasture.

9.3.6. Exchange of ammonia between plants and atmosphere

Emission of NH_3 to and reabsorption of NH_3 from the atmosphere by vegetation and soil has been observed in the field (Denmead et al. 1976, 1978), and absorption by plants has been found in controlled laboratory experiments at high NH_3 concentrations (Hutchinson et al. 1972, Porter et al. 1972, Cowling and Lockyer 1981). Deposition of NH_3 in rain and its absorption by wet vegetation are both well documented. There is a critical partial pressure of NH_3 in the leaves of plants for each species, temperature and condition of tissue (N status and age). If the external concentration of atmospheric NH_3 exceeds this value, absorption will occur, while below it NH_3 will be emitted (Farquhar et al. 1980). In heavily grazed or fertilized pastures, it is likely that NH_3 absorption by plant leaves occurs in significant amounts (Cowling and Lockyer 1981).

The possibilities of NH_3 and other gases being emitted from plants to the atmosphere have been reviewed by Wetselaar and Farquhar (1980) (see also Farquhar et al., this book). Opportunities are thought to be greatest after anthesis, but only small losses of NH_3 were detected from wheat plants at this stage by Hooker et al. (1980). Losses reached only about 0.1% of the total nitrogen content of the plant. Much larger losses appear to occur by other means during maturation of annual grasses (Lapins and Watson 1970).

9.4. The special aspects of nitrogen economy in pastures which facilitate gaseous losses

At this point we review the question of why gaseous N losses from pastures are likely to be different in nature and extent from those over a cultivated site under crop, or those in a natural ecosystem which is less disturbed by Man and his grazing animals.

9.4.1. Nitrogen input

In productive pastures there is a high input of N by symbiotic fixation or from fertilizer since the production of meat, milk and wool depends on the maintenance of a high intake of herbage protein (Simpson and Stobbs 1981). The range of inputs might be 100–500 kg N/ha/year if biological and anthropogenic forms of N are totalled for pastures of areas receiving >500 mm of effective rainfall in the temperate to tropical zones, assuming no other nutritional or toxicity limitations. As mentioned earlier the inputs by biological fixation are not well quantified for grazed field situations, but the total inputs to the pastures as just defined would exceed those for most arable crops.

Furthermore, the biological N input to pastures through symbiotic fixation involves an input directly into vegetation of N which is then recycled, after plant senescence, trampling or consumption and digestion by animals, so that it returns to the soil/atmosphere interface. Fertilizers when applied to pastures, are also usually broadcast on the surface, whereas in cultivated annual crops, mineralization of soil N and application of N fertilizers take place mainly below the surface. Thus the opportunities for losses of gaseous N produced at the soil/atmosphere interface are greater in grazed pasture systems.

9.4.2. Biological activity

The gradual accumulation of organic N which takes place close to the surface of pasture soils is controlled by climate and the input of C, N, P and S. The ratios of these elements in the accumulated organic matter tends to be fairly constant, e.g. 15:1:0.13:0.07 for C:N:S:P, as found by Williams and Steinbergs (1958) for a range of Australian soils. This organic accumulation leads to a high biological activity involving plant rhizospheres, microbes and free enzymes very close to the soil/atmosphere interface.

Losses of gaseous N can be related to this high biological or enzymic activity. For instance, Simpson (1968a), by applying ^{15}N-labelled urea to the surface of pasture soil cores of varying urease activity, found that the loss of ^{15}N was logarithmically proportional to the urease activity. The high urease activity at the surface did not change with seasons when the soils were brought to the same moisture content. Thus, 101–116 μg urea-N/g soil was hydrolysed from an addition of 200 μg N/g in 3 hr by the surface 2.5 cm layer, regardless of time of sampling. However, the subjacent layer would hydrolyse only 10–46 μg N/g in the same period of incubation.

Daniel *et al.* (1980) have demonstrated that certain strains of the nitrogen-fixing symbiotic bacteria, *Rhizobium*, when in free-living form, can also denitrify. Rhizobia abound in the organic surface layers of pasture soils and are now reported to convert nitrate to nitrite, N_2O or N_2, depending on the strain. This may contribute to the amounts of N_2O evolved from grassland soils (Denmead *et al.* 1979).

Organic accumulations in pasture soils generally give rise to an increasing denitrification potential. There is a long-established relationship between the organic carbon content of soils and their denitrification capacity (Allison 1973).

9.4.3. Concentrations of nitrogen voided in animal excreta

Grazing animals void high concentrations of N (up to 15 g N/l) in soluble form as urine and since each urination covers only 0.4–0.8 m^2 for cattle or about 0.03 m^2 for sheep (Simpson and Stobbs 1981) a high heterogeneity is produced in pastures. Individual patches thus receive 30–50 g N/m^2 (300–500 kg N/ha). Carbonaceous compounds which might otherwise cause the immobilization of N, so inhibiting the volatilization of nitrogenous gases, are voided elsewhere in faeces, possibly on particular campsite areas of the pasture (Hilder 1966). This leads to an unstable situation, giving rise to accumulations of ammonium, nitrite and nitrate in the urine patch, and thus to a high potential for emissions of NH_3, N_2O, N_2 and NO.

9.4.4. Vegetative cover

The presence of a perennial cover of vegetation with its persistently active root absorption promotes recycling of mineralized N which might otherwise be leached into the subsoil and groundwater. In this way, N is continually brought back to the soil surface (Wetselaar and Norman 1960, Simpson 1961) as vegetation senesces, decomposes or is consumed by animals and the excreta are returned. With each recycling of soluble N a new opportunity is created for its conversion to gaseous forms. Since the vegetative canopy is constantly being disturbed by grazing animals, the reabsorption of nitrogenous gases is disrupted. Foliage and senescent vegetative material are a replenishing source of urease and other NH_3-producing enzymes which assist in creating high concentrations of NH_3 near the soil surface (McGarity and Hoult 1971, Denmead *et al.* 1976).

9.4.5. Climatic effects

Land is utilised for pasture in many different climatic zones and the areas which are fully irrigated are very small. Thus plants and soils are subjected to wide ranges of temperature and moisture stresses. Rapid drying and intermittent wetting promote losses of NH_3 (Freney et al. 1981), and N_2O (Denmead et al. 1979) from soils. Senescence of plant shoots and roots induced by seasonal changes can lead to N loss from herbage (Lapins and Watson 1970, Wetselaar and Farquhar 1980) and nitrate accumulation in the soil (Simpson 1962) followed by denitrification in the rhizosphere (Woldendorp 1962). Large losses of NH_3, and some of N_2O, have been reported from pastures under warm-dry conditions (Galbally et al. 1980, Ball and Keeney 1981, Harper et al., this book). Perhaps this is an indication that gaseous N losses as a whole are more important in warm or hot, intermittently dry regions than in cooler, moist environments. However, far more work is required to substantiate such projections.

9.4.6. Wind erosion

Wind erosion, removing N in dust, can occur frequently during droughts in warm or semi-arid environments. The heavy defoliation, denudation and trampling which occur during droughts cause exposure of pulverised organic matter on the surface of pasture soils where it is easily removed. This is not a true gaseous loss but a removal of fine particulate matter containing organic N. Ammonium and nitrate N can accumulate very close to the soil surface during droughts (Simpson 1962) so that these N forms could also be removed.

If the surface organic matter is not completely removed from an area of pasture it tends to accumulate in small heaps in depressions and around obstacles under the combined action of wind and subsequent heavy rain on dry surfaces. This aggravates the opportunities for true gaseous losses since sites with high N concentrations are consistently identifiable as potential sites for N loss (Simpson et al. 1974). The effects of drought seasons on N loss have not been well documented, but Vallis (1972), in Queensland, has reported decreases in soil organic N during dry seasons.

9.4.7. Burning of vegetation

Burning of dry vegetation at the end of the dry season, to promote the regeneration of palatable green material after rain is a common practice in some pastoral regions, particularly in the tropics. It is also a source of substantial gaseous N loss. Not only is the carbon lost, with its potential for immobilizing the labile forms of inorganic N, but the combustion gases also contain some 80–90% of the plant N as NH_3, NO_x and N_2O (Crutzen *et al.* 1979). These are also removed in particulate matter. The charred bare soil surface which remains is a site of rapid nitrification on remoistening after its partial sterilisation by heat (Birch 1960), which promotes subsequent denitrification.

9.5. Manageable factors controlling the rates of gaseous nitrogen emissions from pastures

The many soil, environmental and biological factors which control the rates of gaseous N efflux from soils and plants have been discussed in earlier chapters. All that must be done here is to list the principal factors briefly to assist an understanding of the ways in which pasture management might offset gaseous losses of N.

9.5.1. Soil chemical factors

(i) The supply and availability of other nutrients, especially P, S and Mo, affects N_2 fixation, plant growth, N uptake and hence the potential for N cycling including gaseous losses.

(ii) The chemical nature of the soil surface, particularly its cation exchange capacity as modified by organic matter content and pH, is important in NH_3 retention. If more ammonium ion is held as an adsorbed cation, the concentration of NH_3 in the soil solution is decreased and possibilities of NH_3 efflux to the atmosphere are reduced (cf. Gasser 1964, Simpson 1981, Freney *et al.*, this book).

(iii) The pH at the soil surface affects the concentration of free NH_3 in solution and thus the rate of loss. Within urine patches however, the rise in pH is rapid regardless of original soil pH. The buffering capacity of the soil is much more important than the original pH (Avnimelech and Laher 1977). At low pH, nitrite formed during reactions at a higher pH, e.g. in a urine patch, becomes unstable and a potential source of NO, N_2O or N_2.

These soil surface properties can be affected by pasture management and fertilizer practice (e.g. Simpson *et al.* 1974).

9.5.2. Physical or environmental factors

(i) Soil moisture content and evaporation rate have strong effects on NH_3 emission, nitrification, denitrification and hence on the efflux of N_2O and N_2.

(ii) Temperature regulates the processes of formation, diffusion and vaporisation which control losses of NH_3, N_2O and other gases.

(iii) The wind velocity close to the soil surface has a dominant influence on gaseous losses (Denmead *et al.* 1982) by its direct effect of removing the gases and also by lowering their partial pressures near to the surface so that a net upward diffusion is promoted. Other effects of wind velocity probably occur indirectly via effects on the rate of evaporation of soil water and on surface temperature.

The type of vegetative canopy in the pasture, its height and structure, and the presence of a litter mulch, determine the effective windspeeds and temperature variations near the ground. These factors, as modified by grazing and irrigation, thus control the rate of NH_3 efflux and reabsorption of NH_3 within the canopy (Denmead *et al.* 1976, 1982).

9.5.3. Biological factors

(i) Urease activity and microbial N transformations in the surface soil determine the rates of formation and accumulation of ammonium or other labile N forms. These rates are closely linked to the organic C and N concentrations near the surface, which generally increase with the age of pastures.

(ii) By removing organic matter from the soil surface, the soil fauna assist its incorporation away from the soil/atmosphere interface. The consumer fauna such as dung beetles, worms and termites can affect reaction rates in dung patches and surface litter. The effect of these activities on the gaseous efflux of N have not been studied directly, although this could be one facet of the well documented improvements in soil fertility associated with an active soil fauna. Generally, because of their burying activities, the soil fauna could be expected to reduce gaseous losses, in the short term at least.

(iii) The growth rate and density of the vegetation determine the rate of N uptake and its immobilization into less labile forms. In a rapidly growing pasture, N fixed by the legume *Rhizobium* symbioses is rapidly immobilized by plants and rhizosphere organisms. It can be transferred gradually from plant to plant and retained in the vegetation or the surface organic matter.

(iv) When the N content of the herbage is high, the N content of the excreta from grazing animals – particularly the urine – also tends to be high so that opportunities for gaseous N losses are increased both from the excreta and by direct emissions from the animals.

Some of these effects are unavoidably controlled by seasonal fluctuations, and pastureage, as well as by management.

9.5.4. Pasture management effects

(i) The stocking rate of the pasture affects the defoliation of the vegetative canopy and the amounts of excreta voided per unit area. Grazing in the dormant or dry seasons may be less conducive to gaseous losses than grazing during flushes of growth, because the N contents of herbage and excreta are lower in the former case and thus less favourable for losses. The net effect of high grazing pressures tends towards lower N accumulation in the pasture, but pasture systems appear to be well buffered against such changes and the effects are not always obvious (Brockman and Wolton 1963, Simpson *et al.* 1974).

The more recent studies of Ball (1979) do suggest, however, that some intensively grazed pasture systems and environments may be more conducive to N losses. Different grazing systems, e.g. rotational grazing versus set stocking, large paddocks versus small subdivisions, probably affect gaseous losses through the differing degrees of concentration of excreta and defoliation in one area at one time (Hilder 1966). Rotational grazing of small paddock systems allows less concentration of excreta in one site.

(ii) The type of pasture species sown affects the input of N by symbiotic fixation, some legumes being more dominant and persistent than others. Deep rooted, perennial grasses and those which grow actively throughout the year, will in general be efficient at suppressing the accumulation of mineral N in the soil and the movement of nitrate. Annual pastures which senesce in summer or species which are dormant for long periods will, on the other hand, allow accumulations of soil ammonium and nitrate with (presumably) associated losses of N to the atmosphere from soil and herbage (Simpson 1962).

(iii) When N fertilizer is used on pastures, the chemical and physical form of the fertilizer can have important effects on gaseous N losses. Surface broadcasting of solid prilled fertilizer is the normal method of applying fertilizer to pasture because incorporation below the surface often causes substantial damage to the pasture. With this method, urea fertilizer can, according to weather conditions before and after application, be a source of substantial NH_3 loss (Simpson 1968b, Harper *et al.*, this book). Urea is more susceptible to NH_3 loss than most other forms of solid N fertilizer e.g. ammonium nitrate, ammonium sulfate, but the N from these sources can still be lost through nitrification-denitrification reactions. Similar principles apply to the use of dairy effluents and farmyard manure on pastures. In the intensive agricultural systems of western Europe and North America, pastures are often the sites of application of both animal production effluents and processed sewage effluents (Beauchamp, this book).

(iv) Irrigation of pastures, by promoting the year-round growth of perennial plants, may have the effect of reducing gaseous N losses but the subject has not been properly studied. Since in the long term the turnover and plant uptake of N are increased, the effects could be complex. Interactions between irrigation frequency and fertilizer management may be expected, and the effects on the fluxes of the various gases could be quite different.

(v) Drought strategies: If complete defoliation of pastures during drought could be avoided, then gaseous N losses, run off, wind erosion and associated losses might all be reduced.

(vi) Zero grazing: The harvesting and carrying off of forage from pastures, to be fed to animals elsewhere, could avoid some of the gaseous N losses associated with normal grazing practices. The effluents from the animal feedlots can be spread evenly back onto the pasture, so avoiding the concentrated urine patches and dunged areas which normally occur. Perhaps gaseous N losses are less under such systems but energetically and economically, this procedure is expensive and is not likely to be an acceptable alternative under most circumstances.

9.6. Conclusions

We have reviewed the complexity of processes and situations surrounding the emission of volatile N compounds from grazed pastures. There is a great deal of evidence coming to light (largely in the last 10 years) that the efflux of N from pastures is important in extent and varied in its nature, involving NH_3, N_2O, NO, N_2 and probably amines. At present, the picture is only just beginning to unfold, and undoubtedly quite different results will emerge from different regions and different pasture systems.

How can we judge the importance of these emissions?

(a) There is the economic cost of the lost N.

(b) There is the cost of environmental pollution, in terms of aerosol smogs of ammonium sulfate or NO_x and reactions of N_2O in the ozone layer of the stratosphere.

At this stage of our knowledge, (b) is most difficult to assess. In determining (a) we must invoke some system of comparison involving the productivity of the pasture. In productive pastures N_2 is continually being absorbed and converted via NH_3 to plant proteins, then animal proteins via NH_3 again. The rate of biological N_2 fixation appears to be controlled by physical environment, plant genotype and strain of *Rhizobium* present, defoliation, competition and the increasing availability of soil N as the pasture becomes older. All these factors and others, combine to produce a buffered system (Simpson *et al.* 1974) in which gaseous N losses to the atmosphere may be an integral part. It follows that a deliberate attempt to reduce gaseous losses could induce an appropriate response in the pasture system resulting in a reduction of biological N_2 fixation over a period of a few growing seasons.

If we attempt to assess gaseous losses of N to the atmosphere against the input of biologically fixed N_2, we have a double problem because neither the gaseous losses nor the rate of fixation is simple to determine. Difficulties in measuring N_2 fixation are due to the uncertainties and extrapolations involved in the acetylene reduction technique and to possible inaccuracies in the field use of the ^{15}N dilution technique (Rennie *et al.* 1978, Knowles 1981). It may be possible, on a small scale, to check the results of the above techniques using the natural discrimination of ^{15}N against ^{14}N which occurs during N_2 fixation (Amarger *et al.* 1979) if a sufficiently sensitive mass spectrometer is available. Thus, until more data are available and better techniques are developed, the true economic and agronomic significance of gaseous N losses in pastures remains somewhat obscure.

9.7. References

Agricultural Research Council. 1965 The nutrient requirements of farm livestock, No. 2 Ruminants. H.M.S.O., London.

Allison, F.E. 1966 The fate of nitrogen applied to soils. Adv. Agron. 18, 219–258.

Allison, F.E. 1973 Soil Organic Matter and its Role in Crop Production. Elsevier Scientific Publishing Co., Amsterdam. 673 pp.

Amarger, N., Marriotti, A., Marriotti, F., Durr, J.C., Bourguignon, C. and Lagacherie, B. 1979 Estimate of symbiotically fixed nitrogen in field-grown soybeans using variations in ^{15}N natural abundance. Plant Soil 52, 269–280.

Avnimelech, Y. and Laher, M. 1977 Ammonia volatilization from soils: equilibrium considerations. Soil Sci. Soc. Am. J. 41, 1080–1084.

Baber, H.L. and Wilson, A.T. 1972 Nitrate pollution of groundwater in the Waikato region. J. N.Z. Inst. Chem. 36, 179–183.

Ball, R. 1979 Nitrogen relationships in grazed and cut grass-clover systems. Ph. D. Thesis, Massey University, N.Z. 217 pp.

Ball, P.R. and Keeney, D.R. 1981 Nitrogen losses from urine-affected areas of a New Zealand pasture, under contrasting seasonal conditions. Proc. 14th Int. Grassl. Congr. Lexington (in press).

Bergersen, F.J. 1970 The quantitative relationship between nitrogen fixation and the acetylene-reduction assay. Aust. J. Biol. Sci. 23, 1015–1025.

Bergersen, F.J., Turner, G.L. and Appleby, C.A. 1973 Studies of the physiological role of leghaemoglobin in soybean root nodules. Biochim. Biophys. Acta 292, 271–282.

Biggar, J.W. 1978 Spatial variability of nitrogen in soils. In: Neilsen, D.R. and MacDonald, J.G. (eds.), Nitrogen in the Environment, Vol. 1, pp. 201–211. Academic Press, New York.

Birch, H.F. 1960 Nitrification in soils after different periods of dryness. Plant Soil 12, 81–96.

Brockman, J.S. and Wolton, K.M. 1963 The use of nitrogen in grass/white clover swards. J. Br. Grassl. Soc. 18, 7–13.

Burris, R.H. 1974 Methodology, In: Quispel, A. (ed.), The Biology of Nitrogen Fixation, pp. 9–33. North Holland, Amsterdam.

Catchpoole, V.R. 1975 Pathways for losses of fertiliser nitrogen from a Rhodes grass pasture in south-eastern Queensland. Aust. J. Agric. Res. 26, 259–268.

Crutzen, P.J., Heidt, L.E., Krasnec, J.P., Pollock, W.H. and Sieler, W. 1979 Biomass burning as a source of atmospheric gases CO, H_2, N_2O, NO, CH_3Cl and COS. Nature (Lond.) 282, 253–256.

Cowling, D.W. and Lockyer, D.R. 1981 Increased growth of ryegrass exposed to ammonia. Nature (Lond.) 292, 337–338.

Daniel, R.M., Steele, K.W. and Limmer, A.W. 1980 Denitrification by rhizobia. A possible factor contributing to nitrogen losses from soils. N.Z. Agric. Sci. 14, 109–112.

Denmead, O.T., Freney, J.R. and Simpson, J.R. 1976 A closed ammonia cycle within a plant canopy. Soil Biol. Biochem. 8, 161–164.

Denmead, O.T., Freney, J.R. and Simpson, J.R. 1979 Studies on nitrous oxide emission from a grass sward. Soil. Sci. Soc. Am. J. 43, 726–728.

Denmead, O.T., Freney, J.R. and Simpson, J.R. 1982 Dynamics of ammonia volatilisation during furrow irrigation of maize. Soil. Sci. Soc. Am. J. 46, 149–155.

Denmead, O.T., Nulsen, R. and Thurtell, G.W. 1978 Ammonia exchange over a corn crop. Soil Sci. Soc. Am. J. 42, 840–842.

Denmead, O.T., Simpson, J.R. and Freney, J.R. 1974 Ammonia flux into the atmosphere from a grazed pasture. Science (Wash. D.C.) 185, 609–610.

Dixon, R.O.D. 1972 Hydrogenase in legume root nodule bacteroids: occurrence and properties. Arch. Mikrobiol. 85, 193–201.

Doak, B.W. 1952 Some chemical changes in the nitrogenous constituents of urine when voided on pasture. J. Agric. Sci. 42, 162–171.

Donald, C.M. and Williams, C.H. 1954 Fertility and productivity of a podzolic soil as influenced by subterranean clover (*Trifolium subterraneum* L.) and superphosphate. Aust. J. Agric. Res. 5, 664–687.

During, C. 1972 Fertilisers and soils in New Zealand farming. New Zealand Department of Agriculture Bulletin No. 409, 312 pp.

Farquhar, G.D., Firth, P.M., Wetselaar, R. and Weir, B. 1980 On the gaseous exchange of ammonia between leaves and the environment. Determination of the ammonia compensation point. Plant Physiol. 66, 710–714.

Floate, M.J.S. and Torrance, C.J.W. 1970 Decomposition of the organic materials from hill soils and pastures. 1. Incubation method for studying the mineralisation of carbon, nitrogen and phosphorus. J. Sci. Food Agric. 21, 116–120.

Freney, J.R., Simpson, J.R. and Denmead, O.T. 1981 Ammonia volatilization In: Clark F.E. and Rosswall T. (eds.), Terrestrial Nitrogen Cycles. Ecol. Bull. 33, 291–302.

Fried, M. and Broeshart, H. 1975 An independent measure of the amount of nitrogen fixed by a legume crop. Plant Soil 43, 707–711

Fried, M. and Middleboe, V. 1977 Measurement of amount of nitrogen fixed by a legume crop. Plant Soil 47, 713–715.

Galbally, I.E. and Roy, C.R. 1978 Loss of fixed nitrogen from soils by nitric oxide exhalation. Nature (Lond.) 275, 734–735.

Galbally, I.E., Freney, J.R., Denmead, O.T. and Roy, C.R. 1980 Processes controlling the nitrogen cycle in the atmosphere over Australia. In: Trudinger, P.A., Walter, M.A. and Ralph, B.J. (eds.), Biogeochemistry of Ancient and Modern Environments, pp. 319–325. Australian Academy of Science, Canberra.

Gasser, J.K.R. 1964 Some factors affecting losses of ammonia from urea and ammonium sulphate applied to soils. J. Soil Sci. 15, 258–272.

Gilliam, J.W., Dasberg, S., Lund, L.J. and Focht, D.D. 1978 Denitrification in four Californian soils: Effect of soil profile characteristics. Soil Sci. Soc. Am. J. 42, 61–66.

Harmsen, G.W. and Kolenbrander, G.J. 1965 Soil inorganic nitrogen. In: Bartholomew, M.V. and Clark, F.E. (eds.), Soil Nitrogen. pp. 43–92. American Society of Agronomy, Madison.

Healy, T.V., McKay, H.A.C., Pilbeam, A. and Scargill, D. 1970 Ammonia and ammonium sulfate in the troposphere over the United Kingdom. J. Geophys. Res. 75, 2317–2321.

Henzell, E.F. 1970 Problems in comparing the nitrogen economies of legume-based and nitrogen fertilized pasture systems. Proc. 11th Int. Grassl. Congr., Surfers Paradise, pp. 112–120.

Henzell, E.F. 1972 Loss of nitrogen from a nitrogen-fertilized pasture. J. Aust. Inst. Agric. Sci. 38, 309–310.

Henzell, E.F. and Norris, D.O. 1962 Processes by which nitrogen is added to the soil-plant system. In: A Review of Nitrogen in the Tropics with Particular Reference to Pastures. Commonw. Bur. Pastures Field Crops Hurley, Berkshire, Bull. 46, 1–18.

Henzell, E.F. and Ross, P.J. 1973 The nitrogen cycle of pasture ecosystems. In: Butler, G.W. and Bailey, R.W. (eds.), Chemistry and Biochemistry of Herbage. Vol. 2, pp. 227–246. Academic Press, London.

Henzell, E.F., Fergus, I.F. and Martin, A.E. 1966 Accumulation of soil nitrogen and carbon under a *Desmodium uncinatum* pasture. Aust. J. Exp. Agric. Anim. Husb. 6, 157–160.

Henzell, E.F., Martin, A.E. and Ross, P.J. 1970 Recovery of fertilizer nitrogen by Rhodes grass. Proc. 11th Int. Grassl. Congr., Surfers Paradise. pp. 411–416.

Hilder, E.J. 1966 Distribution of excreta by sheep at pasture. 10th Int. Grassl. Congr., Helsinki, pp. 977–981.

Hoglund, J.H., Crush, J.R., Brock, J.L. and Ball, R. 1979 Nitrogen fixation in pasture. XII: General discussion. N.Z. J. Exp. Agric. 7, 45–51.

Hooker, M.L., Lander, D.H., Peterson, G.A. and Daigger, L.A. 1980 Gaseous N losses from winter wheat. Agron. J. 72, 789–792.

Hutchinson, G.L., Millington, R.J. and Peters, D.B. 1972 Atmospheric ammonia: absorption by plant leaves. Science (Wash. D.C.) 175, 771–772.

Jackman, R.H. 1960 Organic matter stability and nutrient availability in Taupo pumice. N.Z. J. Agric. Res. 3, 6–23.

Jackman, R.H. 1964 Accumulation of organic matter in some New Zealand soils under permanent pasture II. Rates of mineralisation of organic matter and the supply of available nutrients. N.Z. J. Agric. Res. 7, 472–479.

Kohn, G.D., Osborne, G.J., Batten, G.D. Smith, A.N. and Lill, W.J. 1977 The effect of topdressed superphosphate on changes in nitrogen:carbon:sulphur:phosphorus and pH on a Red Earth soil during a long term grazing experiment. Aust. J. Soil Res. 15, 147–158.

Kolenbrander, G.J. 1960 Calculation of parameters for the evaluation of the leaching of salts under field conditions, illustrated by nitrate. 'Soil-Water-Plant', Proceedings of the Int. Commission of Hydrological Research TNO No 15, 69–73.

Knowles, R. 1981 The measurement of nitrogen fixation. In: Gibson A.H. and Newton W.E. (eds.), Current Perspectives in Nitrogen Fixation, pp. 327–333. Australian Academy of Science, Canberra.

Lapins, P. and Watson, E.R. 1970 Loss of nitrogen from maturing plants. Aust. J. Exp. Agric. Anim. Husb. 10, 599–603.

McAuliffe, C., Chamblee, D.S., Uribe-Arango, H. and Woodhouse, W.W. 1958 Influence of inorganic nitrogen on nitrogen fixation by legumes as revealed by [15]N. Agron. J. 50, 334–337.

McGarity, J.W. and Hoult, E.H. 1971. The plant component as a factor in ammonia volatilisation from pasture swards. J. Br. Grassl. Soc. 26, 31–34.

McGarity, J.W. and Rajaratnam, J.A. 1973 Apparatus for the measurement of losses of nitrogen as gas from the field and simulated field environments. Soil Biol. Biochem. 5, 121–131.

Mosier, R.A., Andre, C.E. and Viets, F.G. Jr. 1973 Identification of aliphatic amines volatilized from cattle feedyard. Environ. Sci. Technol. 7, 642–644.

O'Connor, K.F. 1974 Nitrogen in agrobiosystems and its environmental significance. N.Z. Agric. Sci. 8, 137–148.

Owens, L.D. 1960 Nitrogen movement and transformation in soils evaluated by a lysimeter study utilising isotopic nitrogen. Soil Sci. Soc. Am. Proc. 24, 372–376.

Porter, L.K., Viets, F.G. Jr. and Hutchinson, G.L. 1972 Air containing nitrogen-15 ammonia. Foliar absorption by corn seedlings. Science (Wash. D.C.) 175, 759–761.

Rennie, R.J., Rennie, D.A. and Fried, M. 1978 Concepts of ^{15}N usage in dinitrogen fixation studies. In: Isotopes in Biological Dinitrogen Fixation, pp. 107–133. IAEA, Vienna.

Rinaudo, G., Hamad-Feres, I. and Dommergues, Y.R. 1972 Nitrogen fixation in the rice rhizosphere. Methods of measurement and practices suggested to enhance the process. In: Ayanaba, A. and Dart, P.J. (eds.), Biological Nitrogen Fixation in Farming Systems of the Tropics, pp. 313–322, John Wiley and Sons, New York.

Russell, J.S. and Harvey, D.L. 1959 Changes in the nitrogen content of the Mobilong clay as influenced by land use. Aust. J. Agric. Res. 10, 637–650.

Schubert, K.R. and Evans, H.J. 1976 Hydrogen evolution: A major factor affecting the efficiency of nitrogen fixation in nodulated symbionts. Proc. Nat Acad. Sci. USA 73, 1207–1211.

Scott, R.S. 1977 Effects of animals on pasture production. II. Pasture production and N and K requirements of cattle and sheep pastures measured under a common method of defoliation. NZ J. Agric. Res. 20, 31–36.

Sears, P.D. 1953 Pasture growth and soil fertility. 1. The influence of red and white clovers, superphosphate, lime and sheep grazing, on pasture yields and botanical composition. NZ J. Sci. Technol 35A, Supplement 1, 1–29.

Sears, P.D., and Evans, L.T. 1953 Pasture growth and soil fertility. III. The influence of red and white clovers, superphosphate, lime and dung and urine on soil composition and on earthworm and grass-grub populations. N.Z. J. Sci. Technol 35A, Supplement 1, 42–52.

Sears, P.D., Goodall, V.C., Jackman, R.H. and Robinson, G.S. 1965 Pasture growth and soil fertility. VII. The influence of grasses, white clovers, fertilisers, and the return of herbage clippings on pasture production of an impoverished soil. N.Z. J. Agric. Res. 8, 270–283.

Simpson, J.R. 1961 The effects of several agricultural treatments on the nitrogen status of a red earth in Uganda. East Afr. Agric. For. J. 26, 156–163.

Simpson, J.R. 1962 Mineral nitrogen fluctuations in soils under improved pasture in southern New South Wales. Aust. J. Agric. Res. 13, 1059–1072.

Simpson, J.R. 1968a Losses of urea nitrogen from the surface of pasture soils. Trans. 9th Int. Congr. Soil Sci. Vol. 2, pp. 459–466.

Simpson, J.R. 1968b Comparison of the efficiencies of several nitrogen fertilisers applied to pasture in autumn and winter. Aust. J. Exp. Agric. Anim. Husb. 8, 301–308.

Simpson, J.R. 1976 Transfer of nitrogen from three pasture legumes under periodic defoliation in a field environment. Aust. J. Exp. Agric. Anim. Husb. 16, 863–870.

Simpson, J.R. 1981 A modelling approach to nitrogen cycling in agro-ecosystems. In: Wetselaar, R., Simpson, J.R. and Rosswall, T. (eds.), Nitrogen cycling in South East Asian Wet Monsoonal Ecosystems, pp. 174–179. Australian Academy of Science, Canberra.

Simpson, J.R. and Stobbs, T.H. 1981 Nitrogen supply and animal production from pastures. In: Morley, F.H.W. (ed.), Grazing Animals, pp. 261–287. Elsevier, Amsterdam.

Simpson, J.R., Bromfield, S.M. and Jones, O.L. 1974 Effects of management on soil fertility under pasture. 3. Changes in total soil nitrogen, carbon, phosphorus and exchangeable cations. Aust. J. Exp. Agric. Anim. Husb. 14, 487–494.

Sinclair, A.G., 1975 Measurement of atmospheric nitrogen fixation in legume-based pasture turfs using the acetylene reduction assay. N.Z. J. Agric. Res. 18, 189–195.

Söderlund, R. and Svensson, B.H. 1976 The global nitrogen cycle. In: Svensson, B.H. and Söderlund, R. (eds.), Nitrogen, Phosphorus and Sulphur-Global Cycles. SCOPE Report 7. Ecol. Bull. 22, 23–73.

Steele, K.W. 1982a Nitrogen in New Zealand grassland soils. In: Lynch, P.B. (ed.), The Place of Nitrogen Fertiliser in New Zealand Agriculture, N.Z. Institute of Agricultural Science (in press).

Steele, K.W. 1982b Quantitative measurements of nitrogen turnover in pasture systems with particular reference to the role of [15]N. FAO/IAEA, Advisory Group Meeting on Nuclear Techniques in Improving Pasture Management Vienna, Austria (in press).

Steele, K.W., Saunders, W.M.H. and Wilson, A.T. 1980a Transformation of ammonium and nitrate fertilisers in two soils of low and high nitrification activity. N.Z. J. Agric. Res. 23, 305–312.

236

Steele, K.W., Wilson, A.T. and Saunders, W.M.H. 1980b Nitrification activity in New Zealand grassland soils N.Z. J. Agric. Res. 23, 249–256.

Stutte, C.A. and Wieland, R.T. 1978 Gaseous nitrogen loss and transpiration of several crop and weed species. Crop. Sci. 18, 887–889.

Tough, H.J. and Crush, J.R. 1979 Effect of grade of acetylene on ethylene production by white clover (*Trifolium repens* L.) during acetylene reduction assays of nitrogen fixation. N.Z. J. Agric. Res. 22, 581–583.

Vallis, I. 1972 Soil nitrogen changes under continuously grazed legume-grass pastures in subtropical coastal Queensland. Aust. J. Exp. Agric. Anim. Husb. 12, 495–501.

Vallis, I., Henzell, E.F. and Evans, R.T. 1977 Uptake of soil nitrogen by legumes in mixed swards. Aust. J. Agric. Res. 28, 413–425.

Vallis, I., Harper, L.A., Catchpoole, V.R. and Weier, K.L. 1982 Volatilization of ammonia from urine patches in a subtropical pasture. Aust. J. Agric. Res. 33, 97–107.

Walker, T.W. and Adams, A.F.R. 1958 Competition for sulphur in a grass-clover association. Plant Soil 9, 353–366.

Walker, T.W., and Adams, A.F.R. 1959 Studies on soil organic matter: 2. Influence of increased leaching at various stages of weathering on levels of carbon, nitrogen, sulfur and organic and total phosphorus. Soil Sci. 87, 1–10.

Walker, T.W., Thapa, B.K. and Adams, A.F.R. 1959 Studies on soil organic matter: 3. Accumulation of carbon, nitrogen, sulphur, organic and total phosphorus in improved grassland soils. Soil Sci. 87, 135–140.

Weeda, W.C. 1970 The effect of fertiliser nitrogen on the production of irrigated pasture with and without clover. N.Z. J. Agric. Res. 13, 896–908.

Weier, K.L. and MacRae, I.C. 1981 Nitrogenase activity in grass-bacteria associations in Northern Australia. In: Gibson, A.H. and Newton, W.E. (eds.), Current Perspectives in Nitrogen Fiscation. p. 491. Australian Academy of Science, Canberra.

Wetselaar, R. and Farquhar, G.D. 1980 Nitrogen losses from tops of plants. Adv. Agron. 33, 263–302.

Wetselaar, R. and Norman, M.J.T. 1960 Recovery of available soil nitrogen by annual fodder crops at Katherine, Northern Territory. Aust. J. Agric. Res. 11, 693–704.

Whitehead, D.C. 1970 The role of nitrogen in grassland productivity. Commonw. Bur. Pastures Field Crops Hurley, Berkshire, Bull. 48, 202 pp.

Williams, C.H. and Steinbergs, A. 1958 Sulphur and phosphorus in some eastern Australian soils. Aust. J. Agric. Res. 9, 483–491.

Woldendorp, J.W. 1962 The quantitative influence of the rhizosphere on denitrification. Plant Soil 17, 267–270.

Woodmansee, R.G., Vallis, I. and Mott, J.J. 1981 Grassland nitrogen. In: Clark, F.E. and Rosswall, T. (eds.), Terrestrial Nitrogen Cycles. Ecol. Bull. 33, 443–462.

10. Fate of fertilizer nitrogen applied to wetland rice

E.T. CRASWELL and P.L.G. VLEK

10.1. Introduction

Dramatic increases in the production and use of fertilizer N in rice-growing countries have occurred during the past decade. In developing Asia, fertilizer N production has increased from 3.6×10^6 t in 1970 to an estimated 18×10^6 t in 1982/83, of which 85% will be urea (Stangel 1979). Expanded use of N fertilizer is being combined with fertilizer-responsive rice varieties and increased areas under irrigation in an all-out effort to produce enough food for the vast Asian population. Unfortunately, N fertilizer, which is already a costly input for the rice farmer, is becoming more expensive in response to the rising cost of the oil-based feedstocks used in fertilizer production. It is therefore imperative that N fertilizer is used efficiently.

Rice utilizes conventionally broadcast N fertilizers very inefficiently. Mitsui (1954) has estimated that rice commonly recovers only 30–40% of applied fertilizer N, whereas upland crops recover 50–60%. The other 60–70% of the N applied to rice is subject to gaseous losses through nitrification-denitrification and ammonia volatilization, or to losses in water through leaching and runoff (see reviews by Broadbent 1978, 1979, Craswell and Vlek 1979a, Patnaik and Rao 1979). Even some of the N absorbed by the rice plant might be lost as a gas from the plant foliage (Wetselaar and Farquhar 1980). Nitrogen losses are not only extremely costly but also may be environmentally detrimental. Further-more, broadcasting fertilizer N may disturb the natural biological systems which fix N_2 in wetland rice soils, causing negative net N balances in wetland rice-soil systems (Watanabe et al. 1981).

Much research has been initiated to develop more efficient N fertilizer products or practices for wetland rice (Prasad and De Datta 1979). Split broadcast applications (incorporated where possible) and the use of slow-release, deep placement, and nitrification inhibitors all have some effect in reducing gaseous losses. However, the relative value of each of these approaches has been shown to vary widely from site to site and season to season in experiments conducted by international networks formed to evaluate improved fertilizers for rice (Yamada et al. 1979, IRRI 1979). Progress in this research is being hampered by the lack of an adequate understanding of the fate of fertilizer N in different wetland rice soils. Such knowledge would provide a rational basis not only for the interpretation of the divergent results from field experiments with improved fertilizers but also for the development of other innovative approaches to improving fertilizer efficiency.

The transformations of fertilizer N in wetland soils are unique because of the influence of a layer of free water above the soil. The floodwater blocks the entry of oxygen into the soil and thus creates reduced conditions in much of the plough layer. Furthermore, fertilizer broadcast on the surface of the soil dissolves in the floodwater which becomes a unique environment for a number of N transformations such as NH_3 volatilization, immobilization, particularly by algae, and nitrification-denitrification. Another particular characteristic of wetland soils is that the floodwater layer exerts a pressure which can promote runoff, seepage, and leaching losses of fertilizer N.

The dominating influence of the water regime on the fate of fertilizer N makes it important to consider briefly the hydrology of rice-growing soils. The hydrology of wetland soils varies widely depending on landscape position, climate, soil characteristics, and the degree of water control (Moorman and Van Breemen 1978). Even the 42% of the world's rice-growing soils which are irrigated are not necessarily kept continuously flooded during the crop's growth, since direct-seeded rice is usually grown with delayed flooding and mid-season drainage is practiced in some Asian countries. In the vast areas of rainfed rice in monsoonal Asia, which comprise 39% of the world's rice area, the water depth may vary from 0 to 3 m depending upon the site and the stage of the monsoon season (Barker and Herdt 1979). However, most of the information reviewed in this chapter was obtained from experiments with good water control; research on the fate of fertilizer nitrogen applied to rainfed rice is therefore needed urgently (De Datta and Craswell, 1980).

This chapter reviews research on the fate of fertilizer N in flooded rice soils with particular emphasis on recent work on gaseous losses of N. The subject has been reviewed previously by Patrick and Mahapatra (1968), Broadbent (1978, 1979), Craswell and Vlek (1979a), and Patnaik et al. (1979). The general chemical and microbiological properties of submerged soils have been described recently by Ponnamperuma (1978) and Watanabe and Furusaka (1980), respectively.

10.2. Transformations of fertilizer nitrogen after application

Shortly after fertilizer N has been applied to flooded soils, a number of transformations are possible, as shown in Fig. 10.1. Some of these reactions, e.g., urea hydrolysis, fuel the processes leading to gaseous N loss while others, such as leaching, runoff, and immobilization, compete with and reduce the magnitude of gaseous emissions.

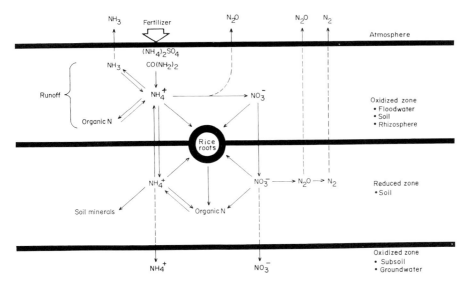

Fig. 10.1. Fate of fertilizer N in wetland soil.

10.2.1. Nitrogen immobilization

Fertilizer N applied to soils enters the soil inorganic N pool and can then undergo biological interchange with the soil organic pool (Jansson 1958, Broadbent 1979). Whether mineralization or immobilization of the soil N predominates depends on the N requirement of the soil heterotrophic microflora, but some of the added fertilizer N invariably is subject to net immobilization. In flooded soils, the extent of this immobilization is less than that in upland soils because anaerobic microbes have a relatively low N requirement for organic matter decomposition (Acharya 1935, Broadbent and Nakashima 1970).

Net immobilization of fertilizer N measured using ^{15}N is strongly influenced by fertilizer placement. Craswell and Vlek (1979b) and Byrnes *et al.* (1979) found that the net immobilization of broadcast and incorporated urea and deep point-placed urea supergranules was 62% and 39%, respectively, of the ^{15}N remaining in the soil: plant system at harvest. Apparently, the volume of soil with which the fertilizer ^{15}N could interchange was much less for the point-placed fertilizer which therefore underwent less immobilization. Besides, surface broadcast urea is particularly subject to immobilization by algae growing in the floodwater (Mitsui 1954). Up to 40% of applied urea-N may be immobilized by algae growing in the floodwater during the first week after fertilizer application (Vlek *et al.* 1980). The algal N would be subsequently remineralized; recent ^{15}N experiments show that 37–52% of added algal N can be mineralized and utilized by a single rice crop (Wilson *et al.* 1980).

Kai and Wada (1979) and Broadbent (1978, 1979) have reviewed available information on the effects of the application of high C:N ratio straw on the immobilization of fertilizer N. Much of the information available has been obtained from laboratory or greenhouse experiments in temperate rice-growing areas. Field research is needed in the tropics using straw in its natural condition.

10.2.2. Runoff and leaching

Physical movement of N away from the site of application in percolating or flowing water is a loss process which competes with gaseous N losses. Nevertheless, the competition may be only temporary and the transported N may be converted to gaseous forms at a later stage.

Surface outflow of water from rice fields is variable, depending on the site, season, degree of water control, and water management. Takamura *et al.* (1977), Singh *et al.* (1978), and Bilal *et al.* (1979) reported runoff losses of 4–16 kg N/ha, 19–30 kg N/ha, and 19–30 kg N/ha under normal fertilization practices in Japan, the Philippines, and California, respectively. Thus, runoff losses of fertilizer N are significant and probably are a major disincentive to fertilizer use by farmers in rainfed rice-growing areas (Patnaik and Rao 1979). These losses are not only costly to the farmer but may also be a serious source of pollution of lakes. Broadcast applications of urea may be particularly susceptible to runoff losses because urea is only weakly adsorbed by the soil.

Few field studies of leaching losses of fertilizer N have been reported, probably because of the difficulty of measuring leaching losses under undisturbed field conditions. Koshino (1975) reported losses of 3–25% of applied fertilizer N in his review of field lysimeter studies conducted between 1928 and 1971 in Japan. Leaching losses depended on the rate of water percolation which is influenced by the soil texture and land preparation. Bilal *et al.* (1979) and Lin *et al.* (1975) reported field leaching losses of 4–30% from ammonium sulfate; whereas Rao (1977) reported 17% loss from urea.

The magnitude of leaching losses depends on the method of fertilizer application. Pande and Adak (1971) showed that losses of split-applied fertilizer were only 11–33%; whereas those of basally applied fertilizer N were 45–65%. Vlek *et al.* (1980) recently showed that deep point-placed urea supergranules are particularly susceptible to leaching losses compared with broadcast urea. This finding explains some of the variation in the agronomic effectiveness of urea supergranules (Tejeda *et al.* 1980).

10.2.3. Urea hydrolysis

Urea hydrolysis is rapid in both flooded and nonflooded soils (Delaune and Patrick 1970) resulting in the formation of ammonium and bicarbonate ions.

$$CO(NH_2)_2 + H^+ + 2H_2O \rightarrow 2NH_4^+ + HCO_3^-$$

The ammoniacal N and alkalinity so produced create an environment which is conducive to NH_3 volatilization. The soil urease activity is located intra-cellularly in the soil biomass and extracellularly in the soil solution or on soil colloids. The contribution of each to the soil urease activity varies with time and soil type, but at least a portion of the soil urease activity is rather constant (Pettit et al. 1976). The fraction of the stable urease in soils depends on the soil's ability to protect urease against microbial degradation or other in-activating processes, which differs from soil to soil (Zantua and Bremner 1977). Factors reportedly affecting the urease activity in soils are temperature (Von Talsky and Klunker 1967, Pettit et al. 1976), soil moisture (Sankhayan and Shukla 1976), soil pH (Kistiakowski and Shaw 1953, Delaune and Patrick 1970), and urea concentration (Von Talsky and Klunker 1967, Ayanaba and Kang 1976).

There have been relatively few studies of the urease activity in flooded soils. The urease activity in flooded soils varies widely from soil to soil, and changes with time in a particular soil (Sahrawat 1980, Vlek et al. 1980). Delaune and Patrick (1970) found the urease activity in the soil to be about 20 times higher than that in the floodwater. Similar results were reported by Sahrawat (1980). Vlek et al. (1980) showed that the urease activity in floodwater was largely associated with suspended colloids or biomass. Using ^{15}N techniques, they showed that urea in the floodwater was rapidly hydrolyzed, presumably at the soil-water interface. Following hydrolysis, a substantial fraction ($\sim 50\%$) of the floodwater N was lost to the atmosphere, while the remaining N was adsorbed by the surface soil or immobilized by algae. Since many fertilizer practices in rice-growing areas cause high urea concentrations in floodwater (Craswell et al. 1981), it is not surprising that NH_3 volatilization is a cause for concern.

10.3. Crop recovery of fertilizer nitrogen

Since plant uptake competes against all of the other transformations illustrated in Fig. 10.1, the effectiveness of the plant in absorbing fertilizer N critically influences the rate of the various loss processes. The rice plant can use ammonium-, nitrate-, and even urea-N (Mitsui and Kurihara 1962, Fried et al. 1965), but the timing of fertilizer N availability is a key factor influencing plant uptake since the rice plant does not have a well-developed root system until

Table 10.1. Recovery by rice of ^{15}N-labelled fertilizers

Type of experiment	Soil or location	Fertilizer Material[a]	Management	^{15}N recovery in plant (%)	References
Pot		AS	Basal	38	Patnaik (1965)
			2/3 planting 1/3 boot	40	
Field	Sri Lanka	AS	Surface	11	Nagarajah and Al-Abbas (1967)
			5 cm deep	20	
Field	Philippines	AS, U	Surface	28	De Datta et al. (1968)
			Deep	68	
			Split	34	
Pot	Philippines	AS, U	Surface	50	De Datta et al. (1968)
			Deep	66	
			Split	47	
Field	Maahas	AS	Basal	35	De Datta et al. (1969)
			Best split	45	
Field	Maligaya	AS	Basal	18	
			Best split	47	
Field	Thailand	AS	Basal	17–42	Koyama et al. (1973)
			Split	40	
			Deep	18	
			Surface	8	
			Nitrification inhibitor	12	
			Flowering	77	
Field	U.S.A.	AS		17–23	Patrick et al. (1974)
Field	India	AS, U	Basal	11	Upadhya et al. (1974)
			Panicle initiation	27	

243

Table 10.1. Continued.

Type of experiment	Soil or location	Fertilizer Material[a]	Management	¹⁵N recovery in plant (%)	References
Field	India	AS, U	Surface	18	Khind and Datta (1975)
			Incorporated	29	
			Deep	38	
			Heading	37	
Field	U.S.A.	AS	Deep	48	Reddy and Patrick (1976)
			Early	38–51	
			Midseason	33	
			Split	35–61	
Field	Japan		Surface	23	Murayama (1977)
			Incorporated	48	
			Basal	46	
			Split	47	
			Deep placement	63	
			Basal	7	
			Heading	49	
Field	China	ABC	Surface	11–31	Li and Chen (1980)
			Incorporated	18–50	
			Deep-point placement	41–79	
Field	Philippines	U	Split	25–34	Savant *et al.* (1982)
			Deep-point placement	50–61	

[a] AS = ammonium sulfate; ABC = ammonium bicarbonate; U = urea

later in its growth (Tanaka *et al.* 1959). Unfortunately, N taken up late in the plant's growth is not very efficiently utilized to increase grain yield.

The isotope ^{15}N has been used extensively to determine the recovery by rice of fertilizer N. Hauck and Bremner (1976) and Jansson (1971) have pointed out the problems of interpreting ^{15}N data on crop fertilizer recovery. Nevertheless, many ^{15}N studies of crop recovery of fertilizer N have been conducted as shown in Table 10.1.

Most of these ^{15}N experiments have been conducted using microplots in the field. The data show that surface applications of fertilizer N at transplanting are poorly recovered by the rice crop (7–38%). Such applications produce high concentrations of soluble N in the floodwater where other N transformations can compete against the plant for uptake. Split or delayed applications produce higher recoveries (35–61% or 18–68%, respectively) by giving the rice plant a better chance to compete. The International Atomic Energy Agency (IAEA 1970, 1978) has coordinated a series of experiments which have demonstrated the advantages of split or delayed application of N at a number of sites throughout Asia.

Most of the data in Table 10.1 have been obtained using ammonium sulfate rather than urea, which is now the main source of N available to rice farmers (Stangel 1979). Recent research by Savant *et al.* (1982) shows that only 25–34% of split-applied urea was recovered by tropical wetland rice, and research reported by IAEA (1970, 1978) shows that urea is generally an inferior source of N to ammonium sulfate. Ammonium bicarbonate, which is widely used in China, showed an average recovery of only 22% which is much lower than recoveries of ammonium sulfate (Li and Chen 1980). While generalizations about the recovery by rice of broadcast fertilizer N could be misleading, the data show that crop responses are often poor.

Incorporation of the broadcast N applications into the soil generally improves the crop recovery, but Table 10.1 shows that deep placement of the fertilizer increases N recovery even further. The effectiveness of deep placement in improving ^{15}N recovery was widely demonstrated by IAEA (1970, 1978). Deep-point placement of ammonium bicarbonate 'pills' (Li and Chen 1980) or urea supergranules (Savant *et al.* 1982) is particularly effective. Savant *et al.* (1982) showed that virtually none of the point-placed urea was absorbed during the first 20–30 days after application; the length of this lag period increased as the lateral and vertical distance from the plant to the placement site increased.

The data on crop recovery of fertilizer N indicate the extent to which the plant was able to compete with various other transformations of N but do not indicate which particular transformation was responsible for low plant recovery. Nevertheless, many authors have interpreted the increased crop recovery due to deep placement as being caused by a reduction in nitrification-denitrification losses. This interpretation does not take into account the evidence reviewed later, that deep placement reduces NH_3 volatilization losses.

Nitrogen can also be lost from the aerial parts of rice plants; Wetselaar and Farquhar (1980), who recently reviewed this topic, quote losses of 47–48 kg N/ha from the tops of rice plants. Heavily fertilized rice crops lost the greatest amount of N, and most of this was lost during the period between anthesis and crop maturity. Tanaka and Navasero (1964) suggested that N was lost due to leaching of the plant by rain and dew, but evidence is mounting that direct gaseous losses can occur as NH_3 and nitrogen oxides that are evolved from the plant foliage (Wetselaar and Farquhar 1980).

10.4. ^{15}N balance in the plant-soil system

The results of experiments in which the recovery of ^{15}N in both the rice plant and the soil was determined and used to calculate the losses of fertilizer N are shown in Table 10.2. In contrast to the crop recoveries shown in Table 10.2, the ^{15}N balance data were largely obtained under greenhouse conditions, possibly because of the difficulty of conducting ^{15}N balance experiments in the field. Since losses are measured indirectly by the ^{15}N balance method, careful techniques of sampling, subsampling, and ^{15}N analysis must be employed in these experiments to avoid apparent losses caused by poor technique. In this regard, it is essential that the ability of the techniques used to recover the applied ^{15}N at zero time be tested experimentally (Craswell and Martin 1975).

Surface broadcast applications of urea, ammonium sulfate, or ammonium bicarbonate at transplanting cause extensive (up to 64%) N losses. Losses of ammonium sulfate (12–40%) were less than losses of urea (19–50%) in the three studies in which direct comparisons of these two sources were made, but losses of broadcast ammonium bicarbonate were the most extensive (64%). These trends implicate NH_3 volatilization losses as a major loss mechanism since the susceptibility of urea and ammonium bicarbonate to NH_3 loss would be greater than that of ammonium sulfate (Vlek and Craswell 1981).

In the only study in which delayed application was compared with basal broadcast application (Yoshida and Padre 1977), losses were considerably reduced by delaying the application. Where losses from fallow soils have been compared with planted soils (Manguiat and Broadbent 1977b, Fillery and Vlek 1982), plant growth has reduced losses considerably, showing the extent to which the plant can compete against the N loss mechanisms.

Deep placement of ammonium and amide fertilizers decreased losses in most experiments. Deep-point placement was particularly effective in reducing losses (to <5%) although no comparisons of deep-point placement with other methods of deep placement have been published. The low losses when fertilizer is deep placed have been attributed to a reduction in nitrification-denitrifi-

Table 10.2. ¹⁵N Balance experiments with rice

Type of experiment	Soil or location	Fertilizer Material[a]	Management	¹⁵N loss (%)	References
Pot	Thailand (pH 4.5) Philippines (pH 6.0) UAR (pH 7.1)	AC	Incorporated	9 29 56	Merzari and Broeshart (1967)
Pot	6 soils	AS	Surface Deep	35 18	Aleksic et al. (1968)
Pot		AS	Broadcast Band Incorporated Split	16 7 13 13	Broadbent and Mikkelsen (1968)
		U	Broadcast Band Incorporated Split	19 12 20 24	
Field tanks	Australia	AS	Surface Deep	51 37	Wetselaar et al. (1973)
		SN	Surface Deep	66 96	
Pot		AS	Continuous flooding Intermittent flooding Percolation	30 39 28	Maeda and Onikura (1976)
Field	U.S.A.	AS	Broadcast Deep	22 25	Patrick and Reddy (1976a)
Pot		AS	Planted Fallow	36 43	Manguiat and Broadbent (1977b)
Pot		AS	Continuous flooding Intermittent flooding	30 30	Manguiat and Broadbent (1977a)

Table 10.2. Continued.

Type of experiment	Soil or location	Fertilizer Material[a]	Management	^{15}N loss (%)	References
Pot	Maahas clay	AS	Broadcast	40	Yoshida and Padre (1977)
			Incorporated	16	
			Delayed	15	
Pot		AS		49	Kudeyarov (1977)
		CN		50	
Field	Japan	AS	Split application	18	Koyama et al. (1977)
		AC		10	
		*AN		20	
		A*N		72	
		*A*N		47	
		U		47	
Pot		AS		12	Craswell and Vlek (1979b)
		U		41	Byrnes et al. (1979)
		SGU		5	
		SCSGU		0	
Field	India	U		49	Krishnappa and Shinde (1980)
Field	China	U		50	Chen and Zhu (1981)
		AS		40	
		ABC		64	
Pot		U	Continuous flooding	16	Fillery and Vlek (1982)
			Intermittent flooding	18	

[a] AC = ammonium chloride, AS = ammonium sulfate, ABC = ammonium bicarbonate, U = urea, SGU = supergranule urea, SCSGU = sulfur-coated supergranule urea (10 cm deep), SN = sodium nitrate, CN = calcium nitrate, and AN = ammonium nitrate
* Indicates position of ^{15}N label

cation (Aleksic *et al.* 1968), but deep placement also reduces NH_3 volatilization losses (Mikkelsen *et al.* 1978, Vlek and Craswell 1979).

The data in Table 10.2 give an indication of the magnitude of fertilizer N losses, but they do not provide information on the mechanisms of the losses. Nevertheless, the main loss mechanisms responsible for the deficits reported in Table 10.2 are probably gaseous losses caused by NH_3 volatilization and nitrification-denitrification because leaching and runoff losses have been either measured or prevented in most of the experiments reported. Maeda and Onikura (1976) measured leaching losses of only 2% in their pot systems. Koyama *et al.* (1977) did, however, suggest that deep percolation, which was neither measured nor prevented in their field study, may have been responsible for part of the losses which they measured.

Further research is needed in which the fate of fertilizer N is traced in the whole soil:plant:water:atmosphere system. Such studies have been reported for other crops such as wheat (Craswell and Martin 1975), but the only experiments with rice were reported by Datta *et al.* (1971) and Mandal and Datta (1975) whose gas lysimeters did not include the atmosphere surrounding the plant so that nitrogenous gases could have been lost through the plant itself (Yoshida and Broadbent 1975, Wetselaar and Farquhar 1980). Nevertheless, Datta *et al.* (1971) recovered their added ^{15}N almost completely in their system; they measured a loss as N_2 of 24% of applied ammonium sulfate.

10.5. Denitrification

The differentiation of oxidized and reduced soil layers in flooded soils was first reported by Pearsall (1938). The thickness of the oxidized layer varies from 0 to 2–3 cm in thickness, depending on the oxygen concentration in the floodwater (Howeler and Bouldin 1971). Phuc *et al.* (1975) showed that the dissolved oxygen content of floodwater varies from 50% of the saturation index at night to as high as 200% during the day when algal photosynthesis is at a maximum. The oxidized rhizosphere of the rice plant has been included with the surface oxidized layer in Fig. 10.1, but the extent and biochemical significance of the rhizosphere is not yet clearly understood (Bouldin 1966). Watanabe and Furusaka (1980), for example, decided that no conclusive evidence is available on the stimulation of the aerobic chemoautotrophic bacteria, such as the nitrifiers, in the oxidized zone of the rhizosphere.

Shioiri (1941) first recognized that ammonium could be oxidized to nitrate in the aerobic surface layer of submerged soils and that the nitrate could diffuse into the underlying anaerobic soil layer and be lost by denitrification. Since this pioneering work, a large volume of literature has been produced on this phenomenon (see reviews by Mitsui 1977, Watanabe and Mitsui 1979).

Attempts have been made to use the knowledge available to synthesize mathematical models of the nitrification-denitrification process. Reddy et al. (1976), for example, concluded from their model that the rates of ammonium diffusion and nitrification determined the rate of N loss. Focht (1979), on the other hand, suggested from his model that oxygen diffusion is the most important factor regulating nitrification-denitrification, although he did not take into account the oxygen produced by algae in the floodwater, which may be very significant under some conditions (Harrison and Aiyer 1913, Phuc et al. 1975).

Because autotrophic micro-organisms are involved, nitrification is sensitive to a number of ecological and environmental factors, particularly the pH and ammonium concentration (Focht 1979). Watanabe and Furusaka (1980) reviewed Japanese evidence showing an enrichment of nitrifying organisms in the surface-oxidized soil layer. Denitrification, on the other hand, involves the diverse heterotrophic microflora which responds largely to changes in temperature and the amount of available organic substrate (MacRae et al. 1968, Burford and Bremner 1975). In the reduced soil layer, the redox potential is usually lower than the -300 mV below which the thermodynamic potential for nitrate reduction is high (Ponnamperuma 1978). Ottow (1981) has, however, recently shown that a low redox potential is not sufficient condition for nitrate reduction and that the enzyme nitrate reductase must be present for denitrification to occur.

Much of the evidence on the significance of the denitrification process is circumstantial since the gaseous products of denitrification have not been measured directly in most experiments. Thus, losses have often been attributed to denitrification when other loss mechanisms, in particular NH_3 volatilization, may have been responsible. The main gaseous products of nitrification-denitrification are N_2 and N_2O. Emissions of N_2 are extremely difficult to measure against a background of 78% N_2 in the atmosphere. Nevertheless, the production of N_2 can be measured using ^{15}N-labelled fertilizer or by using an artificial atmosphere such as helium.

Broadbent and Tusneem (1971), who measured the $^{15}N_2$ produced from labelled ammonium, demonstrated under controlled conditions that oxygen must be present for ammonium to be lost through nitrification-denitrification. Kosuge (1979) measured N_2 losses of 12 kg N/ha in ^{15}N experiments in which 96% of the added ^{15}N was recovered in the soil, plant, and atmosphere. Patrick and Reddy (1976b) measured denitrification directly as N_2 and N_2O in laboratory incubations. Only small quantities of N_2O were produced, possibly because the N_2O was reabsorbed and further reduced to N_2 in the closed systems which Patrick and Reddy used; Garcia (1975) has demonstrated that submerged soils have a high potential to reduce N_2O to N_2.

Focht (1979) stated that wetland soils seldom evolve any N_2O. However,

Denmead *et al.* (1979) recently found that some N_2O was evolved from a rice-growing soil during the denitrification of nitrate which had accumulated during the fallow period before flooding. When nitrate was added later in the cropping season, no N_2O was evolved unless an extraneous energy source was added to the soil. More recently, Freney *et al.* (1981) detected N_2O evolution which commenced 32 hours after ammonium sulfate was broadcast on the surface of a wetland soil. The evolution continued for 4 days but amounted to a total of only 0.1% of the fertilizer applied. Craswell and Hartantyo (unpublished data) subsequently showed that N_2O fluxes were higher from broadcast urea than from broadcast ammonium sulfate but that the fluxes were small and never exceeded 100 ng $N/m^2/sec$.

The N_2O fluxes measured after broadcast applications of ammonium or amide fertilizers demonstrate that nitrification-denitrification occurs in the wetland soils in the field. However, whether the N_2O was emitted during the nitrification step, as has been shown with aerobic soils (Bremner and Blackmer 1978, Freney *et al.* 1978), or during denitrification is uncertain. Furthermore, the N_2O data do not indicate the overall significance of losses due to nitrification-denitrification since N_2, the major product of the process, was not measured. Measurements of N_2 fluxes from denitrification under field conditions are urgently needed.

Nitrification-denitrification has been thought to be especially important in soils subject to intermittent flooding; nitrate formed by nitrification during the aerobic phase is subject to denitrification during the flooded or anaerobic phase of the cycle (Patrick and Wyatt 1964). Ammonium N added to fallow soils before flooding seems to be particularly susceptible to loss by this mechanism, and losses of up to 63% have been measured from soils undergoing short but frequent cycles of wetting and drying in the laboratory (Reddy and Patrick 1975). The potential for loss by this mechanism can be estimated by measuring the concentration of nitrate in the soil at the end of the aerobic phase of the cycle. Using this approach, Ponnamperuma (1978) estimated that an average of 26 kg N/ha would be lost from wetland rice soils in the Philippines when nitrate N, accumulated in the soil between the harvesting and replanting of rice crops, is lost during subsequent flooding.

While nitrification and subsequent denitrification seem to be important in fallow soils, a number of greenhouse studies have suggested that this may not be an important loss mechanism in soils planted to rice (Manguiat and Broadbent 1977a, Craswell and Vlek 1979b, Fillery and Vlek 1982). Fillery and Vlek (1982) employed a wide range of intermittent flooding times and durations but found negligible losses and concluded that the rice plant could compete effectively for inorganic N in the soil, thus reducing losses. Data from one of their experiments are summarized in Table 10.3. Field research on the practical significance of this loss mechanism is presently underway to verify the

Table 10.3. Recovery of ^{15}N from a silt loam after incorporation of urea as influenced by water regime and presence of plants

Water regime	Treatment	Total recovery of ^{15}N labelled urea (%)	Plant recovery of ^{15}N labelled urea (%)
Continuously	Planted	92	49
flooded	Fallow	61	—
Intermittently	Planted	91	42
flooded[a]	Fallow	53	—

[a] Pots were flooded for 10 days after transplanting (DAT) and fertilizer application. A 10 day drying period was imposed 14 DAT; the pots were then reflooded. A second 10 day drying period followed 36 DAT. Each pot was flooded to 3 cm between drying periods and before harvest 60 DAT
Source: Fillery and Vlek 1982

relevance of these findings to the fate of fertilizer N applied to rainfed rice which is grown over vast areas of South and Southeast Asia.

10.6. Ammonia volatilization

The various aspects of N transformations in paddy soils have been studied at least for half a century (Harrison 1930, Harrison and Aiyer 1913). Since that time the issue of the relative importance of NH_3 volatilization has been controversial. A number of Indian scientists maintained that NH_3 volatilization was a major component of the loss of N from flooded soil (Sreenivasan and Subrahmanyan 1935, Gupta 1955). Reports from Japan by Iwata and Okuda (1937) and Mitsui *et al.* (1954) claimed that, with the exception of strongly alkaline soils (pH >8.5), NH_3 volatilization was negligible. More recent research on NH_3 volatilization generally tends to downplay the importance of NH_3 loss from flooded soils (Okuda *et al.* 1960, Delaune and Patrick 1970, MacRae and Ancajas 1970, Ventura and Yoshida 1977, Wetselaar *et al.* 1977, Mikkelsen *et al.* 1978, Freney *et al.* 1981). In contrast, the findings by Ernst and Massey (1960), Bouldin and Alimagno (1976), Vlek and Craswell (1979), and Sahrawat (1980) suggest that NH_3 loss from flooded soil can be significant.

It would be futile to attempt to tabulate and compare the reported losses of NH_3 due to the difficulty in assessing the validity of the measurements. Moreover, even for valid measurements it remains difficult to determine for what conditions such data can be considered representative.

252

10.6.1. Floodwater chemistry

The main problems in comparing results from volatilization studies are the complexity of the process and the multitude of factors influencing it. The many factors that influence NH_3 volatilization from flooded soils are environmental, chemical, and biological in nature. The NH_3 volatilization process can be regarded as a chain of events, the overall rate of which can be controlled at any point in the chain.

Fig. 10.2 schematically represents the process and some of the key factors affecting the magnitude of various N fluxes. Floodwater ammonium may originate from the soil N pool or from extraneous sources such as fertilizer, organic amendments, and crop residues. The presence of algae may increase or decrease the floodwater ammonium level, dependent on whether the algal biomass is decaying or growing (Vlek *et al.* 1980).

It is widely recognized that the concentration of ammoniacal N in the floodwater depends on the nature of the soil (particularly texture and permeability), fertilizer rate, source, and mode of application, and floodwater depth (Mikkelsen *et al.* 1978, Vlek and Craswell 1979). Whether incorporated or not, a large fraction of surface-applied N is generally found in the floodwater. One way to prevent these high ammoniacal N levels from developing is by deep placement of the fertilizer in the puddled soil.

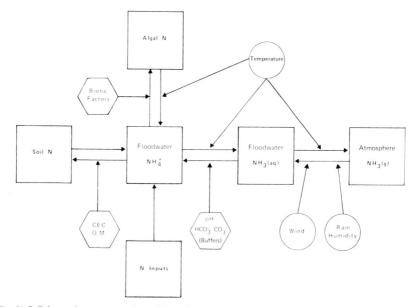

Fig. 10.2. Schematic representation of the origin and fate of ammonium in floodwater and some of the key factors influencing its transformation (Vlek and Craswell 1981).

The chemical dynamics of NH_3 loss from floodwater was described by Vlek and Stumpe (1978) as a consecutive reaction with an opposing step while NH_3 volatilization *per se* follows first-order reaction kinetics (Folkman and Wachs 1973, Moeller and Vlek 1982). The rate of NH_3 volatilization is thus directly related to the concentration of aqueous NH_3 [NH_3 (aq)] and therefore to the concentration of ammoniacal N, pH, and temperature of the floodwater, regardless of the source of ammonium ions.

Mikkelsen and De Datta (1979) derived a complicated set of equations and concluded that the pressure of NH_3 in solution is a function of pH, the concentrations of ammonium, bicarbonate, and carbonate, and the pressure of CO_2 in solution. However, as is shown below, the partial pressure of NH_3 in solution, $p(NH_3)$, is in fact uniquely determined by the ammoniacal N concentration (AN) and pH of that solution at a particular temperature. The ammoniacal N concentration in solution is equal to the sum of the various species:

$$AN = NH_4^+ + NH_3(aq) + NH_4L \tag{1}$$

Ignoring ion-pair formation (NH_4L) which is insignificant at the pH levels commonly found in floodwater (Vlek and Craswell 1981) and activity corrections and substituting

$$\frac{NH_3(aq) \cdot H^+}{NH_4^+} = K \ , \ pK^\circ = 9.24, \ \text{(for T = 25}^\circ \text{C)} \tag{2}$$

yields:

$$NH_3(aq) = K \frac{[AN - NH_3(aq)]}{H^+} = \frac{K \cdot AN}{H^+ + K} \tag{3}$$

and thus

$$p(NH_3) = \frac{NH_3(aq)}{K_H} = \frac{K \cdot AN}{K_H(H^+ + K)}, \ pK^\circ_H = 1.77 \tag{4}$$

Conversion of ammonium ions to aqueous NH_3 is an extremely rapid (first-order) process with a rate constant of 24.6/sec (Emerson *et al.* 1960) and is thus rarely limiting to the overall volatilization process. The concentration of aqueous NH_3 changes in direct proportion with ammoniacal N and increases about tenfold per unit increase in pH up to pH ≈ 9 (Vlek and Stumpe 1978). At a given total ammoniacal N concentration and pH, the aqueous NH_3 concentration increases approximately linearly with increasing temperature (Beutier and Renon 1978), resulting in a nearly fourfold increase with a change in temperature from 10° to 40° C.

Following the hydrolysis of urea at the soil-water interface, the floodwater receives some of the ammonium and carbonate which is not adsorbed by the soil and becomes a weak ammonium bicarbonate solution. Because of its inherent buffering capacity, this solution will tend to maintain a pH of about 8. The dilute ammonium bicarbonate solution is an ideal growth medium for green algae, though generally algal development is delayed for a few days. The extent of algal biomass production will depend on soil characteristics and nutrient availability as well as climatic factors such as temperature and light. Algal biomass production in Senegal paddy fields was found to reach as high as 5 t/ha of wet material (Reynaud and Roger 1978), whereas, Watanabe *et al.* (1977) report algal biomass production in excess of 20 t/ha. Photosynthetic and respiratory activities of such algal populations can cause significant diurnal fluctuation in the floodwater CO_2 level and, consequently, in the floodwater pH (Bouldin and Alimagno 1976).

In order that NH_3 volatilization can proceed buffering substances in the form of alkalinity must be present in the system to neutralize the production of H^+ ions that accompanies the production of aqueous NH_3 from ammonium (Vlek and Stumpe 1978). This is represented by equation (5).

$$NH_4^+ \rightarrow NH_3 + H^+ \tag{5}$$

The equilibrium constant is such that appreciable amounts of NH_3 can be formed only at pH >7.5. In order to keep the pH high enough for volatilization to continue, the protons must be removed and the solution buffered at high pH. The only proton acceptor capable of that and present in appreciable quantities in the surface water of rice fields is bicarbonate, HCO_3^-:

$$HCO_3^- + H^+ \rightarrow H_2O + CO_2. \tag{6}$$

Combining reactions (5) and (6) gives

$$NH_4^+ + HCO_3^- \rightarrow NH_3 + H_2O + CO_2. \tag{7}$$

The stoichiometry of reaction (7) is such that equivalent amounts of ammonium and bicarbonate are consumed in the reaction. Thus, the amount of bicarbonate in the surface water sets an upper limit to the amount of ammonium that can possibly be volatilized.

The world's average river water contains about 1 meq HCO_3^-/l while rainwater contains virtually none. Consequently, rainfed paddy water exhibits essentially no buffering capacity; large fluctuations in pH result even at relatively low algal activity. On the other hand, approximately 0.1 meq C/l needs to be photosynthesized by algae to raise the pH of river water from 7.5 to 9.0. To accomplish the same pH change in floodwater derived from river water but fertilized with ammonium sulfate or urea will require immobilization of substantial quantities of carbon (Vlek and Craswell 1981), particularly for urea which serves as an effective buffer following hydrolysis. Extraneous sources of CO_2 such as alkaline irrigation water or $CaCO_3$ in the underlying topsoil may also prevent large diurnal fluctuations in floodwater pH due to algae.

In general, NH_3 volatilization from urea-fertilized soils far exceeds that from ammonium sulfate-fertilized soils (Vlek and Craswell 1979) because urea, upon hydrolysis, creates a solution ideally suited for volatilization; i.e., the solution has a high alkalinity and pH (Vlek and Stumpe 1978). Ammonia volatilization losses from ammonium sulfate will be substantial only if the soil-floodwater system can provide the alkalinity necessary to buffer the system at a high enough pH to sustain the volatilization process. Such conditions may be found in areas with calcareous soils or where soils are irrigated with alkaline well waters (e.g., the International Rice Research Institute [IRRI] farm).

10.6.2. Environmental conditions

From a chemical standpoint, NH_3 volatilization is thus largely determined by the floodwater dynamics with respect to pH, alkalinity, and ammoniacal N. The net result of these three factors on the dynamics of aqueous NH_3 in the floodwater can be considered the NH_3 volatilization potential, while actual NH_3 volatilization loss rates will further be determined by environmental factors such as temperature, wind speed, and rainfall conditions. Some of these factors were studied and quantified for water impoundments of non-agricultural nature (Stratton and Asce 1969, Folkman and Wachs 1973, Bouldin and Alimagno 1976, Weiler 1979). The effects of changing ammoniacal N concentration, pH, temperature, and wind speed were evaluated simultaneously using an NH_3 diffusion model (Bouwmeester and Vlek 1981a).

The importance of wind on the rate of NH_3 volatilization has often been ignored by agricultural scientists and has led to a general underestimation of this loss mechanism. The importance of air movement in the NH_3 volatilization process is easily appreciated from simple calculations, as demonstrated by Vlek and Craswell (1981). Such calculations show that if NH_3 volatilization measurements are made under enclosures at low flow rates, the rate of NH_3 loss is controlled by the rate of airflow through the enclosures. On the other hand, if flow rates are high, the rate of NH_3 loss is no longer affected by the rate of airflow, but becomes entirely dependent on the rate of efflux of NH_3 from the liquid phase. Unfortunately, this phenomenon has not been recognized by many scientists in the past and has led to drastic underestimation of the magnitude of NH_3 loss due to inadequate airflow (Watkins et al. 1972, Kissel et al. 1977, Vlek and Stumpe 1978).

In a more detailed analysis, the NH_3 transfer from paddy water to the atmosphere was found to be a diffusion-controlled process (Bouwmeester and Vlek 1981a). The diffusion model developed by Bouwmeester and Vlek (1981a) distinguishes the various rate-controlling factors: (a) the reaction rate of equation (5), (b) the transfer resistance in the liquid phase, and (c) the transfer resistance in the gas phase. For low windspeeds the volatilization rates are very small, and the gas phase resistance dominates. However, with increasing windspeed the volatilization rates increase, and the liquid phase resistance becomes more significant due to depletion of NH_3 in the surface film of the liquid phase. This shift from gas phase resistance to liquid phase resistance is more evident for high pH-values.

Finally, the dimensions of the rice paddy may influence the volatilization rate of NH_3 from the floodwater, although the effect of fetch was found to be of little consequence within the size range of most rice paddies (Bouwmeester and Vlek 1981a).

10.6.3. Measurements of ammonia volatilization

Techniques for measuring fertilizer losses of NH_3 are manifold and controversial, varying from closed systems with or without air turbulence to micrometeorological and aerodynamic techniques in undisturbed fields. To some extent, the differences in methodology employed can be explained by the differences in the objectives of the measurement programs. For instance, for the purpose of studying actual losses of ammonia from the field, micrometeorological and aerodynamic techniques are preferable (Denmead and McIlroy 1971, Denmead et al. 1977). On the other hand, for studying the effect of environmental and N management factors on NH_3 volatilization rates, wind tunnel techniques are most appropriate (Bouwmeester and Vlek 1981a, b). To assess the effect of fertilizer management on the *potential* for NH_3 volatilization, measurements under small enclosures can be employed provided the system is set up so that the gas phase resistance does not dominate the volatilization process (Craswell and Vlek 1979a).

Each of the methodologies has merits and demerits. All these techniques provide information only for the place and time of the experiment. Due to the stochastic character of the environmental factors affecting NH_3 loss, such measurements are not easily extrapolated to different environmental conditions. The disturbance of the natural conditions under enclosures, however, causes these measurements to be inaccurate unless care is taken to simulate the gas phase resistance of the undisturbed field throughout the measurement (Bouwmeester and Vlek 1981b). Generally, the rate of airflow in such systems has been low, thus the rate of volatilization is limited (Vlek and Craswell 1981). When airflow rates under enclosures are high, the measurement may, at best, indicate the upper limit of volatilization.

The first attempts to assess NH_3 losses from flooded soils without disturbing the atmospheric environment were probably made by Bouldin and Alimagno (1976) at IRRI. The main approach used in their research was based on the disappearance of mineral N from small plots (20×20 cm) isolated from the bulk paddy by metal sleeves pushed into the soil down to the plowsole. Four such plots were sampled daily down to the plowsole, and the mineral N unaccounted for was assumed lost as NH_3. To further verify the loss mechanism, a similar system was also maintained undercover throughout the study period of 5–10 days; thus, NH_3 volatilization was prevented. Ammonia volatilization losses estimated by these methods appeared alarmingly high (up to 60% of broadcast ammonium sulfate).

Only very recently have efforts been made to apply micrometeorological measurement techniques to flooded soils (Freney et al. 1981). The various micrometeorological methods have been adequately described elsewhere (Denmead and McIlroy 1971). Bouwmeester and Vlek (1981b) evaluated aerodynamic methods of calculating ammonia volatilization rates with the so-called longitudinal flux method as the basis of comparison. The studies were carried out under simulated paddy conditions in a wind-water tunnel. They found that the longitudinal flux method (Denmead et al. 1977) and the aerodynamic methods were in close agreement and suitable for flooded soils. Freney et al. (1981) used a micrometeorological technique to calculate losses from ammonium sulfate applied to fields at IRRI. They reported losses of no more than 10.6% of the applied N, a drastic difference from the magnitude of loss reported for the same fields by Bouldin and Alimagno (1976).

The discrepancy between the two reports may be due to differences in weather conditions during the experiments. Freney et al. (1981) measured losses in the wet season, whereas Bouldin and Alimagno (1976) worked in the dry season. In the wet season, the solar radiation is less, the wind is more variable, and the usage of irrigation water, which at IRRI is alkaline, is less than in the dry season. An enhancement of air turbulence above the microplots used by Bouldin and Alimagno might have been caused by the metal walls protruding from the water surface. This enhanced air turbulence may have resulted in an overestimation of the volatilization rate and, thus, could partially explain this discrepancy. Obviously, further work comparing these methods is warranted to clarify this point. Micrometeorological measurements of NH_3 losses from urea are also needed. These measurements should be accompanied by measurements of changes in the floodwater ammonium content, pH, and temperature since Vlek and Craswell (1979) found that under controlled conditions a good correlation existed between volatilization losses and the aqueous NH_3 concentration of the floodwater sampled between 1000 and 1100 hours each day.

The wide variation in seasonal and agroclimatic conditions under which rice is grown may cause wide variation in measured NH_3 loss. A proper assessment of the relative importance of NH_3 volatilization as compared to other loss mechanisms will require an extensive as well as intensive field measurement program. A single field measurement of the loss of fertilizer N through NH_3 volatilization is therefore inadequate for the assessment of the magnitude of the problem. Hopefully, a relatively simple measurement technique will eventually be devised which, after proper calibration against a micrometeorological technique, can be used for extensive measurements under a wide range of conditions.

Until further field experiments are conducted on loss processes wherein denitrification and NH_3 volatilization are measured directly, the enigma as to which are the important mechanisms of fertilizer N loss from flooded soils will remain unresolved.

10.7. References

Acharya, C.N. 1935 Studies on the anaerobic decomposition of plant materials. III. Comparison of the course of decomposition under anaerobic, aerobic and partially aerobic conditions. Biochem. J. 29, 1116–1120.

Aleksic, Z., Broeshart, H. and Middelboe, V. 1968 Shallow depth placement of $(NH_4)_2SO_4$ in submerged rice soils as related to gaseous losses of fertilizer nitrogen and fertilizer efficiency. Plant Soil 29, 338–342.

Ayanaba, A. and Kang, B.T. 1976 Urea transformation in some tropical soils. Soil Biol. Biochem. 8, 313–316.

Barker, R. and Herdt, R.W. 1979 Rainfed lowland rice as a research priority – An economist's view. IRRI Res. Pap. Ser. 26, 1–50.

Beutier, D. and Renon, H. 1978 Representation of ammonia-hydrogen sulfide-water, ammonia-carbon dioxide-water, and ammonia-sulfur dioxide-water, vapor-liquid equilibriums. Ind. Eng. Chem. Proc., Design and Development 17, 220–230.

Bilal, I.M., Henderson, D.W. and Tanji, K.K. 1979 N Losses in irrigation return flows from flooded rice plots fertilized with ammonium sulfate. Agron. J. 71, 279–284.

Bouldin, D.R. 1966 Speculation on the oxidation-reduction status of the rhizosphere of rice roots in submerged soils. FAO/IAEA Tech. Rep. 65, 128–139.

Bouldin, D.R. and Alimagno, B.V. 1976 NH_3 volatilization losses from IRRI paddies following broadcast applications of fertilizer nitrogen (Terminal report of Bouldin, D.R., as visiting scientist at IRRI).

Bouwmeester, R.J.B. and Vlek, P.L.G. 1981a Rate control of ammonia volatilization from rice paddies. Atmos. Environ. 15, 131–140.

Bouwmeester, R.J.B. and Vlek, P.L.G. 1981b Wind-tunnel simulation and assessment of ammonia volatilization from ponded water. Agron. J. 73, 546–552.

Bremner, J.M. and Blackmer, A.M. 1978 Nitrous oxide: emission during nitrification of fertilizer nitrogen. Science (Wash. D.C.) 199, 295–296.

Broadbent, F.E. 1978 Nitrogen transformation in flooded soils. In: Soils and Rice, pp. 543–559. IRRI, Los Baños.

Broadbent, F.E. 1979 Mineralization of organic nitrogen in paddy soils. In: Nitrogen and Rice, pp. 105–118. IRRI, Los Baños.

Broadbent, F.E. and Mikkelsen, D.S. 1968 Influence of placement on uptake of tagged nitrogen by rice. Agron. J. 60, 674–677.

Broadbent, F.E. and Nakashima, T. 1970 Nitrogen immobilization in flooded soils. Soil Sci. Soc. Am. Proc. 34, 218–221.

Broadbent, F.E. and Tusneem, M.E. 1971 Losses of nitrogen from some flooded soils in tracer experiments. Soil Sci. Soc. Am. J. 35, 922–926.

Burford, J.R. and Bremner, J.M. 1975 Relationships between the denitrification capacities of soils and total water-soluble and readily-decomposable soil organic matter. Soil Biol. Biochem. 7, 389–394.

Byrnes, B.H., Vlek, P.L.G. and Craswell, E.T. 1979 The promise and problems of urea supergranules for rice production. In: Proceedings Final Review Meeting INPUTS Project, pp. 75–83. East-West Center, Honolulu.

Chen, R.Y. and Zhu, Z.L. 1981 The fate of fertilizer nitrogen in paddy soils. In: Proceedings Symposium on Paddy Soil, pp. 597–602. Institute of Soil Science, Academia Sinica, Nanjing.

Craswell, E.T. and Martin, A.E. 1975 Isotope studies of the nitrogen balance in a cracking clay. I. Recovery of added nitrogen from soil and wheat in the glasshouse and gas lysimeter. Aust. J. Soil Res. 13, 43–52.

Craswell, E.T. and Vlek, P.L.G. 1979a Fate of fertilizer nitrogen applied to wetland rice. In: Nitrogen and Rice, pp. 175–192. IRRI, Los Baños.

Craswell, E.T. and Vlek, P.L.G. 1979b Greenhouse evaluation of nitrogen fertilizers for rice. Soil Sci. Soc. Am. J. 44, 1184–1188.

Craswell, E.T., De Datta, S.K., Obcemea, W.N. and Hartantyo, M. 1981 Time and mode of nitrogen fertilizer application to tropical wetland rice. Fert. Res. 2, 247–259.

Datta, N.P., Banerjee, N.K. and Prasad Rao, D.M.V. 1971 A new technique for study of nitrogen balance sheet and an evaluation of nitrophosphate using ^{15}N under submerged conditions of growing paddy. In: International Symposium on Soil Fertility Evaluation. Vol. 1, pp. 631–638. Indian Society of Soil Science, New Delhi.

De Datta, S.K. and Craswell, E.T. 1980 Nitrogen fertility and fertilizer management. In: Rice Research Strategies for the Future, pp. 671–701. IRRI, Los Baños.

De Datta, S.K., Magnaye, C.P. and Magbanua, J.T. 1969 Response of rice varieties to time of nitrogen application in the tropics. In: Proceedings of the Symposium on Tropical Agricultural Research, pp. 73–87. Tokyo.

De Datta, S.K., Magnaye, C.P. and Moomaw, J.C. 1968. Efficiency of fertilizer nitrogen (^{15}N-labelled) for flooded rice. 9th Int. Congr. Soil Sci. Trans. 4, 67–76.

Delaune, R.D. and Patrick, W.H., Jr. 1970 Urea conversion to ammonium in waterlogged soils. Soil Sci. Soc. Am. Proc. 34, 603–607.

Denmead, O.T. and McIlroy, I.C. 1971 Measurement of carbon dioxide exchange in the field. In: Sestak, Z., Catsky, J. and Jarvis, P.G. (eds.), Plant Photosynthetic Production, Manual of Methods, pp. 467–516. W. Junk, The Hague.

Denmead, O.T., Freney, J.R. and Simpson, J.R. 1979 Nitrous oxide emission during denitrification in a flooded field. Soil Sci. Soc. Am. J. 43, 716–718.

Denmead, O.T., Simpson, J.R. and Freney, J.R. 1977 The direct field measurement of ammonia emission after injection of anhydrous ammonia. Soil Sci. Soc. Am. J. 41, 1001–1004.

Emerson, M.T., Grunwald, E. and Kromhout, R.A. 1960 Diffusion control in reaction of ammonium ion in aqueous acid. J. Chem. Phys. 83, 547–555.

Ernst, J.W. and Massey, H.F. 1960 The effect of several factors on volatilization of ammonia formed from urea in the soil. Soil Sci. Soc. Am. Proc. 24, 87–90.

Fillery, I.R.P. and Vlek, P.L.G. 1982 The significance of denitrification of applied nitrogen in fallow and cropped rice soils under different flooding regimes. Plant Soil 65, 153–169.

Focht, D.D. 1979 Microbial kinetics of nitrogen losses in flooded soils. In: Nitrogen and Rice, pp. 119–134. IRRI, Los Baños.

Folkman, Y. and Wachs, A.M. 1973 Nitrogen removal through ammonia release from holding ponds. In: Jenkins, S.H. (ed.), Advances in Water Pollution Research. Pergamon Press, New York.

Freney, J.R., Denmead, O.T. and Simpson, J.R. 1978 Soil as a source or sink for atmospheric nitrous oxide. Nature (Lond.) 273, 530–532.

Freney, J.R., Denmead, O.T., Watanabe, I. and Craswell, E.T. 1981 Ammonia and nitrous oxide losses following applications of ammonium sulphate to flooded rice. Aust. J. Agric. Res. 32, 37–45.

Fried, M., Zsoldos, F., Vose, P.B. and Shatokhin, I.L. 1965 Characterizing the nitrate and ammonium uptake process of rice roots by use of ^{15}N-labelled ammonium nitrate. Physiol. Plant 18, 313–320.

Garcia, J.L. 1975 Evaluation de la denitrification dans le rizieres par la methode de reduction de N_2O. Soil Biol. Biochem. 7, 251–256.

Gupta, S.P. 1955 Loss of nitrogen in the form of ammonia from waterlogged paddy soil. J. Indian Soc. Soil Sci. 3, 29–32.

Harrison, W.H. 1930 The principles of heavy manuring. Jour. Bd. Agr. Brit. Guyana 6, 37–40, 71–77.

Harrison, W.H. and Aiyer, P.A.S. 1913 The gasses of swamp rice soils: Their composition and relationship to the crop. Mem. Dept. Agric. India, Chem. Ser. III. 3, 65–106.

Hauck, R.D. and Bremner, J.M. 1976 Use of tracers for soil and fertilizer nitrogen research. Adv. Agron. 28, 219–266.

Howeler, R.H. and Bouldin, D.R. 1971 The diffusion and consumption of oxygen in submerged soils. Soil Sci. Soc. Am. Proc. 35, 202–208.

IAEA. 1970 Rice Fertilization: A six-year isotope study on nitrogen and phosphorus fertilizer utilization. IAEA Tech. Rep. Ser. 108, 1–184.

IAEA. 1978 Isotope Studies on Rice Fertilization. IAEA Tech. Rep. Ser. 181, 1–131.

IRRI. 1979 Summary report on the first and second international trials on nitrogen fertilizer efficiency in rice. 1975–1978. International Network on Fertilizer Efficiency in Rice (INFER), Los Baños.

Iwata, T. and Okuda, A. 1937 Ammonia volatilization from soil (in Japanese, English summary). J. Sci. Soil Manure (Japan) 11, 185–187.

Jansson, S.L. 1958 Tracer studies on nitrogen transformations in soil with special attention to mineralization-immobilization relationships. K. Lantbrukhogsk Ann. 24, 101–361.

Jansson, S.L. 1971 Use of ^{15}N in studies of soil nitrogen. In: McLaren, A.D. and Skujins, J. (eds.), Soil Biochemistry. Vol. 2, pp. 129–166. Marcel Dekker, New York.

Jenkinson, D.S. 1976 The effects of biocidal treatments on metabolism in soil IV. The decomposition of fumigated organisms in soil. Soil Biol. Biochem. 8, 203–208.

Kai, H. and Wada, K. 1979 Chemical and biological immobilization of nitrogen in paddy soils. In: Nitrogen and Rice, pp. 157–174. IRRI, Los Baños.

Khind, C.S. and Datta, N.P. 1975 Effect of method and timing of nitrogen application on yield and fertilizer nitrogen utilization by lowland rice. J. Indian Soc. Soil Sci. 23, 442–446.

Kissel, D.E., Brewer, H.L. and Arkin, G.F. 1977 Design and test of a field sampler for ammonia volatilization. Soil Sci. Soc. Am. J. 41, 1133–1138.

Kistiakowski, G.B. and Shaw, W.H.R. 1953 On the mechanism of the inhibition of urease. J. Am. Chem. Soc. 75, 866–871.

Koshino, M. 1975 Incoming and outgoing fertilizer nutrients in cropped lands. In: Proc. of the Int. Con. of Scientists on the Human Environment. Kyoto, Japan, November 17–26, 1975.

Kosuge, N. 1979 Denitrification in paddy soils, Nogyo Oyobi Engei 54, 495–500.

Koyama, T., Chamnek, C. and Niamrischand, N. 1973 Nitrogen application technology for tropical rice as determined by field experiments using nitrogen-15 tracer technique. Japan Tech. Bull. 3, 1–79. Trop. Agri. Res. Cent., Tokyo.

Koyama, T., Shibuya, M. Tokuyasu, T., Shimomura, T. and Ide, K. 1977 Balance sheet and residual effects of fertilizer nitrogen in Saga paddy field. In: Proceedings of the International Seminar on Soil Environment and Fertility Management in Intensive Agriculture (SEFMIA), pp. 289–296. Society of the Science of Soil and Manure, Tokyo.

Krishnappa, A.M. and Shinde, J.E. 1980 Fate of 15N-labelled urea fertilizer under conditions of tropical flooded-rice culture. In: Soil Nitrogen as Fertilizer or Pollutant, pp. 127–144. IAEA, Vienna.

Kudeyarov, V.N. 1977 The fate of nitrogen fertilizers in paddy soils and recovery of nitrogen by rice. In: Proceedings of the International Seminar on Soil Environment and Fertility Management in Intensive Agriculture (SEFMIA), pp. 269–274. Society of the Science of Soil and Manure, Tokyo.

Li, C.K. and Chen, R.Y. 1980 Ammonium bicarbonate used as a nitrogen fertilizer in China. Fert. Res. 1, 125–136.

Lin, P.L., Chen, C.T. and Liao, C.S. 1975 A lysimeter experiment investigation: the effects of applying various fertilizers on the losses of plant nutrients and certain properties of the rice paddy soil (in Chinese). Mem. Coll. Agric. Natl. Taiwan Univ. 4, 51–68.

MacRae, I.C. and Ancajas, R. 1970 Volatilization of ammonia from submerged tropical soils. Plant Soil 33, 97–103.

MacRae, I.C., Ancajas, R.R. and Salandanan, S. 1968 The fate of nitrate in some tropical soils following submergence. Soil Sci. 105, 327–334.

Maeda, K. and Onikura, Y. 1976 Fertilizer nitrogen balance in soil variously irrigated in a pot experiment with rice plants (in Japanese). J. Sci. Soil Manure (Japan) 47, 99–105.

Mandal, S.R. and Datta, N.P. 1975 Direct quantitative estimation of nitrogen gas loss under submerged rice crops – a study with ^{15}nitrogen. Indian Agric. 19, 127–134.

Manguiat, I.J. and Broadbent, F.E. 1977a Recoveries of tagged N (15N-labelled) under some management practices for lowland rice. Philipp. Agric. 60, 367–377.

Manguiat, I.J. and Broadbent, F.E. 1977b ^{15}N studies on nitrogen losses and transformations of residual nitrogen in a flooded soil-plant system. Philipp. Agric. 60, 354–366.

Merzari, A.H. and Broeshart, H. 1967 The utilization by rice of nitrogen from ammonium fertilizers as affected by fertilizer placement and microbiological activity. In: International Atomic Energy Agency Isotope Studies of the Nitrogen Chain, pp. 79–87. IAEA/FAO, Vienna.

Mikkelsen, D.S. and De Datta, S.K. 1979 Ammonia volatilization from wetland rice soils. In: Nitrogen and Rice, pp. 135–156. IRRI, Los Baños.

Mikkelsen, D.S., De Datta, S.K. and Obcemea, W.N. 1978 Ammonia volatilization losses from flooded rice soils. Soil Sci. Soc. Am. J. 42, 725–730.

Mitsui, S. 1954 Inorganic Nutrition, Fertilization, and Soil Amelioration for Lowland Rice. Yokendo Ltd., Tokyo.

Mitsui, S. 1977 Recognition of the importance of denitrification and its impact on various improved and mechanized applications of nitrogen to rice plants. In: Proceedings of the International Symposium on Soil Environment and Fertility Management in Intensive Agriculture (SEFMIA), pp. 259–268. Society of the Science of Soil and Manure, Tokyo.

Mitsui, S. and Kurihara, K. 1962 The intake and utilization of carbon by plant roots from ^{14}C-labelled urea, IV. Absorption of intact urea molecule and its metabolism in plants. Soil Sci. Plant Nutr. 8, 219–225.

Mitsui, S., Ozaki, K. and Moriyama, M. 1954 On the volatilization of the ammonia transformed from urea. J. Sci. Soil Manure (Japan) 25, 17–19.

Moeller, M.B. and Vlek, P.L.G. 1982 The chemical dynamics of ammonia volatilization from aqueous solution. Atmos. Environ. 16, 709–717.

Moorman, F.R. and Van Breemen, N. 1978 Rice: Soil, Water, Land. IRRI, Los Baños.

Murayama, N. 1977 Changes in the amount and efficiency of chemical fertilizer applied to rice in Japan. In: Proceedings of the International Symposium on Soil Environment and Fertility Management in Intensive Agriculture (SEFMIA), pp. 126–131. Society of the Science of Soil and Manure, Tokyo.

262

Murayama, N. 1979 The importance of nitrogen for rice production. In: Nitrogen and Rice, pp. 5–23. IRRI, Los Baños.

Nagarajah, S. and Al-Abbas, A.H. 1967 Nitrogen and phosphorus fertilizer placement studies on rice using N^{15} and P^{32}. Trop. Agric. 121, 89–103.

Okuda, A., Takahashi, E. and Yoshida, M. 1960 The volatilization of ammonia transformed from urea applied in the upland and waterlogged conditions. J. Jap. Soil Fert. Sci. 31, 273–278.

Ottow, J.C.G. 1981 Mechanism of bacterial iron-reduction in flooded soils. In: Proceedings Symposium on Paddy Soil, pp. 330–343. Institute of Soil Science, Academia Sinica, Nanjing.

Pande, H.K. and Adak, N.K. 1971 Leaching loss of nitrogen in submerged rice cultivation. Exp. Agric. 7, 329–336.

Patnaik, S. 1965 N15 tracer studies on the utilization of fertilizer nitrogen by rice in relation to time of application. Proc. Indian Acad. Sci., Sect. B. 61, 31–39.

Patnaik, S. and Rao, M.V. 1979 Sources of nitrogen for rice production. In: Nitrogen and Rice, pp. 25–43. IRRI, Los Baños.

Patnaik, S., Mohanty, S.K. and Dash, R.N. 1979 Isotope technology as applied to studies of soil fertility, nutrient availability and fertilizer use on flooded rice soils. In: Isotopes and Radiation in Research on Soil-Plant Relationships, pp. 583–605. IAEA, Vienna.

Patrick, W.H., Jr. and Mahapatra, C. 1968 Transformations and availability to rice of nitrogen and phosphorus in waterlogged soils. Adv. Agron. 20, 323–359.

Patrick, W.H., Jr. and Reddy, K.R. 1976a Fate of fertilizer nitrogen in a flooded rice soil. Soil Sci. Soc. Am. J. 40, 678–681.

Patrick, W.H., Jr. and Reddy, K.R. 1976b Nitrification-denitrification reactions in flooded soils and water bottoms: dependence on oxygen supply and ammonium diffusion. J. Environ. Qual. 5, 469–472.

Patrick, W.H., Jr. and Wyatt, R. 1964 Soil nitrogen loss as a result of alternate submergence and drying. Soil Sci. Soc. Am. Proc. 28, 647–653.

Patrick, W.H., Jr., Delaune, R.D. and Peterson, F.J. 1974 Nitrogen utilization by rice using ^{15}N-depleted ammonium sulfate. Agron. J. 66, 819–820.

Pearsall, W.H. 1938 The soil complex in relation to plant communities. I. Oxidation-reduction potentials in soils. J. Ecol. 26, 180–193.

Pettit, N.M., Smith, A.R.J., Freedman, R.B. and Burns, R.G. 1976 Soil urease: activity stability and kinetic properties. Soil Biol. Biochem. 8, 479–484.

Phuc, N., Tanabe, K. and Kuroda, M. 1975 Variation of dissolved oxygen in the flooding water of the paddy fields. J. Fac. Agr., Kyushu Univ. 20, 47–60.

Ponnamperuma, F.N. 1978 Electrochemical changes in submerged soils and the growth of rice. In: Soils and Rice, pp. 421–441. IRRI, Los Baños.

Prasad, R. and De Datta, S.K. 1979 Increasing fertilizer nitrogen efficiency in wetland rice. In: Nitrogen and Rice, pp. 465–484. IRRI, Los Baños.

Rao, C.U.M. 1977 Development of indigenous slow-release nitrogen sources for rice in relation to water management. Ph.D. Dissertation, O.U.A.T., Bhubaneswar, India.

Reddy, K.R. and Patrick, W.H., Jr. 1975 Effect of alternate anaerobic conditions on redox potential, organic matter decomposition, and nitrogen loss in a flooded soil. Soil Biol. Biochem. 7, 87–94.

Reddy, K.R. and Patrick, W.H., Jr. 1976 Yield and nitrogen utilization by rice as affected by method and time of application of labelled nitrogen. Agron. J. 68, 965–969.

Reddy, K.R., Patrick, W.H., Jr. and Phillips, R.E. 1976 Ammonium diffusion as a factor in nitrogen loss from flooded soils. Soil Sci. Soc. Am. J. 40, 528–533.

Reynaud, P.A. and Roger, P.A. 1978 N_2-fixing biomass in Senegal rice fields. Ecol. Bull. 26, 148–157.

Sahrawat, K.L. 1980 Ammonia volatilization losses in some tropical flooded rice soils under field conditions. Il Riso (Milan) 29, 21–27.

Sankhayan, S.D. and Shukla, V.C. 1976 Rates of urea hydrolysis in five soils of India. Geoderma 16, 171–178.

Savant, N.K., De Datta, S.K. and Craswell, E.T. 1982 Distribution patterns of ammonium nitrogen and ^{15}N uptake by rice after deep placement of urea supergranules in wetland soil. Soil Sci. Soc. of Am. J. 46, 567–573.

Shioiri, M. 1941 Denitrification in paddy soil. Kagaku (Tokyo) 11, 24–30.

Singh, V.P., Wickham, T.H. and Corpuz, I.T. 1978 Nitrogen movement to Laguna Lake through drainage from rice fields. In: Proceedings of the International Conference on Water Pollution Control in Developing Countries, pp. 141–145. Asian Institute of Technology, Bangkok.

Sreenivasan, A. and Subrahmanyan, V. 1935 Biochemistry of waterlogged soils. IV. Carbon and nitrogen transformations. J. Agric. Sci. 25, 6–21.

Stangel, P.J. 1979 Nitrogen Requirement and Adequacy of Supply for Rice Production. In: Nitrogen and Rice, pp. 45–69. IRRI, Los Baños.

Stratton, F.E. and Asce, A.M. 1969 Nitrogen losses from alkaline water impoundments. J. Sanit. Eng. Div., Proc. Am. Soc. Civ. Eng. 95, 223–231.

Takamura, Y., Tabuchi, T. and Kubota, H. 1977 Behaviour and balance of applied nitrogen and phosphorus under rice field conditions. In: Proceedings of the International Symposium on Soil and Fertility Management in Intensive Agriculture (SEFMIA), pp. 342–349. Society of the Science of Soil and Manure, Tokyo.

Tanaka, A. and Navasero, A.S. 1964 Loss of nitrogen from the rice plant through rain or dew. Soil Sci. Plant Nutr. 10, 36–39.

Tanaka, A., Patnaik, S. and Abichandani, C.T. 1959 Studies on the nutrition of rice plant (*O. sativa* L.). Part III. Partial efficiency of nitrogen absorbed at different growth stages in relation to yield of rice. Proc. Indian Acad. Sci. Sect. B 49, 207–216.

Tejeda, H.R., De Datta, S.K. and Craswell, E.T. 1980 Efficiency of nitrogen fertilizers on rice in INSFFER experiments as affected by site characteristics. Agron. Abstr. p. 44.

Upadhya, G.S., Datta, N.P. and Deb, D.L. 1974 Note on the effect of selected drainage practices on yield of rice and the efficiency of nitrogen use. Indian J. Agric. Sci. 43, 888–889.

Ventura, W.B. and Yoshida, T. 1977 Ammonia volatilization from a flooded tropical soil. Plant Soil 46, 521–531.

Vlek, P.L.G. and Craswell, E.T. 1979 Effect of nitrogen source and management on ammonia volatilization losses from flooded rice-soil systems. Soil Sci. Soc. Am. J. 43, 352–358.

Vlek, P.L.G. and Craswell, E.T. 1981 Ammonia volatilization from flooded soils. Fert. Res. 2, 227–245.

Vlek, P.L.G. and Stumpe, J.M. 1978 Effect of solution chemistry and environmental conditions on ammonia volatilization losses from aqueous systems. Soil Sci. Soc. Am. J. 42, 416–421.

Vlek, P.L.G., Stumpe, J.M. and Byrnes, B.H. 1980 Urease activity and inhibition in flooded soil systems. Fert. Res. 1, 191–202.

Von Talsky, G. and Klunker, G. 1967 Hydrolyse von Harnstoff durch Urease. Hoppe-Seyler's Z Physiol. Chem. 348, 1372–1376.

Watanabe, I. and Furusaka, C. 1980 Microbial ecology of flooded rice soils. Adv. Microbial. Ecol. 4, 125–168.

Watanabe, I. and Mitsui, S. 1979 Denitrification loss of fertilizer nitrogen in paddy soils – its recognition and impact. IRRI Res. Pap. Ser. 37, 1–10.

Watanabe, I., Craswell, E.T. and App, A.A. 1981 Nitrogen cycling in wetland rice soils in East and South-East Asia. In: Wetselaar, R., Simpson, J.R. and Rosswall, T. (eds.), Nitrogen Cycling in South-East Asian Wet Monsoonal Ecosystems, pp. 4–17. Australian Academy of Science, Canberra.

Watanabe, I., Lee, K.K., Alimagno, B.V., Sato, M., Del Rosario, D.C. and De Guzman, M.R.

1977 Biological nitrogen fixation in paddy field studies by in situ acetylene-reduction assays. IRRI Res. Pap. Ser. 3, 1–16.

Watkins, S.H., Strand, R.F., DeBell, D. S. and Esch, J., Jr. 1972 Factors influencing ammonia losses from urea applied to Northwestern forest soils. Soil Sci. Soc. Am. Proc. 36, 354–357.

Weiler, R.R. 1979 Rate of loss of ammonia from water to the atmosphere. J. Fish Res. Board Can. 36, 685–689.

Wetselaar, R. and Farquhar, G.D. 1980 Nitrogen losses from tops of plants. Adv. Agron. 33, 263–302.

Wetselaar, R., Jakobsen, P. and Chaplin, G.R. 1973 Nitrogen balance in crop systems in tropical Australia. Soil Biol. Biochem. 5, 35–40.

Wetselaar, R., Shaw, T., Firth, P., Oupathum, J. and Thitipoca, H. 1977 Ammonia volatilization losses from variously placed ammonium sulphate under lowland rice field conditions in Central Thailand. In: Proceedings of the International Symposium on Soil Environment and Fertility Management in Intensive Agriculture (SEFMIA), pp. 282–288. Society of the Science of Soil and Manure, Tokyo.

Wilson, J.T., Eskew, D.L. and Habte, M. 1980 Recovery of nitrogen by rice from blue-green algae added in a flooded soil. Soil Sci. Soc. Am. J. 44, 1330–1331.

Yamada, Y., Ahmed, S., Alcantara, A. and Khan, N.H. 1979 Nitrogen efficiency under flooded paddy conditions: A Review of INPUTS study I. In: Proceedings Third Review Meeting INPUTS Project, pp. 39–74. East-West Center, Honolulu.

Yoshida, T. and Broadbent, F.E. 1975 Movement of atmospheric nitrogen in rice plants. Soil Sci. 120, 288–291.

Yoshida, T. and Padre, B.C., Jr. 1977 Transformations of soil and fertilizer nitrogen in paddy soil and their availability to rice plants. Plant Soil 47, 113–123.

Zantua, M.I. and Bremner, J.M. 1977 Stability of urease in soils. Soil Biol. Biochem. 9, 135–140.

11. The fate of nitrogen compounds in the atmosphere

I.E. GALBALLY and C.R. ROY

11.1. Introduction

The atmosphere plays an important role in the redistribution of nitrogen (N) compounds in terrestrial ecosystems. In some cases the atmosphere plays a passive role, acting only as a medium for transport, e.g. for ammonia (NH_3) release and uptake, whereas in other cases one or more chemical transformations take place in the atmosphere before each cycling N atom returns to a terrestrial ecosystem.

In this chapter we review the role of the atmosphere in redistributing and transforming N compounds. Table 11.1 presents a brief review of typical concentrations of some N compounds in the atmosphere. The observed concentrations of these species depend on the strength and proximity of the source, the extent of mixing in the atmosphere and the effectiveness of removal mechanisms: e.g. the concentration of nitric oxide (NO) and nitrogen dioxide (NO_2) in the urban atmosphere is high due to the dominating influence of combustion processes, but low in the rural atmosphere. Atmospheric NH_3

Table 11.1. Typical concentrations of nitrogen compounds in the atmosphere

Constituent	Urban	Rural	Oceanic
Gases			
N_2O (ppbv)[a]	305	300 ± 4	300 ± 4
NH_3 (ppbv)	1–20	1–5	0.1–1
$NO + NO_2$ (ppbv)	20–500	1–10	0.05–0.2
HNO_3 (ppbv)	1–20	0.1–5	0.01–0.07
Particulates			
NH_4^+ (ppbm)[b]	~1	0.3–3	0.1–0.4
NO_3^- (ppbm)	1–18	0.1–0.6	~0.2
Rainwater			
NH_4^- (gN/m^3)	0.1–1	0.1–1	0.001–0.08
NO_3^- (gN/m^3)	0.03–0.3	0.03–0.3	0.01–0.2

[a] 10^{-9} volume/volume
[b] 10^{-9} mass/mass

concentrations vary markedly and some of this variation is due to the dual propensities of the landscape to act as a net source or sink for NH_3 depending on the moisture status of the soil (Denmead *et al.* 1978). Apparent variations may also occur as a result of different calibration standards. This is a constant problem in many fields where trace amounts of a compound are being analysed.

11.2. Nitrous oxide

At present most nitrous oxide (N_2O) appears to be produced in soils by the action of microorganisms on other nitrogeneous compounds but an increasing amount of N_2O is produced by combustion processes (Pierotti and Rasmussen 1976).

Measurements made during the last five years show the N_2O concentration to be remarkably constant in the atmosphere (Roy 1979, Weiss 1981). The most recent absolute calibrations of N_2O show the atmospheric concentration to be 300 ± 4 ppb (Connell *et al.* 1980, Goldan *et al.* 1981, Weiss 1981). The measurements of Weiss (1981) show the tropospheric N_2O concentration to be increasing by $\sim 0.2\%$/year. So far there are no other observations confirming this increase. Weiss (1981) suggests that this increase is due to N_2O from fossil fuel combustion and agricultural activity.

The remarkable constancy of N_2O in the atmosphere is attributed to the size of the atmospheric reservoir of N_2O, currently calculated to be 1500 TgN, compared with the suggested release rates, globally averaged at about 10 TgN/year. Hence local emissions cause little fluctuation in the concentration of N_2O.

There can be no doubt that soil has the capacity to act as a sink for N_2O as many soil organisms can reduce N_2O to dinitrogen (N_2) (Payne 1973). However the reduction of N_2O to N_2 by soil microorganisms is inhibited by nitrate (Blackmer and Bremner 1978) and it is most unlikely that soils will act as a sink for N_2O whenever they contain nitrate or produce nitrate from ammonium or soil organic matter (Freney *et al.* 1978).

The only known process in the atmosphere for removal of N_2O is by photodissociation by ultraviolet solar radiation and chemical reaction in the stratosphere (McElroy and McConnell 1971, Nicolet and Peetermans 1972).

The loss of N_2O in the stratosphere proceeds via two reactions, the second of which has two pathways

$$N_2O + h\nu \rightarrow N_2 + O(^1D)$$

(for wavelengths less than 260 nm)

and

$$N_2O + O(^1D) \rightarrow N_2 + O_2$$

$$\rightarrow 2NO.$$

Calculations by Johnston *et al.* (1979) and Levy *et al.* (1979) suggest that these mechanisms are responsible for removing 10 ± 3 TgN/year of N_2O from the atmosphere. The error is based on the uncertainty in solar radiation flux, chemical reaction rates and the distribution of N_2O in the atmosphere. As the total atmospheric burden of N_2O is about 1500 TgN its atmospheric lifetime is around 150 years.

Thus N_2O in the stratosphere is converted to N_2 and NO; according to present knowledge (Nicolet and Peetermans 1972) more than 90% of the N_2O will be transformed in the stratosphere to N_2, the remaining fraction is converted to NO. Some of this NO is converted to NO_2 and nitric acid (HNO_3) and all three of these substances are ultimately returned to the troposphere. The net source of NO_x ($NO + NO_2$) plus HNO_3 to the troposphere is estimated to be 0.5 to 1 TgN/year (~ 0.01 kg N/ha/year) (Levy *et al.* 1980, Logan *et al.* 1981). This NO_x plus HNO_3 presumably ends up as nitrate in rain water, but the input is small compared with the intense nitrate cycle near the earth's surface. Thus any N_2O that is produced in an area, or extra N_2O produced due to modification of soil processes will, after release to the atmosphere, be evenly distributed throughout the global atmosphere.

11.3. Ammonia and ammonium

Ammonia is present in the atmosphere as a gas and in the form of ammonium in water droplets and solid particles. The concentrations of NH_3 and ammonium vary widely in the atmosphere due to the inhomogeneity of the sources and sinks (e.g. biological sources and precipitation scavenging).

Particulate ammonium in the atmosphere is generally ammonium sulfate or bisulfate. The vapour pressure of NH_3 over sulfuric acid-sulfate aerosol is very low compared with atmospheric gaseous NH_3 levels (Scott and Cattel 1979). Hence the particulate ammonium concentration is predominately determined by the presence of acid sulfate particles and could be expected to be relatively uncorrelated with gaseous NH_3 concentrations. This may not apply in those urban atmospheres where sulfuric acid concentrations are high and where gaseous NH_3 concentrations may be depleted in a shallow polluted layer.

In mid latitude land sites recent NH_3 measurements vary between 0.2 and 36 ppbv and the mean value is estimated at 3 ppbv with a possible range of 1–5 ppbv (Denmead *et al.* 1978, Tuazon *et al.* 1978, Galbally *et al.* 1980, Hoell *et al.* 1980, Lenhard and Gravenhorst 1980, Abbas and Tanner 1981, Tjepkema *et al.* 1981, Tuazon *et al.* 1981).

The only observations that have been reported for NH_3 in the planetary boundary layer over land in the tropics are those of Lodge *et al.* (1974) giving an annual average value of 15 ppbv. Observations by Georgii and Gravenhorst (1977) over the Sargasso Sea and the Caribbean of 8 to 10 ppbv NH_3 suggest that the high values reported by Lodge *et al.* (1974) are widespread over the ocean near land as well as on land. No similar maximum of NH_3 concentrations is observed in air over the tropical waters of the Pacific Ocean away from land sources (Tsunogai 1971).

Over the oceans the observations appear to vary between the hemispheres (Tsunogai 1971, Georgii and Gravenhorst 1977, Ayers and Gras 1980). A value of 1 ppbv (range 0.5 to 2) is representative of the region north of 30° S and a value of 0.1 (range 0.05 to 0.5 ppbv) for the region 90° S to 30° S.

There are many more measurements of particulate ammonium concentrations than of NH_3 concentrations in the atmosphere (Tsunogai 1971, Brosset *et al.* 1975, Georgii and Gravenhorst 1977, Heintzenberg *et al.* 1979, Galbally *et al.* 1980, Lenhard and Gravenhorst 1980, Tjepkema *et al.* 1981, and particularly Huebert and Lazrus 1980b). Over land north of 30° S typical average values are 1 ppbm (range 0.5 to 2) ammonium in the planetary boundary layer and 0.2 ppbm (range 0.1 to 0.4 ppbm) south of 30° S. These latter values are representative also over all ocean areas.

The primary sources of atmospheric NH_3 appear to be animal urine, the microbial decomposition of organic material on the earth's surface (Söderlund and Svensson 1976), and metabolic processes in terrestrial plants (Farquhar *et al.* 1980) but as with N_2O, large emissions of NH_3 can be expected to occur sporadically from applications of nitrogen fertilizers.

Ammonia is also emitted to the atmosphere from combustion processes, most notably from the burning of coal (Robinson and Robbins 1972).

Denmead *et al.* (1976) found that the NH_3 emissions from the soil surface in an ungrazed grass-clover pasture were at rates comparable with grazed pastures. However most of the NH_3 released from the soil surface was absorbed by the plants above. Denmead *et al.* (1978) show a similar complex pattern of NH_3 uptake and release in a growing corn crop. The release of NH_3 from the soil in these cases is presumably due to microbial decomposition of old plant material (Dawson 1977).

Studies of plants (Farquhar *et al.* 1979, 1980) land surfaces (Dawson 1977) and oceans (Georgii and Gravenhorst 1977) have all shown that depending on whether the partial pressure of NH_3 in the atmosphere exceeds or is less than some equilibrium partial pressure of NH_3 for the underlying surface then the surface acts as a source or sink for atmospheric NH_3.

Farquhar *et al.* (1979, 1980, this book) have provided a key for understanding this behaviour. They have identified a finite partial pressure of NH_3, $p(NH_3)$, at which the net flux between the atmosphere and the plant is zero.

This partial pressure called the NH_3 compensation point, γ, is thought to correspond to the equilibrium $p(NH_3)$ in the intercellular cavity due to the many reactions involving NH_3 and ammonium in the plant cells. The molar flux density of NH_3 between plants and the atmosphere, F, can be expressed as

$$F = \frac{g}{P} \left[p(NH_3)_{atmos.} - \gamma \right]$$

where g is the conductance for diffusion of the gas through the boundary layer of the leaf and the stomata and P is atmospheric pressure. When $p(NH_3)_{atmos.} > \gamma$, NH_3 is taken up by the plant whereas when $p(NH_3)_{atmos.} < \gamma$, NH_3 is released from the plant to the atmosphere. This compensation point has a positive temperature dependence and for the four species so far measured, has values of $2-6 \times 10^{-9}$ atmosphere at NTP (Farquhar *et al.* 1980). Similar equilibria have been described for the soil and oceans.

There are probably adjoining areas where the underlying surfaces are alternately NH_3 rich and poor. Here a sequence of emission by the NH_3 rich underlying surface, transport by the atmosphere and subsequently uptake by the surface in the adjoining area makes up part of a local atmospheric NH_3 cycle.

The complete cycle of NH_3/NH_4^+ flow through the atmosphere can be summarised by the following scheme:

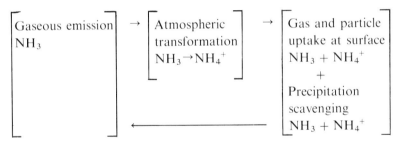

The concept of dry deposition of NH_3 (i.e. irreversible uptake at the surface) for land and ocean surfaces used in other studies (e.g. Söderlund and Svensson 1976) can no longer be considered appropriate, because of the possibility of re-emission of NH_3 from plant, soil and water surfaces.

The conversion of NH_3 to ammonium takes place due to the presence of aqueous sulfuric acid aerosol formed in the atmosphere and also through

$$NH_3 \text{ (gas)} + \text{cloud droplet} \rightarrow NH_3 \text{ (aqueous)} \rightarrow NH_4^+.$$

This uptake of gaseous NH_3 accelerates many of the oxidation processes that produce acidity in cloud water.

It is generally accepted that the sub-micrometre sulfate aerosol originates largely via nucleation from supersaturated sulfuric acid vapour. Nuclei are first formed with radii of order 0.001 μm, but grow rapidly to >0.01 μm by coagulation and condensation, or until their growth ceases when they are removed by washout or deposition processes. The production of sulfate aerosol may be represented by:

$$S\,(\text{gas}) \xrightarrow[\text{multiple steps}]{\text{oxidant}} H_2SO_4\,(\text{gas}) \xrightarrow{\text{water vapour}} H_2SO_4 \text{ (aqueous aerosol)}$$

where S (gas) is usually sulfur dioxide (SO_2), especially in urban or industrial areas, but can also be hydrogen sulfide, organic sulfide, carbon disulfide and carbonyl sulfide in non-urban or maritime atmospheres (see for example, Sze and Ko 1980). The most likely oxidising species in both urban and remote atmospheres is the hydroxyl radical, with other radicals such as hydroperoxyl and methylperoxyl contributing to some less extent (Altshuller 1979, Moller 1980).

The newly formed aqueous sulfuric acid aerosol quickly absorbs NH_3 (Huntzicker et al. 1980). This is the origin of the ubiquitous 'ammonium sulfate' background aerosol well-known to cloud physicists (Twomey 1977). This process can be explained simply in terms of chemical equilibria. The saturation vapour pressure of sulfuric acid at tropospheric temperatures and 75% relative humidity is very low, approximately 10^{-20} and 10^{-17} atmospheres at $0°$ C and $25°$ C respectively (Ayers et al. 1980). Thus sulfuric acid will always be at its saturation vapour pressure in the gaseous phase and virtually all the sulfuric acid will be in the aerosol phase.

The equilibrium gaseous partial pressure of NH_3, $p(NH_3)_{\text{equilib.}}$ over condensed acid sulfate varies from 10^{-21} atmosphere over 99:1 $H_2SO_4:NH_4HSO_4$ to 10^{-15} atmosphere at 99:1 $(NH_4)_2SO_4:NH_4HSO_4$ at $0°$ C in dry air. At $25°$ C in dry air $p(NH_3)_{\text{equilib.}}$ is 10^{-14} atmosphere over 99:1 $(NH_4)_2SO_4:NH_4HSO_4$ (Scott and Cattel 1979). Tang (1980) has estimated $p(NH_3)_{\text{equilib.}}$ over aqueous aerosol composed of H_2SO_4, NH_3 and HNO_3 along with their ionic products. At $25°$ C, 85% relative humidity and a molar ratio of $NH_4^+:SO_4^{2-}$ of 1:1, $p(NH_3)_{\text{equilib.}}$ is approximately 10^{-11} atmospheres. $p(NH_3)_{\text{equilib.}}$ is much lower for lower ratios, i.e. more acidic or less ammoniated solutions as would be expected. Hence when NH_3 is present at partial pressures around 10^{-13} atmospheres or greater, it will react with any condensed acid sulfate removing gas phase NH_3 by forming NH_4HSO_4 and $(NH_4)_2SO_4$ aerosol.

Measured NH_3 atmospheric partial pressures of $\sim 10^{-10}$ atmosphere (maritime, Ayers and Gras 1980) to greater than 10^{-9} atmosphere (continental, Georgii and Lenhard 1978) are adequate to ammoniate sulfuric acid to NH_4HSO_4 and $(NH_4)_2SO_4$ under most tropospheric conditions. The result is that gaseous NH_3 is converted to ammonium in aerosol or aqueous droplets.

There are now two competing processes regulating $p(NH_3)_{\text{atmos.}}$ Plants at the earth's surface (in as far as our present knowledge is correct) will attempt to maintain a $p(NH_3)_{\text{atmos.}}$ of a few times 10^{-9} atmosphere due to their internal chemistry, whereas sulfuric acid present will equilibrate forming NH_4HSO_4 and $(NH_4)_2SO_4$ taking up NH_3 until $p(NH_3)_{atmos.}$ is reduced to around 10^{-13} atmosphere or less or until all the acid sulfate is converted to $(NH_4)_2SO_4$. Thus the addition of sulfuric acid, or other sulfur compounds that are oxidised to sulfuric acid in the atmosphere effectively removes NH_3 from the gaseous phase in the atmosphere, preventing its recycling through the quasi-equilibrium with plants. This results in enhanced ammonium content in aerosol. Ammonia which would otherwise be recycled through the biosphere is redistributed as NH_4HSO_4 and $(NH_4)_2SO_4$ aerosol in the atmosphere.

The NH_3 converted to ammonium in aerosol is not available for direct reabsorption by the plants, but is ultimately returned to the biosphere via precipitation scavenging as ammonium in rainwater.

This process is illustrated by observations over the continent of Australia. Ayers (1980) has estimated that 0.2 Tg S/year is cycled through the 'natural' submicrometre sulfate aerosol and, on the assumption of complete ammoniation, these sulfuric acid particles 'fix' ~ 0.2 Tg N/year of NH_3. This compares with a natural NH_3 release to the atmosphere of around 0.7 Tg N/year (Galbally et al. 1980).

The best available estimate of the total anthropogenic SO_2 emission for Australia is 0.7 Tg S/year (Davey 1980). About 0.2 Tg S/year of this release occurs at an isolated smelter at Mt. Isa in northern Australia. This SO_2 can be removed from the atmosphere over Australia by conversion to sulfate, by deposition at the ground and on vegetation, by scavenging by clouds and rain and by advection beyond the continental boundaries. There have been several studies of the plume from the Mt. Isa smelter. Over much of this region of Australia clouds are the exception rather than the rule and dry deposition of SO_2 does not appear to be unusually fast (Milne et al. 1979). The gas phase oxidation rates measured by Roberts and Williams (1979) suggest that one-third to one-half of the emitted SO_2 would be converted to sulfate aerosol within one week. Ayers et al. (1979) showed that 95% of all sulfuric acid particles formed in this plume were ammoniated at 15 or more hours after the release of SO_2 from the smelter ($\gtrsim 370$ km plume travel from the smelter). Thus this anthropogenic SO_2 source of 0.2 Tg S/year causes between 0.05 and 0.1 Tg N/year of NH_3 to be cycled with S through the submicrometre ammonium sulfate aerosol. Similarly we assume that at least 0.2 Tg N/year of NH_3 must be withdrawn from the biosphere and cycled through the aerosol phase by the total anthropogenic release of 0.7 Tg S/year. This ammonium aerosol will ultimately be returned to the earth's surface at some distant location most probably in precipitation.

Lenhard and Gravenhorst (1980) have made measurements over a rural area of West Germany, presumably an NH_3 'source' region and observed a total flux of NH_3 and ammonium from the surface of 60 μg N/m^2/hour (5 kg

N/ha/year) from eleven aircraft flights during summer and winter. These measurements represent an area of about 10^4 km^2. The measured flux represents about 30% of the N content of organic fertilizer and cattle excrements spread on soil in the region neighbouring the measuring area. They found the NH$_3$ was converted to ammonium in the atmosphere with a time constant (for conversion) of about one day. The amount of ammonium deposited in rainwater in the region was less than half the flux to the atmosphere so presumably over this region there is a net outflow of NH$_3$ and subsequent deposition in surrounding regions.

There are many ($\sim 10^2$) locations around the world where annual measurements of ammonium in rainwater have been made. The most recent survey of this data by Bottger et al. (1978) have a global deposition of ammonium on land of 15 ± 7 Tg N/year and on oceans of 6 ± 6Tg N/year. There are large variations in deposition rate from site to site. Data on ammonium in rainwater in south east Asia from references in Galbally and Wetselaar (1981) show that both the concentration and average deposition varies by more than an order of magnitude between different sites in the region.

In the light of Lenhard and Gravenhorst's (1980) work it is reasonable to suppose that ammonium found in rainwater can come from areas 10^2 or 10^3 km away. Ammonia volatilized at a particular site may be reabsorbed by direct uptake nearby, as Denmead et al. (1976) show, but that fraction converted to aerosol travels to more distant locations.

Some NH$_3$ is converted in the atmosphere to NH$_2$ radicals due to reaction with hydroxyl radicals and may be oxidised to NO. However recent calculations by Lenhard and Gravenhorst (1980) show that the conversion to NH$_2$ is only 1% of the observed flux in the lower atmosphere and hence is negligible in these considerations.

11.4. Nitrogen oxides

Recent advances in NO$_x$ (NO + NO$_2$) detector technology has resulted in an increase in the number of reported studies and an improvement in the quality of NO$_x$ measurements.

Nitric oxide is present at very low concentrations in background surface air – often at or near the detection limit for the chemiluminescent detector used for measurements. For surface, or near surface, air McFarland et al. (1979) reported 0.004 ppbv for the tropical Pacific, Galbally and Roy (1981) 0.005 ppbv for marine air in southern Australia and Helas and Warneck (1981) <0.01 ppbv on the west coast of Ireland. In the free troposphere Schiff et al. (1979) found 0.03–0.06 ppbv for maritime and North American continental air and Roy et al. (1980) reported 0.2 ppbv for the upper troposphere over southern Australia.

Nitrogen dioxide concentrations in surface air in non-background locations show large variations due to anthropogenic inputs. Perner and Platt (1979) and Platt and Perner (1980a) reported NO_2 concentrations up to 100 ppbv in a semi-rural location in West Germany and in the Los Angeles basin, while Platt et al. (1979) found 5–10 ppbv in a rural area of West Germany. In rural areas a diurnal variation in atmospheric concentration has been observed. Galbally et al. (1980) and Galbally and Roy (1981) found $\sim$3 ppbv (day) and $\sim$10 ppbv (night) in southern Australia.

Much lower concentrations have been measured for clean maritime air. For the west coast of Ireland Cox (1977) found <0.4 ppbv, Platt and Perner (1980b) <0.1–0.4 ppbv, and Helas and Warneck (1981) 0.1 ± 0.1 ppbv and for the NW coast of Tasmania, Australia, Galbally and Roy (1981) found 0.1–0.2 ppbv.

Measurements of NO_2 made at a height of 3 km in the Rocky Mountains of North America by Noxon (1975, 1978) and Kelly et al. (1980) resulted in mixing ratios of <0.1–0.25 ppbv. However Noxon (1981) reported 0.03 ppbv from optical absorption measurements made at a height of 3 km at Mauna Loa in the mid Pacific. In the upper troposphere above North America, Kley et al. (1981) find NO_x at 0.2 ppbv.

Nitric oxide is produced naturally in soils by chemical and microbial reactions (Steen and Stojanovic 1971) and in the atmosphere by oxidation of NH_3 (Cox et al. 1975), oxidation of N_2 by lightning (Chameides et al. 1977) and meteoroid impact on the upper atmosphere. Of these processes, only lightning and meteoroid impact add to the pool of fixed nitrogen, the other reactions merely recycle fixed nitrogen from one reservoir to another.

Nitric oxide is formed in both waterlogged and aerobic soils and appears to be produced by the decomposition of nitrite. Nitrite is produced as an intermediate in the denitrification of nitrate or during the nitrification of ammonium. However, it normally accumulates in soil only under certain conditions which depend on fertilizer additions, the ammonium level, pH and organic matter content of the soil, temperature, moisture and soil-fertilizer geometry (Van Cleemput et al. 1976).

The NO flux from unamended upland soils has been shown to vary from 0.06 to 0.73×10^{-11} kg $N/m^2/sec$ (Galbally and Roy 1978) but it is to be expected that this flux will increase as a result of nitrite accumulation after addition of ammonium or ammonium-yielding fertilizers. Steen and Stojanovic (1971), for example, observed increased NO loss on addition of urea and ammonium sulphate to a calcareous soil and the loss was related to the amount of nitrogen applied.

Galbally and Roy (1981) report a further nine days measurements from 12 sites on a grazed legume pasture where exhalation rates of 0.1 to 5.0×10^{-11} kg $N/m^2/sec$ were observed; the higher values were associated with high mineral N levels within the soils. Some recent laboratory work on nitrifying bacteria has linked NO and N_2O production by these bacteria (Lipschultz *et al.* 1981). Both the latter work and that of Galbally and Roy (1978, 1981) suggest a global source of NO from soils of 10 to 15 Tg N/year (1 kg N/ha/year over land).

Nitric oxide and NO_2 are freely interchangeable in the atmosphere by reaction with ozone (O_3), a pervasive constituent of the lower atmosphere, and by photodissociation in sunlight. The reactions are

$$NO + O_3 \rightarrow NO_2 + O_2$$
$$NO_2 + h\nu \rightarrow NO + O \quad \text{(for wavelengths less than 410 nm)}.$$

It appears that to some extent NO exhalation from soils is counterbalanced by NO_2 uptake in soils and plants.

Gaseous NO_2 is continually removed from the atmosphere by absorption by plant, soil and water surfaces. The usual way of describing gaseous uptake, U, (e.g. g $N/m^2/sec$) is in terms of an uptake or 'deposition' velocity, Vg, and the concentration of the gas within the air, C, where

$$U = Vg \times C.$$

In using this equation, we assume that for a given surface in a given condition, Vg is, to a first approximation, constant and U varies linearly with C. The equation is obviously an oversimplified description of gaseous transport and uptake in the real world, but in the present context, it provides some useful pointers to the fate of atmospheric nitrogen species. The uptake velocity for NO_2 of grassland and soil is about 0.5×10^{-2} m/sec (Galbally 1974, Galbally and Roy 1981).

Many other studies have confirmed the effectiveness of plant and soil surface in absorbing NO_2 (e.g. Hill 1971, Rogers *et al.* 1979, Elkiey and Ormrod 1981). This uptake of gas by plants is regulated by variations in stomatal conductance, as the ultimate fate of the NO_2 appears to be chemical reaction after passage to the intercellular cavities within the leaves (Hill 1971, Rogers *et al.* 1979, Elkiey and Ormrod 1981). Nitric oxide is taken up much more slowly than NO_2 (Hill 1971, Galbally and Roy 1981).

These processes are illustrated with some data from a grazed pasture in Southern Australia (at Rutherglen, Victoria) (Fig. 11.1.)

Essentially, in a grazed pasture situation, the following apply:
(1) there are generally low background NO emissions with higher emissions at specific sites;
(2) all NO emissions show little day to night variation;
(3) there is continuous (horizontally uniform) uptake of NO and NO_2 by plants and soil;
(4) during the daytime, when there is good vertical mixing bringing down

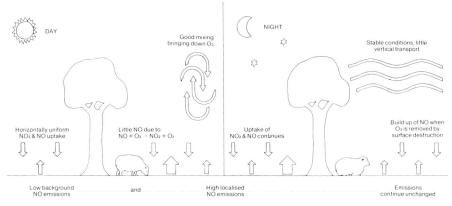

Fig. 11.1. The odd nitrogen cycle over pasture land (Rutherglen, Victoria, April 1979).

O_3 from above the planetary boundary layer, the reaction $NO + O_3 \rightarrow NO_2 + O_2$ is promoted and the NO_x is present almost entirely as NO_2;

(5) at night time, in stable conditions following the supression of turbulence and after the O_3 is removed by surface destruction, there is a buildup of NO in the surface air;

(6) on windy nights, when turbulence persists and O_3 remains present in the surface air, the NO level remains low.

It appears that in this situation the NO_2 uptake, extrapolated to a yearly average, is of the order of 1 kg N/ha/year.

The other major sources of NO and NO_2 are combustion sources, automobiles, furnances, etc. The amount of fixed N from these sources is comparable with that from natural sources (Galbally 1975).

The oxidation of NH_3 could influence the concentration of nitrogen oxides in the atmosphere (Cox *et al.* 1975, Logan *et al.* 1981). The oxidation commences by reaction with hydroxyl radicals:

$$NH_3 + OH \rightarrow NH_2 + H_2O.$$

Subsequent reactions (Kurasawa and Lesclaux 1979, 1980) can include:

$$NH_2 + NO \rightarrow \text{products } (N_2, N_2O)$$

$$NH_2 + NO_2 \rightarrow \text{products } (N_2, N_2O)$$

$$NH_2 + O_3 \rightarrow \text{products } (NH, HNO, NO).$$

Hence the oxidation of ammonia can act as either a source or sink of nitrogen oxides. The reactions consume NO_x if $NO + NO_2 > 0.06$ ppbv or produce NO_x if $NO + NO_2 < 0.06$ ppbv (Logan *et al.* 1981, Cox *et al.* 1975). Most probably both processes occur in the troposphere. In view of the uncertainty in the concentration of $NO + NO_2$ throughout the troposphere we calculate the upper limit of chemical conversion of NH_3 to NO_x as ± 4 Tg N/year, using the observed NH_3 concentrations. This NO_x flux reflects the fact that the chemical conversion may be acting as either a source or sink and at present we cannot identify even the direction of this process.

Nitric oxide is produced in the atmosphere by the heating of the air by the shock waves from lightning strokes. Noxon (1976) has observed nitrogen fixation in the atmosphere during a lightning storm by measuring the absorption spectra of NO_2. The fixation efficiency was estimated at approximately 10^{26} NO_2 molecules per lightning stroke or an annual fixation of several Tg NO_x-N. Recent theoretical works suggest values of 2–4 Tg N/year (Tuck 1976, Hill 1979, Hill et al. 1980, Dawson 1980, Levine et al. 1981). Measurements of tropospheric NO_x profiles by Kley et al. (1981) support this suggested production rate of 2–4 Tg N/year (~ 0.1 kg N/ha/year globally).

The extent to which bushfires fix atmospheric nitrogen, as well as release fixed nitrogen from the plant material used as fuel, is presently not known although elevated nitrogen oxide concentrations are found in bushfire smoke plumes (Evans et al. 1977, Crutzen et al. 1979).

Combustion sources, lightning and bushfires all contribute fixed nitrogen which ultimately through dry deposition and precipitation scavenging passes as an accession of fixed nitrogen to the terrestrial or oceanic biosphere.

Some of the NO and NO_2 in the atmosphere previously released from plant/soil systems, combustion sources, lightning and bushfires is converted to HNO_3. The production of HNO_3 vapour in the atmosphere is thought to proceed in a manner similar to the homogeneous gas phase production of H_2SO_4, that is

$$NO_x \text{ (gas)} \xrightarrow[\text{multiple steps}]{\text{oxidant}} HNO_3 \text{ (gas)}$$

where the major pathway is probably $NO_2 + OH$.

Nitric acid has been measured using both the impregnated filter method and the chemiluminescence technique with apparently consistent results. In the marine boundary layer of the equatorial Pacific Huebert (1980) measured values in the range 0.01–0.07 ppbv with an average of 0.04 ppbv. However, in the continental boundary layer higher and variable values were obtained. Kelly and Stedman (1979) found 0.5–5 ppbv in the presence of 5–25 ppbv NO_x and Appel et al. (1980) found 0.3–1.8 ppbv in the presence of high NO_x concentrations. In remote areas values in the range 0.04–3.6 ppbv have been reported (Huebert and Lazrus 1978, 1980a, Grennfelt 1980). In the free troposphere Huebert and Lazrus (1978) find 0.04–0.4 ppbv with most values <0.15 ppbv. Huebert and Lazrus (1978) found the ratio of HNO_3/NO_3^- to vary from 3:1 to 5:1 in the free troposphere and to be <1 in the boundary layer.

As a result of the relatively high vapour pressure of aqueous HNO_3, the typical observed atmospheric HNO_3 concentrations of $<0.01-1$ ppbv (Huebert and Lazrus 1980a) are insufficient to lead to the formation of new particles composed of HNO_3 by gas to particle conversion. For similar reasons formation of homogeneous ammonium nitrate aerosols is likely to be rare, except in highly polluted regions (Stelson *et al.* 1979). However, aerosol nitrate can occur readily through heterogeneous nucleation (condensation) of HNO_3 vapour on to existing aerosols, leading to surface reactions or dissolution of the acid into solutions of deliquessed aerosols. In either case there is potential for interaction in the aqueous H_2SO_4-NH_3-HNO_3 system. In particular, the partitioning of NH_3 and HNO_3 between the aqueous aerosol phase and the gas phase is determined by the hydronium ion activity in the droplets, in conjunction with Henry's Law. In the case of HNO_3, for example,

$$[HNO_3]_{gas} = K_H [HNO_3]_{aq.}$$

where $[HNO_3]_{aq.}$ is given by

$$[HNO_3]_{aq.} = \frac{a(H^+) \cdot a(NO_3^-)}{K_a}.$$

The a's refer to the appropriate ionic activities, K_H is the Henry's Law constant for HNO_3 and K_a is the dissociation constant of the acid in water. From these equations it can be seen that addition of sulfuric acid to the aerosol, either directly via condensation of sulfuric acid vapour or indirectly as a result of dissolution and subsequent oxidation of SO_2, can lead to an increase in aerosol acidity that causes loss of HNO_3 and/or uptake of NH_3. A similar phenomenon is apparent in the loss of chloride from sea salt particles that encounter elevated levels of SO_2 (Hitchcock *et al.* 1980). The fact that aerosol nitrate resides most often in the large particle fraction also is explicable in these terms (Ayers 1977), as Winkler (1980) has shown that the submicrometre fraction is considerably more acidic in composition than the large particle fraction. More detailed discussions of the equilibria involved may be found in Orel and Seinfeld (1977) and Tang (1980).

The same equilibria apply for the dissolution of HNO_3 in rain droplets. Large particles containing nitrate also are scavenged from the atmosphere by precipitation processes. Both these mechanisms lead to nitrate being incorporated in rainwater. As with NH_3, the wet deposition of nitrate has been measured at many locations around the world. Data from references in Galbally and Wetselaar (1981) show typical nitrate levels in rainwater of 0.03 to $0.3\,gN/m^3$ and typical deposition rates of 0.3 to $3\,kgN/ha/year$ in the south east Asian region.

11.5. Historical records of the atmospheric nitrogen cycle

Historical records of the atmospheric N cycle can be obtained from analyses of nitrate and ammonium in ice cores from Greenland and Antarctica (Parker *et al.* 1977, Busenberg and Langway 1979, Zeller and Parker 1981). The records of ammonium in Greenland ice cores for 1680 AD to the present show occasional large year to year variations (by a factor of 2). When the periods 1680–1790, and 1819–1975 are compared there appears to have been a decrease in ammonium with time. The reasons for this are at present unknown. In the Antarctic (Parker *et al.* 1977) there are large year to year variations of ammonium in ice cores and there appears to be a substantial decrease in ammonium in the cores between 1927 and 1957. The concentrations of ammonium for Greenland and South Pole ice cores are quite similar.

A much longer record for nitrate in Antarctic ice cores extending from 800 AD to the present has been published (Zeller and Parker 1981). It appears that this nitrate originates in the upper atmosphere where it was possibly formed from the oxidation of N_2O. Significant correlations of this nitrate are found with sunspots and the ^{14}C record. Smoothed data varies from 30 μg N/l in 1900 to $\sim 10 \mu g$ N/l in 1700. This suggests that there may have been significant perturbations of the global N_2O cycle on this time scale.

On a shorter time scale there are measurements of ammonium and nitrate available from historical rainfall analyses (Söderlund 1977, Brimblecombe and Pitman 1980). Measurements at Rothamstead (Brimblecombe and Pitman 1980) between 1880 and the present show no change in ammonium but a significant increase in nitrate. In particular the increase in nitrate deposition has occured via a change in the seasonal cycle of nitrate concentration in rain water. Values in winter months have remained constant while deposition in spring has substantially increased. The source of this change is not known.

Data presented by Söderlund (1977) on precipitation analyses at four Scandanavian stations shows a 100% increase in nitrate deposition between 1956 and 1973. This increase is associated with increased fossil fuel combustion.

11.6. Summary

A balance sheet for the atmospheric nitrogen cycle is presented in Table 11.2.

This balance is based on observations made during the last few decades and it should be emphasized that there is considerable uncertainty in these estimates. Some of the contributing terms in this balance are based on one or even no observations.

It is evident from the little historical evidence available that the atmospheric

Table 11.2. The mass and fluxes of nitrogen compounds in the atmosphere

Compound	Atmospheric content (Tg N)	Sources (Tg N/year)	Sinks Tg N/year	Atmospheric lifetime
N_2O	1500 (1400–1600)	Soil Release? Combustion?	Stratosphere 10 (7–13)	150 yrs
NH_3	1.7(0.6–2.7)	Gaseous exchange 50–80 Combustion ~5	Precipitation scavenging 21 (11–31) Balanced by	1 to 2 wks combined
NH_4^+	0.4(0.2-0.8)		gaseous uptake	lifetime
NO_x	0.6(0.2–1.0)	Production by	Precipitation	
HNO_3	0.2(0.1–0.3)	lightning 2–4	scavenging	
NO_3^-	0.1(0.04–0.2)	Inflow from the stratosphere 0.5–1 NH_3 conversion ±4 Release from soils 5–20 Combustion ~20	24 (15–23) Dry deposition 28 (14–56)	1 to 2 wks combined lifetime

nitrogen cycle undergoes significant perturbations on all time scales from years to millenia.

In summary we make the following observations:

(1) Of the N_2O lost from a terrestrial ecosystem 90% is converted to N_2 and the remainder is globally distributed as nitrate in rainwater.

(2) Ammonia in the atmosphere partially results from an attempt by plants and soil processes to maintain an equilibrium concentration consistent with the metabolism and chemistry, thus emission and reabsorption occurs. However, sulfuric acid formed in the atmosphere acts as a sponge for NH_3, leading to its long range transport and subsequent deposition as ammonium in precipitation.

(3) Nitric oxide released by soils is quickly converted to NO_2 by O_3 in the lower atmosphere. This NO_2 can be taken up by the local plant and soil surface or transported long distances by the wind. In the latter case a substantial fraction is converted to HNO_3 by atmospheric reactions and ultimately is returned to the earth's surface as nitrate in rainwater.

11.7. References

Abbas, R. and Tanner, R.L. 1981 Continuous determination of gaseous ammonia in the ambient atmosphere using fluoresence derivization. Atmos. Environ. 15, 277–281.

Altshuller, A.P. 1979 Model predictions of the rates of homogeneous oxidation of sulfur dioxide to sulfate in the troposphere. Atmos. Environ. 13, 1653–1661.

Appel, B.R., Wall, S.M., Tokiwa, Y. and Haik, M. 1980 Simultaneous nitric acid, particulate nitrate and acidity measurements in ambient air. Atmos. Environ. 14, 549–554.

Ayers, G.P. 1977 On the use of nitron for detection of nitrate in individual aerosol particles. Atmos. Environ. 12, 1227–1230.

Ayers, G.P. 1980 Sulfate particles in the troposphere over Australia. In: Freney, J.R. and Nicholson, A.J. (eds.), Sulfur in Australia, pp. 62–65. Australian Academy of Science, Canberra.

Ayers, G.P. and Gras, J.L. 1980 Ammonia gas concentrations over the Southern Ocean. Nature (Lond.) 284, 539–540.

Ayers, G.P., Bigg, E.K. and Turvey, D.E. 1979 Aitken particle and cloud condensation nucleus fluxes in the plume from an isolated industrial source. J. Appl. Meteorol., 18, 449–459.

Ayers, G.P., Gillett, R.W. and Gras, J.L. 1980 On the vapour pressure of sulfuric acid. Geophys. Res. Lett. 7, 433–436.

Blackmer, A.M. and Bremner, J.M. 1978 Inhibitory effect of nitrate on reduction of N_2O to N_2 by soil microorganisms. Soil Biol. Biochem. 10, 187–191.

Bottger, A., Ehhalt, D.H. and Gravenhorst, G. 1978 Atmospharische kreislaufe von stickoxiden und ammoniak. Report Jul-1558, Institut 3: Atmospharische Chemie, Kernforschungsanlage Julich Gmbh.

Brimblecombe, P. and Pitman, J. 1980 Long-term deposit at Rothamsted, southern England. Tellus 32, 261–267.

Brosset, C., Andreasson, K. and Ferm, M. 1975 The nature and possible origin of acid particles observed at the Swedish west coast. Atmos. Environ. 9, 631–642.

Busenberg, E. and Langway Jr., C.C. 1979 Levels of ammonium, sulfate, chloride, calcium, and sodium in snow and ice from southern Greenland, J. Geophys. Res., 84, 1705–1709.

Chameides, W.L., Stedman, D.H., Dickerson, R.R., Rusch , D.W. and Cicerone, R.J. 1977 NO_x production in lightning. J. Atmos. Sci. 34, 143–149.

Connell, P.S., Perry, R.A. and Howard, C.J. 1980 Tunable diode laser measurement of nitrous oxide in air. Geophys. Res. Lett. 7, 1093–1096.

Cox, R.A. 1977 Some measurements of ground level NO, NO_2 and O_3 concentrations at an unpolluted maritime site. Tellus 29, 356–362.

Cox, R.A., Derwent, R.G. and Holt, P.M. 1975 The photo-oxidation of ammonia in the presence of NO and NO_2. Chemosphere 4, 201–205.

Crutzen, P.J., Heidt, L.E., Krasnec, J.P., Pollock, W.H. and Seiler, W. 1979 Biomass burning as a source of atmospheric gases CO, H_2, N_2O, NO, CH_3Cl and COS. Nature (Lond.) 282, 253–256.

Davey, T.R.A. 1980 Anthropogenic sulfur balance for Australia, 1976. In: Freney, J.R. and Nicholson, A.J. (eds.), Sulfur in Australia, pp. 75–87. Australian Academy of Science, Canberra.

Dawson, G.A. 1977 Atmospheric ammonia from undisturbed land, J. Geophys. Res. 82, 3125–3133.

Dawson, G.A. 1980 Nitrogen fixation by lightning. J. Atmos. Sci. 37, 174–178.

Denmead, O.T., Freney, J.R. and Simpson, J.R. 1976 A closed ammonia cycle within a plant canopy. Soil Biol. Biochem. 8, 161–164.

Denmead, O.T., Nulsen, R. and Thurtell, G.W. 1978 Ammonia exchange over a corn crop. Soil Sci. Soc. Am. J. 42, 840–842.

Elkiey, T. and Ormrod, D.P. 1981 Sorption of O_3, SO_2, NO_2 or their mixture by nine *Pao Pratensis* cultivars of differing pollutant sensitivity. Atmos. Environ. 15, 1739–1743.

Evans, L.F., Weeks, I.A., Eccleston, A.J. and Packham, D.R. 1977 Photochemical ozone in smoke from prescribed burning of forests. Environ. Sci. Technol. 11, 896–900.

Farquhar, G.D., Wetselaar, R. and Firth, P.M. 1979 Ammonia volatilization from senescing leaves of Maize. Science 203, 1257–1258.

Farquhar, G.D., Firth, P.M., Wetselaar, R. and Weir, B. 1980 On the Gaseous exchange of ammonia between leaves and the environment: determination of the ammonia compensation point. Plant Physiol. 66, 710–714.

Freney, J.R., Denmead, O.T. and Simpson, J.R. 1978 Soil as a source or sink for atmospheric nitrous oxide. Nature (Lond.) 273, 530–532.

Galbally, I.E. 1974 Gas transfer near the Earth's surface. Adv. Geophys. 18B, 329–339.

Galbally, I.E. 1975 Emission of oxides of nitrogen (NO_x) and ammonia from the Earth's surface. Tellus 27, 67–70.

Galbally, I.E. and Roy, C.R. 1978 Loss of fixed nitrogen from soils by nitric oxide exhalation. Nature (Lond.) 275, 734–735.

Galbally, I.E. and Roy, C.R. 1981 Ozone and nitrogen oxides in the southern hemisphere troposphere. In: London, J. (ed.), Proceedings of the Quadrennial International Ozone Symposium, pp. 431–438. International Ozone Commission.

Galbally, I.E. and Wetselaar, R. 1981 Nitrogen in precipitation in south-east Asia and adjoining areas: A bibliography. In: Wetselaar, R. Simpson, J.R. and Rosswall, T (eds.), Nitrogen Cycling in South-east Asian Wet Monsoonal Ecosystems, pp 195–198. Australian Academy of Science, Canberra.

Galbally, I.E., Freney, J.R., Denmead, O.T. and Roy, C.R. 1980 Processes controlling the nitrogen cycle in the atmosphere over Australia. In: Trudinger, P.A. and Walter, M.R. (eds.), Biochemistry of Ancient and Modern Environments, pp. 319–325. Australian Academy of Science, Canberra.

Georgii, H.W. and Gravenhorst, G. 1977 The ocean as source or sink of reactive trace-gases. Pure Appl. Geophys. 115, 503–511.

Georgii, H.W. and Lenhard, U. 1978 Contribution to the atmospheric NH_3 budget. Pure Appl. Geophys. 116, 385–392.

Goldan, P.D., Kuster, W.C., Schmeltekopf, A.L., Fehsenfeld, F.C. and Albritton, D.L. 1981 Correction of Atmospheric N_2O mixing – Ratio data. J. Geophys. Res. 86, 5385–5386.

Grennfelt, P. 1980 Investigation of gaseous nitrates in an urban and a rural area. Atmos. Environ. 14, 311–316.

Heintzenberg, J., Tragardh, C., Alm, G., Bostrom, R., Granat, L., Hansson, H.CH., Holm, B., Moritz, S., Richter, A and Söderlund, R. 1979 Physical and chemical properties of aerosols under varying European influence. Report AG48, Department of Meteorology, University of Stockholm.

Helas, G. and Warneck, P. 1981 Background NO_x mixing ratios in air masses over the Atlantic ocean. J. Geophys. Res. 86, 7223–7290.

Hill, A.C. 1971 Vegetation: A sink for atmospheric pollutants, J. Air Pollut. Control Assoc. 21, 341–346.

Hill, R.D. 1979 On the production of nitric oxide by lightning. Geophys. Res. Lett. 6, 945–947.

Hill, R.D., Rinker, R.G. and Wilson, H.D. 1980 Atmospheric nitrogen fixation by lightning. J. Atmos. Sci. 37, 179–192.

Hitchcock, D.R., Spiller, L.L. and Wilson, W.E. 1980 Sulfuric acid aerosols and HCl release in coastal atmospheres: Evidence of rapid formation of sulfuric acid particulates. Atmos. Environ. 14, 165–182.

Hoell, J.M., Harward, C.N. and Williams, B.S. 1980 Remote infrared heterodyne radiometer measurements of atmospheric ammonia profiles. Geophys. Res. Lett. 7, 313–316.

Huebert, B.J. 1980 Nitric acid and aerosol nitrate measurements in the equatorial pacific region. Geophys. Res. Lett. 7, 325–328.

Huebert, B.J. and Lazrus, A.L. 1978 Global tropospheric measurements of nitric acid vapor and particulate nitrate. Geophys. Res. Lett. 5, 577–580.

Huebert, B.J. and Lazrus, A.L. 1980a Tropospheric gas-phase and particulate nitrate measurements. J. Geophys. Res. 85, 7322–7328.

Huebert, B.J. and Lazrus, A.L. 1980b Bulk composition of aerosols in the remote troposphere. J. Geophys. Res. 85, 7337–7344.

Huntzicker, J.J., Cary, R.A. and Ling, C-S. 1980 Neutralization of sulfuric acid aerosol by ammonia. Environ. Sci. Technol. 14, 819–824.

Johnston, H.S., Serang, O. and Podolske, J. 1979 Instantaneous global nitrous oxide photochemical rates. J. Geophys. Res. 84, 5077–5082.

Kelly, T.J. and Stedman, D.H. 1979 Measurements of H_2O_2 and HNO_3 in rural air. Geophys. Res. Lett. 6, 375–378.

Kelly, T.J., Stedman, D.H., Ritter, J.A. and Harvey, R.B. 1980 Measurements of oxides of nitrogen and nitric acid in clean air. J. Geophys. Res. 85, 7417–7425.

Kley, D., Drummond, J.W., McFarland, M. and Liu, S.C. 1981 Tropospheric profiles of NO_x. J. Geophys. Res. 86, 3153–3161.

Kurasawa, H. and Lesclaux, R. 1979 Kinetics of the reaction of NH_2 with NO. Chem. Phys. Lett. 66, 602–607.

Kurasawa, H. and Lesclaux, R. 1980 Rate constant for the reaction of NH_3 with ozone in relation with atmospheric processes. Chem. Phys. Lett. 72, 437–442.

Lenhard, U. and Gravenhorst, G. 1980 Evaluation of ammonia fluxes into the free atmosphere over Western Germany. Tellus 32, 48–55.

Levine, J.S., Rogowski, R.S., Gregory, G.L., Howell, W.E. and Fishman, J. 1981 Simultaneous measurements of NO_x, NO and O_3 production in a laboratory discharge: atmospheric implications. Geophys. Res. Lett. 8, 357–360.

Levy, H., Mahlman, J.D. and Moxim, W.J. 1979 A preliminary report on the numerical simulation of the three-dimensional structure and variability of atmospheric N_2O. Geophys. Res. Lett. 6, 155–158.

Levy, H., Mahlman, J.D. and Moxim, W.J. 1980 A stratospheric source of reactive nitrogen in the unpolluted troposphere. Geophys. Res. Lett. 7, 441–444.

Lipschultz, F., Zafiriou, O.C., Wofsy, S.C., McElroy, M.B., Valois, F.W. and Watson, S.W. 1981 Production of NO and N_2O by soil nitrifying bacteria. Nature (Lond.) 294, 641–643.

Lodge, J.P., Jr., Machado, P.A., Pate, J.B., Scheesley, D.C. and Wartburg, A.F. 1974 Atmospheric trace chemistry in the American humid tropics. Tellus 26, 250–253.

Logan, J.A., Prather, M.J., Wofsy, S.C. and McElroy, M.B. 1981 Tropospheric chemistry: a global perspective. J. Geophys. Res. 86, 7210–7254.

McElroy, M.B. and McConnell, J.C. 1971 Nitrous oxide: a natural source of stratospheric NO. J. Atmos. Sci. 28, 1095–1098.

McFarland, M., Kley, D., Drummond, J.W., Schmeltekopf, A.L. and Winkler, R.H. 1979 Nitric oxide measurements in the equatorial Pacific region. Geophys. Res. Lett. 6, 605–608.

Milne, J.W., Roberts, D.B. and Williams, D.J. 1979 The dry deposition of sulphur-dioxide field measurements with a stirred chamber. Atmos. Environ. 13, 373–379.

Moller, D. 1980 Kinetic model of atmospheric SO_2 oxidation based on published data. Atmos. Environ. 14, 1067–1076.

Nicolet, M. and Peetermans, W. 1972 The production of nitric oxide in the stratosphere by oxidation of nitrous oxide. Ann Geophys. 28, 751–762.

Noxon, J.F. 1975 Nitrogen dioxide in the stratosphere and troposphere measured by ground-based absorption spectroscopy. Science (Wash. D.C.) 189, 547–549.

Noxon, J.F. 1976 Atmospheric nitrogen fixation by lightning. Geophys. Res. Lett. 3, 463–465.

Noxon, J.F. 1978 Tropospheric NO_2. J. Geophys. Res. 83, 3051–3057.

Noxon, J.F. 1981 NO_x in the mid-Pacific troposphere, Geophys. Res. Lett. 8, 1223–1226.

Orel, A.E. and Seinfeld, J.H. 1977 Nitrate formation in atmospheric aerosols. Environ. Sci. Tech. 11, 1000–1007.

Parker, B.C., Zeller, E.J., Heiskell, L.E. and Thompson, W.J. 1977 Nitrogen in south polar ice and snow: tool to measure past solar, auroral, and cosmic ray activities. Antarct. J. U.S. 12, 133–134.

Payne, W.J. 1973 Reduction of nitrogenous oxides by microorganisms. Bacteriol. Rev. 37, 409–452.

Perner, D. and Platt, U. 1979 Detection of nitrous acid in the atmosphere by differential optical absorption. Geophys. Res. Lett. 6, 917–920.

Pierotti, D. and Rasmussen, R.A. 1976 Combustion as a source of nitrous oxide in the atmosphere. Geophys. Res. Lett. 3, 265–267.

Platt, U. and Perner, D. 1980a Detection of NO_3 in the polluted troposphere by differential optical absorption. Geophys. Res. Lett. 7, 89–92.

Platt, U. and Perner, D.1980b Direct measurements of atmospheric CH_2O, HNO_2, O_3, NO_2 and SO_2 by differential optical absorption in the near UV. J. Geophys. Res. 85, 7453–7458.

Platt, U., Perner, D. and Patz, H.W. 1979 Simultaneous measurement of atmospheric CH_2O, O_3 and NO_2 by differential optical absorption. J. Geophys. Res. 84, 6329–6335.

Roberts, D.B. and Williams, D.J. 1979 The kinetics of oxidation of sulfur dioxide within the plume from a sulphide smelter in a remote region. Atmos. Environ. 13, 1485–1499.

Robinson, E. and Robbins R.C. 1972 Emissions concentrations and fate of gaseous atmospheric pollutants. In: Strauss W. (ed.), Air Pollution Control, Part 2, pp. 1–93. Wiley Interscience, New York.

Rogers, H.H., Campbell, J.C. and Volk, R.J. 1979 Nitrogen – 15 dioxide uptake and incorporation by phaseolus vulgaris (L.). Science (Wash. D.C.) 206, 333–335.

Roy, C.R. 1979 Atmospheric nitrous oxide in the mid-latitudes of the southern hemisphere. J. Geophys. Res. 84, 3711–3718.

Roy, C.R., Galbally, I.E. and Ridley, B.A. 1980 Measurements of nitric oxide in the stratosphere of the southern hemisphere. Q. J. R. Meteorol. Soc. 106, 887–894.

Schiff, H.I., Pepper, D. and Ridley, B.A. 1979 Tropospheric NO measurements up to 7 km. J. Geophys. Res. 84, 7895–7897.

Scott, W.D. and Cattell, F.C.R. 1979 Vapour pressure of ammonium sulfates. Atmos. Environ. 13, 307–317.

Söderlund, R. 1977 NO_x pollutants and ammonia emissions – A mass balance for the atmosphere over N.W. Europe. Ambio 6, 118–122.

Söderlund, R. and Svensson, B.H. 1976 The global nitrogen cycle. In: Svensson, B.H. and Söderlund, R. (eds.), Nitrogen, Phosphorous and Sulphur – Global Cycles SCOPE Report 7, Ecol. Bull. 22, 23–73.

Steen, W.C. and Stojanovic, B.J. 1971 Nitric oxide volatilization from a calcareous soil and model aqueous solutions. Soil Sci. Soc. Am. Proc. 35, 277–282.

Stelson, A.W., Friedlander, S.K. and Seinfeld, J.H. 1979 A note on the equilibrium relationship between ammonia and nitric acid and particulate ammonium nitrate. Atmos. Environ. 13, 369–371.

Sze, N.D. and Ko, M.K.W. 1980 Photochemistry of COS, CS, CH_3SCH_3 and H_2S: Implications for the atmospheric sulfur cycle. Atmos. Environ. 14, 1223–1239.

Tang, I.N. 1980 On the equilibrium partial pressures of nitric acid and ammonia in the atmosphere. Atmos. Environ. 14, 819–828.

Tjepkema, J.D., Cartica, R.T. and Hemond, H.F. 1981 Atmospheric concentration of ammonia in Massachusetts and deposition to vegetation. Nature (Lond.) 249, 445–446.

Tsunogai, S. 1971 Ammonia in the oceanic atmosphere and the cycle of nitrogen compounds through the atmosphere and hydrosphere. Geochem. J. 5, 57–65.

Tuazon, E.C., Winer, A.M. and Pitts, J.N., Jr. 1981 Trace pollutant concentrations in a multiday smog episode in the California South Coast Air Basin by long path length fourier transform infra red spectroscopy. Environ. Sci. Tech. 15, 1232–1237.

Tuazon, E.C., Graham, R.A., Winer, A.M., Easton, R.R. and Pitts, J.N. 1978 A kilometer pathlength fourier-transform infrared system for the study of trace pollutants in ambient and synthetic atmospheres. Atmos. Environ. 12, 865–875.

Tuck, A.F. 1976 Production of nitrogen oxides by lightning discharges. Q. J. Roy. Meteorol. Soc. 102, 749–755.

Twomey, S. 1977 Atmospheric Aerosols. Elsevier, Amsterdam.

Van Cleemput, O., Patrick, W.H. and McIlhenny, R.C. 1976 Nitrite decomposition in flooded soil under different pH and redox potential conditions. Soil Sci. Soc. Am. J. 40, 55–60.

Weiss, R.F. 1981 The temporal and spatial distribution of tropospheric nitrous oxide. J. Geophys. Res. 86, 7185–7195.

Winkler, P. 1980 Observations on acidity in continental and in marine atmospheric aerosols and precipitation. J. Geophys. Res. 85, 4481–4486.

Zeller, E.J. and Parker, B.C. 1981 Nitrate ion in Antarctic firn as a marker for solar activity. Geophys. Res. Lett. 8, 895–898.

12. Agronomic and technological approaches to minimizing gaseous nitrogen losses from croplands

R.D. HAUCK

12.1. Introduction

Nitrogen is transferred in gaseous forms from the earth's surfaces to the atmosphere through NH_3 (and probably amine) volatilization from soils, waters, and plant leaves; N_2, N_2O, and NO evolution during biological denitrification; N_2, N_2O, NO, and NO_2 evolution during abiological de-nitrification (chemodenitrification); N_2O evolution during nitrification; and NH_3 and NO_x (N oxides) release through combustion. The physical and biological processes (except burning) leading to these gaseous N transfers have been discussed in the preceding chapters of this monograph. Discussed here are some approaches that have been or are being taken to minimize gaseous N losses from cultivated soils. These approaches seek to improve the efficiency of soil and fertilizer N use by crop plants. They include use of chemicals that inhibit biological activity in soils, amendments that alter the physical and/or chemical properties of N fertilizers, and improved crop and soil management practices. Some of the methods directed toward reducing gaseous N loss from cultivated soils also may increase the efficiency of plant use of N by decreasing the movement of inorganic N in solution from the plant rhizosphere, or by regulating the form and amount of N being supplied during any given growth period. Discussion of these latter aspects of N use efficiency is beyond the intended scope of this chapter which focusses on methods of reducing gaseous N losses from fertilized croplands.

12.2. Ammonia volatilization

One-half of the chapters in this monograph are wholly or in part concerned with NH_3 volatilization from soils, soil amendments, waters, and plants, attesting to the importance attached to this phenomenon. Estimates of NH_3-N volatilized annually from the global biosphere range from 18 to 244 Tg/year; comparable estimates of N loss as N_2 and N_2O through denitrification range from 107 to 390 Tg/year (see summary by Hauck and Tanji 1982). Because as much as 236 Tg of NH_3-N may be returned annually from the atmosphere to the earth's surfaces (Söderlund and Svensson 1976), NH_3 volatilization and deposition are important N transfers leading to the redistribution of fixed N over the biosphere. Although the ecological significance of NH_3 transfers between the biosphere and atmosphere has been speculated upon (e.g. Lemon and Van Houtte 1980), the importance of such transfers to the N economy of many ecosystems has yet to be determined. For some agricultural systems, NH_3 volatilization is of considerable economic importance.

12.2.1. Anhydrous and aqua ammonia

Nitrogen may be evolved into the atmosphere during the industrial synthesis and commercial distribution of NH_3, from anhydrous and aqua NH_3 applied to soils, and from NH_3 in irrigation waters.

A. Industrial ammonia production

Almost all N fertilizers are derived from NH_3. World production of NH_3-N during 1980 was estimated to be about 70 Tg, 56.8 Tg of which was assumed to be used in fertilizer manufacture (Harris and Harre 1979). Some NH_3 is lost to the atmosphere throughout the process of NH_3 manufacture, storage, distri-bution, and application to the field. During manufacture, an average of 0.55% of production is lost in old factories that do not remove traces of CO by methanation. Technological improvements in modern factories have reduced NH_3 emissions to about 0.15% of production (Barber 1978). Based on these values, NH_3 emissions in the United States during NH_3 manufacture are estimated to be about 0.025 Tg annually at current levels of production. The global emission of NH_3 during NH_3 synthesis is estimated to be in the range of 0.1 to 0.4 Tg annually. Ammonia can also be lost to the atmosphere during the production of ammonium phosphates, ammonium nitrate, and urea. For example, about 1.5 kg of NH_3 is emitted for each tonne of urea produced. Annual NH_3 emissions during fertilizer N manufacture in the United States at current production levels are estimated to be 0.09 Tg (Barber 1978). Similar values are not available for many countries that manufacture fertilizers, but a global NH_3 emission during processing of NH_3 feedstock of 0.3 Tg/year would be a reasonable estimate. Although these NH_3 emissions to the atmosphere are relatively small compared to NH_3 emissions from other sources, they are N losses connected with agriculture that can be further reduced through improve-ments in production processes. The economic incentives for conserving energy and raw materials, and the pressure of regulations limiting emissions to the atmosphere are continuing to encourage improvements in fertilizer production processes that will result in reduced NH_3 emissions.

The estimates of total NH_3 emissions during fertilizer N manufacture are much lower than the estimated shrinkage, i.e. the difference between NH_3 production and inventory. It is estimated that 10–20% of the NH_3-N produced is lost during handling and transport. Much of this apparent loss may be attributed to inaccuracies in inventory keeping. If the records are presumed to be accurate, the global loss of commercially produced NH_3 may be 7 to 14 Tg/year.

B. *Application*

Anhydrous NH_3 injected into soil as a liquid under pressure immediately vaporizes and expands into the surrounding free pore space. Because it is highly reactive, NH_3 is restrained by soil and soil water from escaping to the atmosphere. Application usually is 10 to 20 cm below the soil surface in a narrow channel made by any one of several types of applicator knives. The applicator is designed to seal the channels immediately after injection. Ammonia losses have been reported as negligible in silt loam soils (Baker *et al.* 1959) to as high as 75% of the NH_3 applied to sandy soils of low cation exchange capacity (Blue and Eno 1954). The extent of NH_3 loss is determined by the amount applied, soil water content, cation exchange capacity, organic matter content, pH, texture, and other physical properties. Ammonia loss immediately after injection usually cannot be correlated with any single factor (Parr and Papendick 1966).

Barber (1978) estimated an average loss during injection of 5% of the NH_3 applied to U.S. soils, corresponding to about 0.2 Tg/year. Currently about 40% of the N applied to U.S. soils is as anhydrous or aqua NH_3. Although reliable data are lacking, probably less than 5% of the fertilizer N used worldwide is applied as NH_3. On this basis, NH_3 emissions during injection into world soils could be about 0.3 to 0.4 Tg/year at current consumption levels. Nitrogen loss from NH_3 applied directly to soils can be reduced through improvement of fertilizer management practices and applicator equipment. Injected NH_3 diffuses into an area ranging in diameter from about 5 cm in silt and clay loam soils to about 15 cm in sandy soils. The extent of diffusion decreases with increasing soil water content. Allowing for drying in the surface 5 cm of soil, the recommended minimum application depths are 10 cm and 20 cm for heavy and light textured soils, respectively. Ideally, NH_3 should be injected when soil water content and tilth allow complete closure of the injection channel. In practice this may not always be possible. For maximum retention, NH_3 should not be injected into excessively wet, heavy soils, cloddy, dry, coarse-textured soils, or soils which are stony or highly eroded. Applicator knives can be front or back swept. In loose, sandy soil a chisel applicator often is used. However, almost all applicators inject the NH_3 forward into the channel which allows some to escape before channel closure. Improvements in applicator design have been made which allow injection to the sides of the channel and/or in the back of the applicator knife or chisel, thereby reducing the free air space into which the NH_3 can vaporize and shortening the time between channel opening and closure.

288

Ammonia in irrigation water also may be volatilized during irrigation or from the drying soil after irrigation. Data summarized by Adams *et al.* (1961) and Warnock (1966) indicate that NH_3 losses during irrigation may be negligible to as much 80% of the N applied in the concentration range of 50 to 110 ppm NH_3-N in the irrigation water. The extent and rate of loss vary with NH_3 concentration, pH of the irrigation water and soil, length of time needed for travel of the NH_3-amended water from the distribution point to the soil, rate of water penetration into soil, air and water turbulence, and ambient temperature during irrigation. Losses of NH_3-N have been observed even from soils which are initially acid but which increase in alkalinity during irrigation and maintain a surface pH >8 for several days after irrigation. No practical means of reducing N loss from NH_3-amended irrigation waters is apparent other than reducing NH_3-N concentrations and irrigating at night when temperatures are lower and wind speeds are negligible. Where NH_3 is added to sprinkler irrigation systems, NH_3-N concentrations lower than 40 ppm generally are recommended. Acidulating the irrigation waters is not an economic means of reducing NH_3 volatilization except, perhaps, where low-cost by-product acid is available (Miyamoto *et al.* 1975). As higher prices for fertilizer N encourage increased N use efficiency, use of N sources other than NH_3 in irrigation waters becomes increasingly more attractive.

Although NH_3 emissions into the global atmosphere during the manufacture and handling of N fertilizers are relatively small, they represent emissions of high concentrations from local areas and are, therefore, of local environmental concern. The application of anhydrous and aqua NH_3 to soils generally results in emissions of lower NH_3 concentrations but these emissions are a direct cost to the farmer. Prospects for minimizing NH_3 emissions during the production and agricultural use of NH_3 are good.

12.2.2. Solid nitrogen fertilizers

Urea and urea-based fertilizers

Urea is becoming the major solid N fertilizer in the United States (about 10% of total N fertilizer used during 1980) and is the dominant N fertilizer in world agriculture (>20 Tg of urea N is applied to world soils, especially in Asia where urea comprises about 80% of the fertilizer N used). Therefore, special attention should be given to the problems associated with use of urea as a fertilizer. The problems result from the rapid hydrolysis of urea to ammonium carbonate in moist soils through soil urease activity, with a concomitant rise in pH and liberation of NH_3. They include seed and seedling damage, nitrite accumulation in toxic amounts or under conditions in which chemodenitrification can occur, and gaseous loss of urea N as NH_3. Of these problems, most research emphasis has been placed on NH_3 evolution from urea that is applied to the soil surface without incorporation soon afterwards. The extent of N loss from surface-applied urea ranges from <1 to $>50\%$ of the N applied, depending on the rate and manner of application, and on a variety of soil characteristics and environmental factors (for reviews, see Gasser 1964, Terman 1965, 1979). Freney *et al.* (1981) recently discussed the chemistry of NH_3 transfer from fertilized soil surfaces to the atmosphere and the effect on NH_3 loss of important variables, including soil buffer capacity, average pH, surface temperature, and water content, and air movement above the soil. Emphasis is placed here on variables that can be manipulated and approaches that are being taken in order to minimize NH_3 loss from surface-applied urea.

Single urea particles or groups of particles hydrolyse in soil to form microsites with a high pH and high ammonium concentration. The pH and ammonium concentration within a microsite is determined by the size, density, and number of particles, the pattern of particle distribution (spacing between adjacent microsites), rate and extent of urea hydrolysis, and other soil and environmental factors that affect ammonium production and diffusion, expansion and/or dilution of the microsite, and subsequent liberation of NH_3 to the atmosphere. The simplest solution to the problem of minimizing NH_3 loss from surface-applied urea is to cover the urea with sufficient soil to trap NH_3 liberated from the microsite (at least 10 to 20 cm, depending on soil pH, texture, cation exchange capacity, and water content, among other factors). Fenn and Kissel (1976) reported that NH_3 loss from calcareous soils decreased with increasing depth of soil incorporation, but was still substantial at a depth of 7.5 cm (20% of the N applied to a silty clay loam was lost). Ammonia loss is negligible from urea that is incorporated into acid soils and can be reduced greatly when urea is mixed or layered below the surface of alkaline soils. However, farmers who surface-apply urea run the risk of losing NH_3 when they delay incorporation for several days after application, or when they fail to maintain a required minimum depth of incorporation when banding urea or discing-in broadcast applications. For instance, uneven terrain will cause application and cultivation equipment to rise above from the soil surface, leaving urea particles uncovered.

Dispersing the fertilizer granules, thereby reducing local concentrations, is another approach to minimizing NH_3 loss from surface-applied urea. Dispersion occurs naturally after heavy rainfall or can be achieved through irrigation. In acid soils, enough water must pass through the urea-soil microsite to prevent high local concentrations of ammonium and high pH from developing within the microsite during urea hydrolysis; i.e. to prevent the buffer capacity of the soil within the microsite from being exceeded. In alkaline soils, sufficient water must be applied to wash the urea well into the soil as a dilute solution. For many soils a lag period of one to several days occurs between urea application and its hydrolysis at a rate sufficient to build up both pH and ammonium concentrations within the fertilizer-soil microsites. Urea can be surface-applied with little risk of NH_3 loss in areas where rainfall patterns can be confidently predicted. The risk of NH_3 loss from urea applied to the soil surface before natural or cultural irrigation increases as the time period between urea application and irrigation increases and as higher temperatures, lower humidity, and air movement increase soil drying. High NH_3 losses may be expected where soil moisture is sufficient to effect urea hydrolysis during the night while daytime conditions cause rapid water loss from the soil surface. Obviously, surface application of urea is to be avoided, if practicable, unless the rate and site of urea hydrolysis can be controlled. However, for some cropping systems, this cannot readily be accomplished. Farmer acceptance of no- or reduced-tillage as a means of conserving energy is leading to rapid expansion of these cultivation practices (according to one U.S. Department of Agriculture prediction, 90% of U.S. cropland will be under reduced-tillage by the year 2000). Commercial equipment is not available for economically incorporating fertilizers beneath the vegetative cover, thereby requiring their surface application. Experiments in the eastern United States on no-tillage maize show substantial yield reductions for surface-applied urea-ammonium nitrate solutions or urea that are not incorporated into soil. Numerous experiments where urea was applied to grassland indicate severe yield reductions attributed to NH_3 loss (for review, see Terman 1979). Split applications of urea do not necessarily increase total N recovery by grasses, especially if later applications are made during hot weather (Burton and DeVane 1952). Because there are no commonly known fertilizer management practices that reduce the risk of NH_3 loss from solid urea under reduced-tillage, technological solutions to this problem are being sought.

One technological approach to reducing problems associated with urea use as a fertilizer is to coat urea particles with some material that slows their rate of dissolution in soils. The main difficulty has been to find effective low-cost coating materials and to develop satisfactory methods for applying them (for reviews, see Hauck and Koshino 1971, Hauck 1972). Progress toward solution of this problem has been made with the development of sulfur-coated urea, which is now in limited commercial production (see Allen 1983). However, for most of the urea which is applied for use by agronomic crops coating is not an economical way to prevent N loss.

A second approach is to mix substances with urea in order to modify the chemical reactions that occur within the fertilizer-soil microsite. Reduced NH_3 loss was obtained for urea amended with superphosphate (Okuda *et al.* 1960, Stephen and Waid 1963, Gasser and Penny 1967), urea added to soil together with phosphoric acid (Stephen and Waid 1963) and urea mixed with acid salts (Low and Piper 1961) or Ca salts that precipitate carbonate within the fertilizer-soil microsite (Mees and Tomlinson 1964, Fenn *et al.* 1981). For a review of the chemistry of these reactions in relation to NH_3 loss from urea, see Terman (1979) and Freney *et al.* (1981). Although salts of alkaline earth metals, especially calcium chloride, that precipitate carbonate reduce NH_3 loss from urea surface-applied to acid and alkaline soils, the amount of salt needed to effect substantial reduction may be so large (e.g. a Ca: urea mole ratio of 1:4 or more) that the increased cost of the amended urea fertilizer may not be commensurate with the benefits obtained (as measured in terms of N savings and crop yield increase). Nevertheless, altering the chemistry of the urea-soil microsite to reduce NH_3 loss is an approach meriting further study. This approach is the basis of work directed toward the production and evaluation of urea-phosphate formulations by the Tennessee Valley Authority (TVA). Urea-phosphate formulations with N:P ratios $<7:1$ show promise of substantially reducing NH_3 loss as compared to unamended urea when surface-applied to soil.

Numerous chemicals have been tested for their potential as urease inhibitors for use with animal feed containing urea or with urea fertilizer. Among the compounds proposed for use with urea fertilizer are pyridine-3-sulfonic acid, *o*-chloro-*p*-aminobenzoic acid, and *γ*-benzene hexachloride (Peterson and Walter 1970), dithiocarbamates (Hyson 1963, Tomlinson 1967), urea derivatives, such as methylurea, dimethylurea, thiourea, and phenylurea (Sor *et al.* 1966, Geissler *et al.* 1970); mercaptans (Gould *et al.* 1978); quinones and polyhydric phenols (Anderson 1969, 1970); salts of metals having atomic weights >50 (Sor 1969) and various halogens, cyanides, urea complexes with amines, coordination complexes of urea, boron- and/or fluorine-containing compounds, and copper and nickel chelates. Of more than 100 compounds tested as soil urease inhibitors, Bremner and Douglas (1971) found that dihydric phenols and quinones were the most effective organic compounds and that silver and mercury salts were the most effective inorganic compounds. Of 34 substituted *p*-benzoquinones tested, Bundy and Bremner (1973) found that those substituted with methyl, chloro, bromo, or fluoro groups were effective urease inhibitors while phenyl-, *t*-butyl-, and hydroxy-substituted *p*-benzoquinones had little or no effect. Review of this rapidly expanding area of research is beyond the scope of this chapter. Much of the literature on soil urease and its inhibition has been reviewed by Bremner and Mulvaney (1978) and Terman (1979). Although considerable information is available from laboratory studies concerning the inhibition of soil urease, few greenhouse and field studies have been made to determine whether chemicals that retard urea hydrolysis in soils can be used to reduce NH_3 loss from surface-applied urea with consequent increase in crop yield. Current information suggests that simple quinones and phenols (Bremner and Mulvaney 1978), 1, 3, 4-thiadiazole-2, 5-dithiol (Gould *et al.* 1978) and phenylphosphorodiamidate (Lang *et al.* 1976) merit further evaluation as inhibitors of urease activity for use with fertilizer urea.

B. Ammonium salts

Little if any NH_3 is liberated from ammonium salts added to unlimed acid soils. These salts form acid microsites in soil in which ammonium remains ionized. Ammonia can be liberated from diammonium phosphate which forms an alkaline microsite in acid soils. In alkaline soils, the liberation of NH_3 increases with decrease in solubility of the reaction products of the ammonium salt with calcium. Thus, Terman and Hunt (1964) reported NH_3 loss from diammonium phosphate and ammonium sulfate which form relatively insoluble calcium salts, but not from ammonium nitrate, which reacts with calcium to form soluble calcium nitrate within the microsite. As indicated earlier, mixing acid-hydrolyzing salts with urea tends to reduce the pH of the fertilizer-soil microsite, thereby reducing NH_3 liberation. Nitrogen loss as NH_3 from surface-applied ammonium fertilizers probably is <0.1 Tg. This estimate is based on world production of 10.1 Tg of N as ammonium fertilizers (ISMA, 1981), $>90\%$ of which are incorporated into soil. To decrease the risk of NH_3 loss from ammonium fertilizers, diammonium phosphate should not be allowed to remain on the surface of calcareous or heavily limed acid soils, nor should ammonium nitrate be surface-applied to drying alkaline soils during hot weather without incorporation.

12.2.3. Nitrogen solutions

The United States is the world's main user of nonpressure N-containing solutions as fertilizers (about 2 Tg of N during 1980). About 0.3 Tg of N as solution fertilizers was used during 1980 in 20 other countries, mainly in Europe. Nitrogen loss from N solutions is estimated to be about 0.35 Tg/year worldwide.

Urea and ammonium nitrate are the main N sources of most nonpressure fluids, especially of clear N solutions. These solutions form an alkaline soil-fertilizer microenvironment within a few hours or days after application, depending on the amount and rate of urea hydrolysis, among other factors. Ammonia may be liberated from these alkaline microsites if they remain undispersed on the soil surface. Nitrogen solutions are applied as sprays or dribbled-on by the gravity feed method. Nitrogen applied as a spray is in the form of many small droplets distributed over a large, discontinuous soil surface area. The same amount of N dribbled-on will be applied in a narrow stream of larger-size drops concentrated over strips of soil surface. Unpublished results from the author's laboratory indicate that NH_3 volatilization from N solutions surface-applied to soil is a complex function of drop size and distribution, composition of the N solution, and soil characteristics. The soil chemistry of N solutions and mixed fluid fertilizers in relation to the manner in which they are applied has received virtually no attention. Yet greenhouse and field tests have demonstrated substantial losses of N as NH_3 from surface application of nonpressure fluids. For example, the relative recoveries of N topdressed on bermudagrass over a 5-year period were: solid ammonium nitrate 100, urea-ammonium nitrate solution 86, and urea 74 (Burton and Jackson 1962). Statistics are not available concerning the percentage of spray $vs.$ dribble-on application of N solutions, nor the extent to which the N is incorporated into the soil after application. In the United States, probably $>75\%$ of N solutions used are spray-applied and not incorporated. Nitrogen solution use probably will increase with increase in area under no-till or minimum tillage. Data are needed on the extent of N loss as NH_3 from N solutions and other nonpressure N fluids, especially as such loss is a function of method of application. Use of urease inhibitors mixed with fluids, and application methods that avoid high local concentrations of ammonium at high pH, should minimize the risk of NH_3 loss.

12.2.4. Manure

Manure is a mixture of feed residues, microorganisms and microbial tissue, and metabolic products such as ammonium salts, urea, and uric acid. Usually, 40 to 60% of the total N of manure is in the form of organic compounds which are readily decomposed within hours to days after excretion. Decomposition yields ammonium salts usually in an inorganic-organic matrix of high pH from which NH_3 evolves. Moist manure continues to lose NH_3 until the pool of soluble organic N is dissipated, decomposition ceases, and/or the equilibrium partial pressure of NH_3 becomes zero (as through neutralization or incorporation into soil).

Recent estimates of the amount of N in manure from poultry, dairy, and slaughter animal populations in the United States range from 5.3 Tg (Yeck et $al.$ 1975) to 10.5 Tg (Taiganides and Stoshine 1971), with an average value (six literature sources) of 6.1 Tg. Reasons for the wide variation among estimates of manure-N in animal populations are discussed by Bouldin and Klausner (1983). Despite the wide range in estimates, it is evident that the amount of N in manure is substantial (probably equivalent to at least one-half of the amount of fertilizer N added annually to U.S. soils). Numerous studies (e.g. see review by Bouldin and Klausner 1983) suggest that on the average about 50% of manure-N is volatilized to the atmosphere as NH_3 or amines between the time of excretion and application to the field, and of the manure applied about 50% of its N is lost from the soil system. In estimating NH_3 fluxes from domestic livestock over N.W. Europe, Söderlund (1977) assumed that 10 and 50% of the N in manure excreted by extensively managed (rangeland) and intensively managed (feedlots) livestock, respectively, was volatilized to the atmosphere as NH_3. The region studied comprised 4×10^{12} m^2, of which 55% was land; the entire region represented about 0.8% of the world's land and water surfaces. Ammonia-N evolution from domestic livestock was estimated at 1.7 Tg/year, as compared with 0.7 Tg of N evolved from annual fertilizer applications to soils of this region. Extrapolating from regional estimates, Söderlund and Svensson (1976) calculated a global flux of NH_3 from domestic animal excretion of 20 to 35 Tg of N/year, an NH_3-N flux that is substantially larger than their estimated NH_3-N flux of 5–12 Tg/year from coal burning. (Comparable values for fertilizer-derived NH_3 were not given.)

Apart from their role in the redistribution of N from one site to another, NH_3 emissions from domestic livestock manures represent a large loss of agricultural N, part of which could be recycled into crop production systems. Obviously, changes in the number of animals produced for human consumption will correspondingly increase or decrease manure production. The effect of such changes on NH_3 emissions will depend in part on the manner in which the animals are raised. There is no obvious practical way to conserve manure N in pastures, unpaved open feed lots, and poultry production systems such as high-rise cage laying operations (Bouldin and Klausner 1983). Because about 50% of the total manure N frequently is lost between time of excretion and application to soil, for some livestock operations, large amounts of N can be conserved through improvements in collection and storage of manures prior to use. Included among the conservation measures are: (1) use of slotted floors with concrete manure collection pits; (2) at least once-daily cleaning; (3) paved feed lots with alleys for flushing excrement; and (4) ponding and stabilization of excrement. However, such manure conservation measures can be too expensive in terms of installation, labour, and energy costs to be adopted by many animal producers. A second problem associated with manure conservation is that in many large livestock operations, feed used in the operation is produced elsewhere. Manure produced in these operations is more a disposal problem than a source of nutrients for feed production. Also, large feed lots may produce more manure than can be loaded into soils of croplands in their immediate vicinity. In mixed animal-grain farming systems, preventing NH_3 loss from manure may prove of economic benefit. For maximum benefit, measures should be taken to shorten the time between manure collection and application to soil (for unprocessed manures), to incorporate the manure into the soil immediately after application, and to apply manure in the amount and at the time needed to satisfy crop requirements either as a sole or supplemental source of N and other nutrients.

Past research on manure use has focussed on ways to derive benefit from, and to minimize adverse environmental effects of manure disposal. Some research has been directed toward the residual value of manures (e.g. Pratt *et al.* 1976). Improved animal production systems have been designed which reduce NH_3 loss during collection and storage of manure and urine. Practical and economical methods are needed for processing, distribution to farms, and application of excrement from these animal production systems. The potential for decreasing NH_3 emissions from manures of domestic livestock is great; whether it is realized depends on the cost/benefit ratio of conserving manure nutrients and on farmer education.

12.2.5. Rice paddies and ponds

About 10% of the N consumed worldwide is applied to rice (Stangel 1979). The expected annual requirement for N during the 1980's is about 9.2 to 10.6 Tg. Although NH_3 volatilization from the floodwater of swamp rice was reported almost 70 years ago (Harrison and Aiyers 1913), whether or not NH_3 loss from floodwater significantly decreases the efficiency of N use in paddy rice culture has yet to be established. Assuming that 75% of the N currently used on rice is applied as urea and that an average of 10% of this N is volatilized as NH_3, current and anticipated annual NH_3-N emissions are speculated to be 0.5 and 0.8 Tg, respectively.

Interest in the mechanism of N loss from paddy has been renewed by recent studies that suggest that as much as 50% of applied N can be lost as NH_3 from alkaline soils (IRRI 1977). The extent of loss is a function of ammonium concentration, buffer capacity and pH of the floodwater, temperature, and wind velocity. Using a model based on measured pH values, Bouldin et al. (1974) reported daily decreases in ammonium concentrations in fertilized ponds ranging from 2 to 32% of the N applied. The model indicated that the major avenue of ammonium depletion was by NH_3 volatilization across the air-water interface. Floodwater pH in paddy can rise after addition of alkaline-hydrolyzing fertilizers, such as urea or diammonium phosphate, by interaction with strongly alkaline soil, or through decrease in CO_2 concentration in a bicarbonate-buffered soil-water system. Consumption of CO_2 by algae during photosynthesis may dramatically increase floodwater pH to values exceeding 8.5 (IRRI 1977). The partial pressure of NH_3 in equilibrium with a solution containing ammonium increases rapidly in the pH range 8.5 to 10, so that NH_3 loss from ponds or paddy floodwaters which rise even temporarily to pH 8.5 or above may be considerable. A review of the chemistry of NH_3 evolution from paddy floodwater is given in chapter 10 of this book.

Approaches to the problem of decreasing fertilizer-derived NH_3 emission from floodwater are directed toward preventing the simultaneous rise in floodwater NH_4OH concentrations and pH. Urea is not only the dominant N fertilizer used in paddy rice culture but also is the solid N source that is most prone to N loss as NH_3. Urea hydrolysis often increases floodwater pH as opposed to ammonium sulfate and ammonium chloride, which being salts of strong acids, often decrease floodwater pH for several days after application.

The effect on floodwater pH exerted by N fertilizers is determined by the source and rate of N applied, pH and depth of floodwater, the water content of soil before flooding, and by other soil and environmental factors. The effect on floodwater pH also is determined by the manner of fertilizer application; this effect is greatest when fertilizer is broadcast directly into floodwater, followed by adding it to a wet soil surface, to dry soil, and least when incorporating into soil. For example, Okuda *et al.* (1960) found high concentrations of ammonium in floodwater followed by NH_3 emissions to the atmosphere when urea was permitted to hydrolyze on the surface of wet soil before flooding, but not when urea was washed into soil before flooding. Placing fertilizer N about 10 cm below the soil surface after puddling has long been known to improve plant uptake of N (e.g. Mitsui 1954), and deep placement has been found to reduce NH_3 volatilization (Mikkelsen *et al.* 1978). Incorporation below the soil surface is relatively a simple operation for direct-seeded rice, but practical means for such incorporation for transplanted rice in puddled soils have not until recently been available. Craswell *et al.* (1981) describe studies with deep, point placement of urea supergranules (spherical particles 1 to 3 g each) at the rate of one per four hills of rice. They found that split applications of prilled urea (the best method of applying N which is readily available to farmers) produced less yield and provided less N to rice plants than the urea supergranules. They also confirmed that split applications in which the initial fertilizer application was incorporated into soil were more efficient than where both applications were broadcast into floodwater.

Another approach to decreasing floodwater ammoniacal N concentrations at high pH is to use soluble N fertilizers coated to slow their rate of dissolution (e.g. sulfur-coated urea) or relatively insoluble materials (e.g. oxamide, isobutylidenediurea). Slow-release N fertilizers have been recommended for use in rice management systems where nitrification followed by denitrification results in significant N loss; they may also reduce NH_3-N loss. For example, Craswell *et al.* (1981) suggested that because floodwater N concentrations were low when sulfur-coated urea is used, N loss as NH_3 is low.

Use of urease inhibitors may be practical under rice management systems where unamended urea has opportunity to hydrolyse before floodwater addition. Phenylphosphorodiamidate (PPDA) shows promise in this regard. In common with other chemicals with potency as urease inhibitors, PPDA is not completely effective in all soils at application rates that could be considered economical. Use of chemicals to suppress algal blooms and thus prevent diurnal rise in floodwater pH does not appear to be attractive.

Assessing the significance of NH_3 volatilization as a N loss mechanism in paddy is complicated by the fact that N loss through denitrification, and perhaps leaching, also may be occuring in the system under study. There seems little question that considerable amounts of NH_3 may evolve from N-fertilized paddy, especially following urea application. However, application of urea to floodwater does not always increase floodwater pH to values above 7.5 to 8. To assess the global flux of NH_3 from fertilized wetlands, statistics are needed for total area under rice culture, proportion of wetlands over alkaline soils (preferably, the areas of soil falling within specific pH ranges), and the pH of irrigation waters correlated with the pH range of the soils to be flooded. Information is also needed on the fertilizer management practices of typical farmers by regions. In the absence of such information, one can only speculate on the extent of NH_3 loss from fertilized paddy. However, one can justifiably assume that means will become available to markedly reduce NH_3 emissions from rice-growing wetlands.

12.2.6. Phyllosphere

Söderlund and Svensson (1976) estimated that wildlife, domestic livestock, humans, and coal burning contribute 26 to 53 Tg/year NH_3-N to the atmosphere. Dry and wet deposition of NH_3-N to global land and water surfaces was estimated to be 87 to 191 Tg/year. Recent evidence (reviewed by Wetselaar and Farquhar 1980, Lemon and Van Houtte 1980) indicated that at least part of the apparent NH_3 imbalance (34 to 105 Tg of N) can be accounted for by taking into account the NH_3 evolved from senescing vegetation and the leaf canopies of healthy plants. The question now arises whether or not NH_3 volatilization from the phyllosphere is an important mechanism of N transfer from the earth to the atmosphere, and if so, what are its effects on global N and C cycles. There is evidence (e.g. Kirkby 1981) that certain plant species, e.g. the *Solanaceae*, which do not efficiently assimilate ammonium, and annuals high in leaf N concentrations, are prone to lose N and NH_3. However, there is as yet little basis for speculating that NH_3 emissions from plants are directly linked to photosynthesis (Stutte *et al.* 1979) or that they can be ameliorated by crop management techniques. The release of NH_3 from and its absorption by living plants is a virgin area of study of great potential importance to our understanding of the N economy of cultivated and undisturbed ecological systems.

12.3. Biological denitrification

Estimates of N loss from field soils via denitrification have been based mainly on indirect measurements; that is, the deficit in a N balance account is presumed to be the denitrification loss. The accuracy of N balance measurements is greatly improved where stable N tracer techniques are used, but the data obtained may be subject to errors of interpretation (Hauck 1978). One source of error may be in the assumption that by accounting for all of the ^{15}N which remains in a soil–plant system, one accounts also for the unlabelled N originally present in the system. Nevertheless, my view is that the ^{15}N deficit approach is the most reliable way to measure denitrification loss from fertilized fields over a period of several weeks or months, provided that N losses by other means and N gains can be measured.

Estimates of denitrification loss from fertilized fields usually range from 10 to 30% of the N applied, which at current consumption levels, corresponds to an annual global N loss of about 6 to 18 Tg. Estimates of the amount of N evolved via denitrification from global terrestrial surfaces range from 107 to 390 Tg/year (see summary by Hauck and Tanji 1982). Fertilizer N may contribute, therefore, as little as 1.5% to as much as 17% of the total terrestrial denitrification loss. The percentage contribution of fertilizer N to total N loss is smaller when denitrification loss from oceans is taken into account, estimated by Söderlund and Svensson (1976) to be 91 Tg N/year. Even though denitrification loss from fertilized fields is only a fraction of the total N loss by this process, the cost to world farmers is considerable. Considering only the cost of N lost and not the cost of reduced crop yields, a reasonable estimate at current world prices for fertilizer N is U.S. 1.5 to 4.5×10^9 dollars annually.

Approaches to reducing denitrification loss involve use of nitrification inhibitors, slow-release N fertilizers, and improved N fertilizer management practices.

12.3.1. Nitrification inhibitors

Control of nitrification is the obvious approach to control of denitrification, because nitrification produces oxidized forms of N, nitrite and nitrate, which can be converted to gaseous N by microbial and chemical reduction processes in soils. Several intensive surveys have been conducted during the past 25 years to find nonphytotoxic chemicals that can selectively inhibit the biological oxidation of ammonium to nitrite in soils. A large number of chemicals, including pyridines, pyrimidines, mercapto compounds, thiazoles, succinimates, and triazine derivatives, have been tested, and several have been patented for use with fertilizers containing ammonium or other reduced forms of N. The chemical characteristics, behaviour in soils, and agronomic effects of several promising chemicals were reviewed about a decade age (Hauck and Koshino 1971, Prasad et al. 1971, Hauck 1972). A more recent review (Hauck 1980) includes the effects on nitrification in soils of numerous agricultural chemicals (fungicides, herbicides, and insecticides) and the potential of substances that release CS_2 in soils as nitrification inhibitors. Many aspects of the potentials and limitations of nitrification inhibitor use, especially in the United States, are discussed in a recent monograph on the subject (Stelly et al. 1980).

Only six chemicals currently are in commercial production for use as nitrification inhibitors with N fertilizers: 2-chloro-6 (trichloromethyl) pyridine (N-SERVE, nitrapyrin), 2-amino-4-chloro-6-methyl pyrimidine ('AM'), sulfathiazole ('ST'), 2-mercapto-benzothiazole (MAST), thiourea, and dicyandiamide. A seventh chemical, 5-ethoxy-3-trichloromethyl-1,2,4,-thiadiazole (DWELL, terrazole) was developed for use as a fungicide but shows promise also as a nitrification inhibitor. By far the largest amount of field data on use of nitrification inhibitors has been obtained with nitrapyrin. The data usually have consisted of measurements of ammonium and nitrate in soil, nitrate movement within the soil profile, and yield response to applied N in the absence and presence of the nitrification inhibitor. Reduction of gaseous N loss via denitrification frequently has been cited as the reason for increased plant use of N applied with nitrification inhibitors, but the occurrence of denitrification was presumed from circumstantial evidence rather than observed from direct measurements of N_2 and/or N_2O. There seems little question that control of nitrification under conditions conducive to denitrification will decrease gaseous loss of N. Generally, yield responses to applied N in the presence of a nitrification inhibitor result from reduced leaching losses on coarse-textured soils and reduced N loss via denitrification from wet, fine-textured soils. Perhaps the greatest potential for economical use of nitrification inhibitors is in direct-seeded flooded rice culture. In this system of rice management, fertilizer N is applied preplant or at the time of seeding to well-aerated soil. Nitrification of ammoniacal-N occurs during a preflood period of usually 2 to 3 weeks which is provided for seedling establishment. After flooding of the paddy, nitrate formed during the preflood period is reduced to gaseous N, thereby decreasing the available N supply and reducing grain yield. Control of nitrification through use of a nitrification inhibitor during the preflood period has been observed in various greenhouse and field experiments to increase rice yields (see review by Prasad *et al.* 1971). In these studies, gaseous N losses were not measured directly but were presumed to be the reason why lower yields were obtained from use of ammonium or ammonium-forming fertilizers in the absence, rather than in the presence, of a nitrification inhibitor. Marked increases in rice yields also were attributed to use of nitrification inhibitor-amended fertilizer N with upland rice in Indian soils subject to alternate wetting and drying (Prasad *et al.* 1970). With irrigated upland rice, N recovery was increased and N loss reduced through use of nitrapyrin (Prasad 1968).

Nitrification can occur in the aerobic zone at the soil–water interface of flooded soil (IRRI 1967), and, even in completely submerged soil, facultative anaerobes may oxidize ammonium to nitrite and nitrate using dissolved oxygen and iron and manganese salts as terminal electron acceptors (Patrick and Mahapatra 1968). Laboratory studies have shown that treatment of flooded soils with nitrapyrin will reduce the small amount of nitrification that may be occurring in them (IRRI 1967, Rajale and Prasad 1970), but use of nitrification inhibitors to suppress nitrification in continuously flooded rice soils is not practical (Hong 1980). Placement of N fertilizers well within the anaerobic zone of continuously flooded soils (deep placement) has long been used to increase crop recovery of applied N. In cases where NH_3 volatilization is thought to be negligible, this increased recovery of applied N by rice usually is ascribed to a reduction in denitrification loss (Shioiri 1941, De Datta 1978). However, in permeable paddy soils, deep placement may increase leaching of ammonium (Prasad *et al.* 1971) especially where urea supergranules (1 g/granule) are used (Juang 1980).

A voluminous literature on the effects of nitrification inhibitors with upland crops suggests that inhibitors can conserve fertilizer N for plant use under conditions where formation of nitrate would lead to loss of N through leaching or denitrification. Because conditions that are conducive to substantial N loss by these means do not always prevail during the growing season, measurable increases in crop yields are not always obtained through use of nitrification inhibitors. Moreover, it is difficult to obtain a reliable assessment of the value of nitrification inhibitors from farmer experience because few farmers have controls against which treatment effects can be measured. In the United States, the usefulness of nitrification inhibitors can be summarized as follows: Agronomists in the Southwest observe that the effectiveness of nitrapyrin in preventing N loss is diminished by its movement away from ammonium during leaching. Under such conditions, delayed N application was more effective in conserving N (Onken 1980). Under the climatic conditions of the Northwest, the potential for conserving N over the winter months through use of nitrification inhibitors is clearly evident but has not been demonstrated under field conditions (Papendick and Engibous 1980). In the semi-arid (Great Plains) region of the Midwest, loss of fertilizer N through leaching and denitrification is considered not to be a major problem. Denitrification losses occur in the irrigated, fine-textured soils of the Midwest but data to demonstrate that use of nitrification inhibitors reduces N loss from these soils are lacking (Hergert and Wiese 1980). In the Eastern Cornbelt, the probability of reducing N losses through use of nitrification inhibitors is high, especially where fertilizer N is applied in the fall for use by spring-planted crops. In field studies conducted between 1973 and 1978, Nelson and Huber (1980) reported that in over 70% of their trials, nitrapyrin or terrazole increased the recovery of fall-applied N by spring-seeded wheat and maize. The benefit from using nitrification inhibitors with spring-applied preplant N was marginal and almost nil for topdressings in the southeastern United States, where the relatively high average winter temperatures militate against use of nitrification inhibitors to conserve fall-applied N. Because soil temperatures can be sufficiently high to sustain nitrification, nitrification inhibitors must resist degradation, remain active, and move with ammonium throughout the four- to six-month period between application and plant use of N if they are to be effective. Thus it is unlikely that the use of nitrification inhibitors with fall-applied N will be beneficial in the Southeast (Touchton and Boswell 1980). Data are lacking for this region, especially regarding nitrification inhibitor use in relation to time of N fertilizer application.

Nitrification inhibitors that are specifically developed for use as a N fertilizer amendment retard the biological oxidation of soil-derived ammonium largely in the fertilizer-inhibitor-soil reaction zone. The bulk of the mineralized soil N is not affected by the inhibitor, nor is any ammonium that is no longer within the zone of effective concentration of the inhibitor. Therefore, one might expect that N loss resulting from the nitrification/denitrification transformations of soil N to be little affected by nitrification inhibitors that are applied with fertilizer N. Soil fumigation for the purpose of inhibiting nitrification through-out the crop rooting profile is not done and probably would not be economically feasible.

Of over 400 papers on the agricultural use of nitrification inhibitors reviewed by the author, none report direct measurements of gaseous N loss via denitrification from cropped soils. Estimates of the extent to which nitrification inhibitors decrease denitrification loss currently must be made by indirect means from yield response data or from the difference in N uptake from inhibitor-amended and unamended fertilizer. Such estimates can reasonably be made only for a few specific local areas, but regional and global estimates cannot justifiably be made from available information.

12.3.2. Slow-release nitrogen fertilizers

Reduction of N loss is one of several advantages cited for use of slow-release N fertilizers (for review, see Army 1963, Hammamoto 1967, Hayase 1967, Hayase et al. 1968, Hauck and Koshino 1971, Prasad et al. 1971, Hauck 1972). Delayed and/or controlled release of fertilizer-derived ammonium limits the supply of substrate for nitrification, thereby indirectly affecting denitrification. Direct measurements of gaseous N loss from slow-release N fertilizers have not been reported, although there is presumptive evidence from greenhouse pot studies with flooded rice that denitrification loss from slow-release N can be con-siderably less than from conventional fertilizer N. For example, N fertilizers were applied to soil under conditions which simulated direct-seeded paddy culture (aerobic preflood period followed by anaerobic, continuous flooding). Urea treated with the nitrification inhibitors nitrapyrin or 'AM', mixtures of urea and thiourea, and the slow-release materials oxamide, isobutyliden-ediurea, and two sulfur-coated urea formulations produced more total dry matter than unamended urea. Apparent recovery of applied N ranged from 60 to 78% for all N sources that were applied to soil immediately before flooding. When incubated under aerobic soil conditions for 20 days before flooding, N recoveries were only 35% for unamended urea, 56 to 61% for nitrification inhibitor-amended urea, and 54 to 62% for the slow-release fertilizers (Hauck and Engelstad, unpublished). In field studies with rice grown under conditions of intermittent flooding (conducive to nitrification/denitrification), sulfur-coated urea and isobutylidenediurea consistently produced more yield than urea when applied preplant (Wells and Shockley 1975).

12.3.3. Timing of nitrogen application

Numerous articles discuss the advantages in terms of N uptake and increased crop yield of delayed or split applications of fertilizer N.

Selective timing of N applications permits plants to compete better with other soil processes for the applied N. Thus, increased plant uptake of N may affect denitrification loss indirectly by removing substrate nitrate. Among the reviews containing information on the effects of N fertilizer placement and timing are those by Olson et al. (1964), Cooke (1964), Scarsbrook (1965), and Olson (1972).

12.3.4. Denitrification inhibitors

Several chemicals have been identified which suppress denitrification activity, including the nitrification inhibitors thiourea, 2-mercapto-benzothiazole, and 2-benzothiazole-sulfane-morpholine, as well as several pesticides (for references, see Hauck 1972). Nitrapyrin also has been observed to slightly retard denitrification activity (Henninger and Bollag 1976). It is questionable whether direct chemical control of denitrifying microorganisms in field soils is a practical approach to reducing N loss. If the denitrification inhibitor is added with fertilizer N, only nitrite and nitrate in the fertilizer-soil reaction zone will be protected against denitrification. Nitrate that moves from this zone and the influence of the inhibitor, is subject to denitrification. Another possibility is to apply the chemical as a soil fumigant, thereby retarding denitrification throughout the upper soil profile. The problem is to find a chemical that specifically suppresses denitrifier activity and that is of high potency, resistant to degradation, nontoxic to beneficial soil organisms, fish, mammals, and crops, safe in the environment, and economical to use. These ideal specifications will be difficult to meet.

12.4. Chemodenitrification

Based on his review of almost 100 articles published between 1871 and 1943, Wilson (1943) suggested that a substantial amount of N not accounted for in fertility experiments involved N loss from soils, plants, and organic residues as a result of nitrite or nitrous acid reactions which form gaseous N. Later, Broadbent and Clark (1965) observed that 'within recent years, it has become increasingly evident that biological or dissimilatory nitrate reduction ... does not adequately explain all losses of gaseous nitrogen from soil' and that N losses 'appear especially pronounced in soils in which nitrites have been observed to accumulate following addition of ammonia or ammonium-releasing materials'. The fact that nitrite is ubiquitous in natural systems, albeit frequently transient and in low concentration, and that it reacts with numerous inorganic and organic substances to form gaseous N suggests that nitrite dismutation reactions could be a major avenue of N loss. Yet field data to substantiate this suggestion are not available, or, as is probably the case, N losses ascribed to biological denitrification may in part be the result of chemodenitrification.

One problem is to assess the importance of chemodenitrification as a mechanism of N loss from croplands. Another is to reduce the extent of loss. With current knowledge, reducing chemodenitrification loss from fertilizer-derived N probably will prove less difficult than making reliable estimates of such loss in productive ecosystems. For N derived from soil or plant residues, assessing N loss through chemodenitrification on a regional basis becomes a formidable task.

Broadbent and Clark (1965) suggested that gaseous N losses resulting from biological denitrification fall into two categories – rapid and extensive losses, and continuous small losses extended over time. Chemodenitrification losses also may fall into similar categories. Those associated with nitrite accumulation after application of fertilizer or green manures may be rapid and intensive. It is possible to reduce such N loss through management practices. Chemodenitrification losses that occur slowly over extended time periods may be losses that cannot readily be reduced.

Preventing the accumulation of nitrite in soil–fertilizer microsites appears to be the most obvious way to reduce chemodenitrification loss. Nitrite can accumulate in acid or alkaline soils from fertilizers such as anhydrous NH_3, urea, and diammonium phosphate which hydrolyze to form alkaline microsites. Materials such as ammonium sulfate, ammonium chloride and monoammonium phosphate will accumulate nitrite only when placed into alkaline or calcareous soils. Whether or not nitrite will accumulate is determined largely by that combination of fertilizer source and rate, soil characteristics, and soil environmental factors which maintain a pH of about 8 within the fertilizer–soil microsite (Hauck and Stephenson 1965). Thus, in acid soils, loss of N through chemodenitrification is more probable from urea than from ammonium sulfate, from high rates of fertilizer N than from low rates, and from band than from broadcast application. However, it should be emphasized that data from field studies are not available to substantiate either that significant fertilizer N is lost through chemodenitrification or that management practices can reduce such loss.

12.5. Nitrification

During the past decade several articles have reported the formation of N_2O during the biological oxidation of ammonium to nitrite. The exact mechanism by which this occurs is not known, but Lees (1963) observed that in pure cultures of *Nitrosomonas*, cyanide induces a condition resembling anaerobiosis under which hydroxylamine (the first reaction product of the nitrification sequence) is oxidized to N_2O and NO. This corroborated the observation of Falcone *et al.* (1962) that cell-free extracts of *Nitrosomonas europaea* produced N_2O from hydroxylamine in the presence of cytochrome *c* under anaerobic conditions. Hooper (1968) and Ritchie and Nicholas (1972) provided evidence that *Nitrosomonas europaea* can produce N_2O by enzymatic reduction of nitrite. Ritchie and Nicholas (1972, 1974) showed that N_2O can be formed from nitrite under anaerobic conditions or during the oxidation of ammonium to nitrite under aerobic conditions, possibly through the decomposition of hyponitrous acid ($H_2N_2O_2$) as suggested earlier by Corbet (1935).

A series of studies (Blackmer and Bremner 1977, Bremner and Blackmer 1978a, b, Bremner *et al.* 1978, Minami *et al.* 1978) established that N_2O can evolve from well-aerated soils of low water content ($< 5\%$). In well aerated soils, N_2O emissions were found not to be significantly correlated with soil nitrate content but were correlated with soil content of nitrifiable N. The amount of N_2O evolved decreased after addition of nitrification inhibitor (nitrapyrin), increased markedly with amount of nitrifiable N but not of nitrate added, and increased after addition of plant residue and nitrogenous organic amendments, the amount of N_2O evolved increasing with decrease in C/N ratio of the amendments and with increase in soil pH. Nitrous oxide emissions from soils amended with ammonium sometimes exceeded those amended with nitrate after waterlogging to promote denitrification. These observations provide evidence that much of the N_2O evolved from well-aerated soils is produced during nitrification rather than denitrification. Since the work summarized above was done with U.S. (Iowa) soils, attention is drawn to the measurement of significant N_2O emissions from Australian soils at low-water contents and under aerobic conditions (Freney *et al.* 1978).

As with chemodenitrification loss, the practical significance to farmers of N loss as N_2O produced during nitrification has yet to be established. In an Iowa field study, Breitenbeck et al. (1980) showed that N_2O emissions induced by applications of urea or $(NH_4)_2SO_4$ markedly exceeded those induced by $Ca(NO_3)_2$ application but represented $<0.2\%$ of the fertilizer N applied. Hutchinson and Mosier (1979) and Cochran et al. (1981) reported N_2O emissions, apparently during nitrification of anhydrous NH_3 applied to western U.S. soils, to be <0.2 and 0.1% of the N applied, respectively. However, Bremner et al. (1981a) found N_2O emissions from three Iowa soils amended with anhydrous NH_3 to range from 12.1 to 19.6 kg of N/ha, representing 4.0 to 6.8% of the N applied. The measurement period was 139 days. During this period, 1.7 to 2.5 kg of N_2O-N/ha evolved from the corresponding unfertilized soils. In another study with anhydrous NH_3-amended soil, 1.66% of the N applied was evolved as N_2O over 264 days (during spring and early summer), with almost no additional emissions thereafter (Bremner et al. 1981b).

Farmer concern about N loss during nitrification of N fertilizers would not be justified on the basis of the limited amount of data that are available, although the results reported by Bremner et al. (1981a) suggest that N_2O emissions during nitrification can be an important avenue of N loss from soils. The use of a nitrification inhibitor (e.g. nitrapyrin) has been observed to markedly reduce N_2O emissions from N fertilizer-amended soil (Bremner et al. 1981b, Bremner and Blackmer 1980), but use of a nitrification inhibitor for this purpose will be economical only where N losses are large enough to warrant the extra cost of amending the fertilizer.

There is current interest in resolving such questions as to whether N_2O that is derived from hydroxylamine is formed through chemical decomposition of hydroxylamine during nitrification or whether it is a product of the reaction between hydroxylamine and nitrite (e.g. Bremner and Blackmer 1981, Bremner et al. 1980). Detailed discussion of the mechanisms of N gas formation during nitrification or as a result of chemodenitrification lies beyond the scope of this article. However, it is relevant to point out that reactions involving nitrite that lead to gaseous N emissions occur in acid soil microsites, while N_2O evolution during nitrification is favored by increase in soil pH. For example, Bremner and Blackmer (1981) reported more than a fivefold greater emission of N_2O during nitrification from soils of pH >8 than from soils within the pH range of 6.0 to 6.5. The reaction conditions leading to nitrite accumulation and dismutation probably are more stringent than those that determine N_2O emissions during nitrification. Nitrite accumulates only in alkaline soil or alkaline soil–fertilizer microsites, from which it must then diffuse or otherwise be moved into acid microsites for further reaction. Unless the conditions within the microsite where nitrite accumulates are such that *Nitrobacter* activity is severely retarded, oxidation of nitrite to nitrate usually occurs before nitrite can move into acid soil microsites. On the other hand, N_2O emissions during nitrification do not have these reaction site restrictions and may occur wherever nitrifiable N is oxidized.

12.6. Perspective

The transfer of N between the earth's atmosphere and its land and water surfaces is of interest to those engaged in crop production and those who assess the ecological and environmental effects of crop production. The interests of each are not mutually exclusive. Indeed, responsible approaches toward increasing the efficiency of N use in crop production consider both interests.

On the basis of current knowledge, N_2 and N_2O evolution during biological denitrification appears to account for the largest percentage of gaseous emissions of N from croplands. Where denitrification loss is estimated by measuring the deficit in a N balance account, values for the amounts of N evolved during denitrification may include the amounts of N_2 and N oxides evolved during chemodenitrification and nitrification and undetected NH_3 emissions. This source of error is especially serious when results of local estimates of denitrification loss are extrapolated to regional and global dimensions.

Improvements in the technology of fertilizer manufacture have already reduced NH_3 emissions. At most, about 1.5 Tg of NH_3–N are now released during NH_3 synthesis, processing into fertilizer, and transport. Prospects for additional reduction in NH_3 emissions are good. The apparent imbalance between NH_3 production and consumption (shrinkage) needs clarification. The apparent shrinkage is five- to tenfold greater than total N emissions during N fertilizer manufacture, which suggests that more effort to secure reliable production and consumption information would be welcomed by those engaged in modelling global N fluxes.

Ammonia emissions from fertilizers applied to soil surfaces or to paddy floodwaters can be reduced through use of urease inhibitors. The development of an effective chemical that can retard urea hydrolysis in soils is a major N research goal of several public and private research organizations. Very large reductions in NH_3 emissions from manure are technologically feasible, the main limitations to reducing such loss being economics and farmer education.

Quantitative speculations on the amount of N evolved during chemo-denitrification and the amount of N_2O evolved during nitrification probably are not justified at this time. Chemodenitrification and N_2O evolution during nitrification may prove to be avenues for significant amounts of gaseous N loss from croplands. If so, methods already are available or are under development that seek to control nitrification, and, therefore, the processes leading to N loss that are the direct or indirect consequences of nitrification.

When considering gaseous fluxes of N between the earth and its atmosphere, it should be kept in mind that croplands which receive fertilizer N comprise only about 11% of the earth's land surface. Croplands thus contribute only a fraction of the total gaseous N emissions to the atmosphere. Processes that lead to gaseous N emissions are also active in unfertilized soils, lakes, rivers, and oceans. However, this fact does not obviate the need to reduce the level of emissions whenever practical. This can be done by increasing the efficiency of agricultural N use by crop plants. Significant improvements in this regard can and undoubtedly will be made. Higher fertilizer and energy costs and the need to maintain the quality of the global environment provide powerful incentives for using agricultural N more efficiently. Progress to date indicates that the prognosis for achieving this objective is good.

12.7. References

Adams, R., Anderson, M.S. and Hulburt, W.C. 1961 Liquid nitrogen fertilizers for direct application. U.S. Dep. Agric., Agric. Res. Serv., Agric. Handbook No. 198, 44 pp.

Allen, S.E. 1983 Slow-release nitrogen fertilizers. In: Hauck, R.D., Beaton, J.D., Goring, C.A.I., Hoeft, R.G., Randall, G.W. and Russel, D.A. (eds.), Nitrogen in Crop Production. American Society of Agronomy, Madison (in press).

Anderson, J.R. 1969 Inhibition of soil urea hydrolysis. Brit. Pat. 1, 142, 245. February 5. 7 pp. (Chem. Abstr. 70, 95902.).

Anderson, J.R. 1970 Inhibition of urea hydrolysis. U.S. Pat. 3, 515, 532. June 2, 3 pp. (Chem. Abstr 73, 65535.).

Army, T.J. 1963 Coated fertilizers for the controlled release of plant nutrients. Agric. Chem. 18, 26–28, 81–82.

Baker, J.H., Peech, M. and Musgrave, R.B. 1959 Determination of application losses of anhydrous ammonia. Agron. J. 51, 361–362.

Barber, J.C. 1978 Ammonia losses in production and agricultural use. Ammonia Plant Saf. 20, 5–10.

Blackmer, A.M. and Bremner, J.M. 1977 Gas chromatographic analysis of soil atmospheres. Soil Sci. Soc. Am. J. 41, 908–912.

Blue, W.G. and Eno, E.F. 1954 Distribution and retention of anhydrous ammonia in sandy soils. Soil Sci. Soc. Am. Proc. 18, 420–424.

Bouldin, D.R. and Klausner, S.D. 1983 Use of nitrogen from manure. In: Hauck, R.D., Beaton, J.D., Goring, C.A.I., Hoeft, R.G., Randall, G.W. and Russel, D.A. (eds.), Nitrogen in Crop Production. American Society of Agronomy, Madison (in press).

Bouldin, D.R., Johnson, R.L., Burda, C. and Kao, C.W. 1974 Losses of inorganic nitrogen from aquatic systems. J. Environ. Qual. 3, 107–114.

Breitenbeck, G.A., Blackmer, A.M. and Bremner, J.M. 1980 Effects of different nitrogen fertilizers on emission of nitrous oxide from soil. Geophys. Res. Lett. 7, 85–88.

Bremner, J.M. and Blackmer, A.M. 1978a Nitrous oxide: emission from soils during nitrification of fertilizer nitrogen. Science (Wash D.C.) 199, 295–296.

Bremner, J.M. and Blackmer, A.M. 1978b Nitrification versus denitrification as a source of atmospheric N$_2$O. 12th Int. Congr. Microbiol. Symp. 152 pp.

Bremner, J.M. and Blackmer, A.M. 1980 Mechanisms of nitrous oxide production in soils. In: Trudinger, P.A., Walter, M.A. and Ralph, B.J. (eds.), Biogeochemistry of Ancient and Modern Environments, pp. 279–291. Australian Academy of Science, Canberra.

Bremner, J.M. and Blackmer, A.M. 1981 Terrestrial nitrification as a source of atmospheric nitrous oxide. In: Delwiche, C.C. (ed.), Denitrification, Nitrification, and Atmospheric Nitrous Oxide, pp. 151–170. John Wiley, New York.

Bremner, J.M. and Douglas, L.A. 1971 Inhibition of urease activity in soils. Soil Biol. Biochem. 3, 297–307.

Bremner, J.M. and Mulvaney, R.L. 1978 Urease activity in soils. In: Burns, R.G. (ed.), Soil Enzymes. pp. 149–196. Academic Press, London.

Bremner, J.M., Blackmer, A.M. and Minami, K. 1978 Effects of organic amendments on fluxes of nitrous oxide between soils and air. Agron. Abstr. p. 21.

Bremner, J.M., Blackmer, A.M. and Waring, S.A. 1980 Formation of nitrous oxide and dinitrogen by chemical decomposition of hydroxylamine in soils. Soil Biol. Biochem. 12, 263–269.

Bremner, J.M., Breitenbeck, G.A. and Blackmer, A.M. 1981a Effect of anhydrous ammonia fertilization on emission of nitrous oxide from soils. J. Environ. Qual. 10, 77–80.

Bremner, J.M., Breitenbeck, G.A. and Blackmer, A.M. 1981b Effect of nitrapyrin on emission of nitrous oxide from soil fertilized with anhydrous ammonia. Geophys. Res. Lett. 8, 353–356.

Broadbent, F.E. and Clark, F.E. 1965 Denitrification. In: Bartholomew, W.V. and Clark, F.E. (eds.), Soil Nitrogen. Agronomy 10, pp. 344–359. American Society of Agronomy, Madison.

Bundy, L.G. and Bremner, J.M. 1973 Effects of substituted p-benzoquinones on urease activity in soils. Soil Biol. Biochem. 5, 847–853.

Burton, G.W. and DeVane, E.H. 1952 Effect of rate and method of applying different sources of nitrogen on the yield and chemical composition of Bermuda grass, Cynodon dactylon, hay, Agron. J. 44, 128–132.

Burton, G.W. and Jackson, J.E. 1962 Effect of rate and frequency of applying six nitrogen sources on Coastal bermudagrass. Agron. J. 54, 40–43.

Cochran, V.L., Elliott, L.F. and Papendick, R.I. 1981 Nitrous oxide emissions from a fallow field fertilized with anhydrous ammonia. Soil Sci. Soc. Am. J. 45, 307–310.

Cooke, G.W. 1964 Nitrogen fertilizers: their place in food production, the forms which are made, and their efficiencies. Proc. Fert. Soc. 80, 1–88.

Corbet, A.S. 1935 The formation of hyponitrous acid as an intermediate compound in the biological or photochemical oxidation of ammonia to nitrous acid. II. Microbiological oxidation. Biochem. J. 29, 1086–1096.

Craswell, E.T., De Datta, S.K., Obcemea, W.N. and Hartantyo, M. 1981 Time and mode of nitrogen fertilizer application to tropical wetland rice. Fert. Res. 2, 247–259.

De Datta, S.K. 1978 Fertilizer management for efficient use in wetland rice soils. In: Soils and Rice, pp. 671–701. International Rice Research Institute, Los Baños.

Falcone, A.B., Shug, A.L. and Nicholas, D.J.D. 1962 Oxidation of hydroxylamine by particles from Nitrosomonas. Biochem. Biophys. Res. Commun. 9, 126–131.

Fenn, L.B. and Kissel, D.E. 1976 The influence of cation exchange capacity and depth of incorporation on ammonia volatilization from ammonium compounds applied to calcareous soils. Soil Sci. Soc. Am. J. 40, 394–398.

Fenn, L.B., Taylor, R.M. and Matocha, J.E. 1981 Ammonia losses from surface-applied nitrogen fertilizer as controlled by soluble calcium and magnesium: general theory. Soil Sci. Soc. Am. J. 45, 777–781.

Freney, J.R., Denmead, O.T. and Simpson, J.R. 1978 Soil as a source or sink for atmospheric nitrous oxide. Nature (Lond.) 273, 530–532.

Freney, J.R., Simpson, J.R. and Denmead, O.T. 1981 Ammonia volatilization. In: Clark, F.E. and Rosswall, T. (eds.), Terrestrial Nitrogen Cycles. Ecol. Bull. 33, 291–302.

Gasser, J.K.R. 1964 Urea as a fertilizer. Soils Fert. 27, 175–180.

Gasser, J.K.R. and Penny, A. 1967 The value of urea nitrate and urea phosphate as nitrogen fertilizers for grass and barley. J. Agric. Sci. 69, 139–146.

Geissler, P.R., Sor, K. and Rosenblatt, T.M. 1970 Fertilizer compositions containing borax and (or) copper sulfate as urease inhibitors U.S. Pat. 3, 523, 018. August 4. 3 pp. (Chem. Abstr. 73, 98051.).

Gould, W.D., Cook, F.D. and Bulat, J.A. 1978 Inhibition of urease activity by heterocyclic sulfur compounds. Soil Sci. Soc. Am. J. 42, 66–72.

Harris, G.T. and Harre, E.A. 1979 World fertilizer situation and outlook – 1978-85. International Fertilizer Development Center Tech. Bull. IFDC-T-13. 27 pp.

Hauck, R.D. 1972 Synthetic slow-release fertilizers and fertilizer amendments. In: Goring, C.A.I. and Hamaker, J.W. (eds.), Organic Chemicals in the Soil Environment. Vol. 2, pp. 633–690. Marcel Dekker, New York.

Hauck, R.D. 1978 Critique – of 'Field trials with isotopically labeled nitrogen fertilizer.' In: Nielsen, D.R. and MacDonald, J.G. (eds.), Nitrogen in the Environment. Vol. 1, pp. 63–77. Academic Press, New York.

Hauck, R.D. 1980 Mode of action of nitrification inhibitors. Meisinger, J.J., Randal, G.W. and Vitosh, M.L. (eds.), Nitrification Inhibitors – Potentials and Limitations. ASA Spec. Publ. No. 38. pp. 19–32. American Society of Agronomy, Madison.

Hauck, R.D. and Koshino, M. 1971 Slow-release and amended fertilizers. In: Olson, R.A., Army, T.J., Hanway, J.J. and Kilmer, V.J. (eds.), Fertilizer Technology and Use, pp. 455–494. Soil Science Society of America, Madison.

Hauck, R.D. and Stephenson, H.F. 1965 Nitrification of nitrogen fertilizers. Effect of nitrogen source, size, and pH of the granule, and concentration. J. Agric. Food Chem. 13, 486–492.

Hauck, R.D. and Tanji, K.K. 1982 Nitrogen transfers and mass balances. In: Stevenson, F.J., Bremner, J.M., Hauck, R.D. and Keeney, D.R. (eds.), Nitrogen in Agricultural Soils. Agronomy 22, pp. 891–926. American Society of Agronomy, Madison.

Hamamoto, J. 1967. Isobutylidene diurea as a slow-acting nitrogen fertilizer and the studies in this field in Japan. Proc. Fert. Soc. 90, 1–64.

Harrison, W.H. and Aiyers, P.A.S. 1913 The gases of swamp rice soils; their composition and relationship to the crop. Mem. Dep. Agric. India Chem. Ser. No. 3, 65–106.

Hayase, T. 1967 On the slowly available nitrogen fertilizers. Part 1. In: Bull. Nat. Inst. Agric. Sci. Ser. B, 18, 129–303 (in Japanese).

Hayase, T., Kurihara, K. and Koshino, M. 1968 Use of controlled release nitrogen fertilizers in paddy fields (II). Meeting, Int. Rice Committee, FAO, Kandy, Ceylon, September 1968.

Henninger, N.M. and Bollag, J.M. 1976 Effect of chemicals used as nitrification inhibitors on the denitrification process. Can. J. Microbiol. 22, 668–672.

Hergert, G.W. and Wiese, R.A. 1980 Performance of nitrification inhibitors in the Midwest (west). In: Stelly, M., Meisinger, J.J., Randal, G.W. and Vitosh, M.L. (eds.), Nitrification Inhibitors – Potentials and Limitations. pp. 89–105. American Society of Agronomy, Madison.

Hong, C.W. 1980 Implications and problems related to recently suggested fertilizer application technologies. In: Increasing Nitrogen Efficiency for Rice Cultivation. Food and Fertilizer Technology Center Book Ser. No. 18, pp. 102–115. FFTC, Taipei.

Hooper, A.B. 1968 A nitrite-reducing enzyme from *Nitrosomonas europaea*. Preliminary characterization with hydroxylamine as electron donor. Biochim. Biophys. Acta 162, 49–65.

Hutchinson, G.L. and Mosier, A.R. 1979 Nitrous oxide emissions from an irrigated cornfield. Science (Wash, D.C.) 205, 1125–1127.

Hyson, A.M. 1963 Urea and dithiocarbamates in fertilizers U.S. Pat. 3, 073, 694. January 15. 2 pp.

(Chem. Abstr. 58, 9590.).

International Rice Research Institute (IRRI). 1967 Annual report, pp. 146–147. IRRI, Los Baños.

International Rice Research Institute (IRRI). 1977 Anual report for 1976. IRRI, Los Baños. 418 pp.

ISMA, Economics Committee. 1981 Fertilizer consumption statistics 1979/80. ISMA Ltd., Paris. 45 pp.

Juang, T.C. 1980 Increasing nitrogen efficiency through deep placement of urea supergranules under tropical and subtropical conditions. In: Increasing Nitrogen Efficiency for Rice Cultivation. Food and Fertilizer Technology Center. Book Ser. No. 18, pp. 83–99. FFTC, Taipei.

Kirkby, E.A. 1981 Plant growth in relation to nitrogen supply. In: Clark, F.E. and Rosswall, T. (eds.), Terrestrial Nitrogen Cycles. Ecol. Bull. 33, 249–267.

Lang, S., Held, P., Hartbrich, H.J., Klepel, M., Rothe, G., Haeckert, H. and Thieme, H. 1976 Reducing the loss of ammonia nitrogen from urea fertilizers. Ger. (East) Pat. 122, 621. October 20. 11 pp. (Chem. Abstr. 87, 4660.).

Lees, H. 1963 Inhibitors of nitrification. In: Hocherster, R.H. and Quastel, J.H. (eds.), Metabolic Inhibitors, Vol. 2, pp. 615–629. Academic Press, New York.

Lemon, E. and Van Houtte, R. 1980 Ammonia exchange at the land surface. Agron. J. 72, 876–883.

Low, A.J. and Piper, F.J. 1961 Urea as a fertilizer. Laboratory and pot-culture studies. J. Agric. Sci. 57, 249–255.

Mees, G.C. and Tomlinson, T.E. 1964 Urea as a fertilizer. Ammonia evolution and brairding of wheat. J. Agric. Sci. 62, 199–205.

Mikkelsen, D.S., De Datta, S.K. and Obcemea, W.N. 1978 Ammonia volatilization losses from flooded rice soils. Soil Sci. Soc. Am. J. 42, 725–730.

Minami, K., Blackmer, A.M. and Bremner, J.M. 1978 Emission of nitrous oxide from well-aerated soils. Agron. Abstr. p. 31.

Mitsui, S. 1954 Inorganic Nutrition, Fertilization, and Soil Amelioration for Lowland Rice. Yokendo Ltd., Tokyo, 107 pp.

Miyamoto, S., Ryan, J. and Stroehlein, J.L. 1975 Sulfuric acid for the treatment of ammoniated irrigation water: I. Reducing ammonia volatilization. Soil Sci. Soc. Am. J. 39, 544–548.

Nelson, D.W. and Huber, D.M. 1980 Performance of nitrification inhibitors in the Midwest (East). In: Stelly, M., Meisinger, J.J., Randal, G.W. and Vitosh, M.L. (eds.), Nitrification Inhibitors – Potentials and Limitations, pp. 75–88. American Society of Agronomy, Madison.

Okuda, A., Takahashi, E. and Yoshida, M. 1960 Volatilization of ammonia from urea under upland and water-logged conditions. Nippon Dojo-Hiroyogaku Zasshi 31, 273–278. (Soils Fert. 24, 2658.) (In Japanese).

Olson, R.A. 1972 Maximizing the utilization efficiency of fertilizer N by soil and crop management. In: Effects of Intensive Fertilizer use on the Human Environment. Soils Bull. No. 16, pp. 34–52. FAO, Rome.

Olson, R.A., Dreier, A.F., Thompson, C., Frank, K. and Grabouski, P.H. 1964 Using fertilizer nitrogen effectively on grain crops. Nebr. Agric. Exp. Stn Bull. SB 479. 42 pp.

Onken, A.B. 1980 Performance of nitrification inhibitors in the Southwest. In: Stelly, M., Meisinger, J.J., Randal, G.W. and Vitosh, M.L. (eds.), Nitrification Inhibitors – Potentials and Limitations, pp. 119–129. American Society of Agronomy, Madison.

Parr, J.F., Jr. and Papendick, P.I. 1966 Retention of ammonia in soils. In: McVickar, M.H., Martin, W.P., Miles, I.E. and Tucker, H.H. (eds.), Agricultural Anhydrous Ammonia, Technology and Use, pp. 213–236. Agricultural Ammonia Institute, Memphis, and American Society of Agronomy, Madison.

Patrick, W.H., Jr. and Mahapatra, I.C. 1968 Transformation and availability to rice of nitrogen and phosphorus in waterlogged soils. Adv. Agron. 20, 323–359.

Papendick, R.I. and Engibous, J.C. 1980 Performance of nitrification inhibitors in the Northwest.

In: Stelly, M., Meisinger, J.J., Randal, G.W. and Vitosh, M.L. (eds.), Nitrification Inhibitors –
Potentials and Limitations. pp. 107–117. American Society of Agronomy, Madison. Wisconsin.

Peterson, A.F. and Walter, C.R., Jr. 1970 Reduction of ammonia loss from the soil by regulating
microbial production of urease enzyme. U.S. Pat. 3, 547, 614. December 15. 3 pp. (Chem. Abstr.
74, 63585.).

Prasad, R. 1968 Dry-matter production and recovery of fertilizer nitrogen by rice as affected by
nitrification retarders 'N-Serve' and 'AM'. Plant Soil 29, 327–330.

Prasad, R., Rajale, G.B. and Lakhdive, B.A. 1970 Effect of time and method of application of urea
and its treatment with nitrification inhibitors on the yield and nitrogen uptake by irrigated
upland rice. Indian J. Agric. Sci. 40, 1118–1127.

Prasad, R., Rajale, G.B. and Lakhdive, B.A. 1971 Nitrification retarders and slow-release nitrogen
fertilizers. Adv. Agron. 23, 337–383.

Pratt, P.F., Davis, S. and Sharpless, R.G. 1976 A four-year field trial with animal manures.
Hilgardia 44, 99–125.

Rajale, G.B. and Prasad, R. 1970 Nitrification/mineralization of urea as affected by nitrification
retarders 'N-Serve' and 'AM'. Curr. Sci. 39, 211–212.

Ritchie, G.A.F. and Nicholas, D.J.D. 1972 Identification of the sources of nitrous oxide produced
by oxidative and reductive processes in *Nitrosomonas europaea*. Biochem. J. 126, 1181–1191.

Ritchie, G.A.F. and Nicholas, D.J.D. 1974 The partial characterization of purified nitrite reductase
and hydroxylamine oxidase from *Nitrosomonas europaea*. Biochem. J. 138, 471–480.

Scarsbrook, C.E. 1965 Nitrogen availability. In: Bartholomew, W.V. and Clark, F.E., (eds.), Soil
Nitrogen. Agronomy 10, pp. 486–502. American Society of Agronomy, Madison.

Shioiri, M. 1941 Denitrification in paddy soil. Kagaku 11, 24–30 (In Japanese).

Söderlund, R. 1977 NO_x pollutants and ammonia emissions – A mass balance for the atmosphere
over NW Europe. Ambio 6, 118–122.

Söderlund, R. and Svensson, B.H. 1976 The global nitrogen cycle. In: Svensson, B.H. and
Söderlund, R. (eds.), Nitrogen, Phosphorus, and Sulfur – Global Cycles. SCOPE Report 7, Ecol.
Bull. 22, 23–73.

Sor, K.M. 1969 Inhibition of urea hydrolysis in fertilizers. Brit. Pat. 1, 157, 400. July 9. 9 pp.
(Chem. Abstr. 71, 69813.).

Sor, K M , Stansbury, R.L. and De Ment, J.D. 1966 Combination of urea with a hydrolysis
inhibitor. U.S. Pat. 3, 232, 740. February 1. 7 pp. (Chem. Abstr. 64, 14918.).

Stangel, P.J. 1979 Nitrogen requirement and adequacy of supply for rice production. In: Nitrogen
and Rice, pp. 45–67. International Rice Research Institute, Los Baños.

Stelly, M., Meisinger, J.J., Randal, G.W. and Vitosh, M.L. 1980 Nitrification Inhibitors – Potential
and Limitations. American Society of Agronomy, Madison. 129 pp.

Stephen, R.C. and Waid, J.S. 1963 Pot experiments on urea as a fertilizer. III. The influence of rate,
form, time, and placement. Plant Soil 10, 184–192.

Stutte, C.A., Weiland, R.T. and Blem, A.R. 1979 Gaseous nitrogen loss from soybean foliage.
Agron. J. 71, 95–97.

Taiganides, E.P. and Stoshine, R.L. 1971 Impact of farm animal production and processing on the
total environment. In: Livestock Waste Management and Pollution Abatement. Proc. Intl.
Symp. on Livestock Wastes. Amer. Soc. Agric. Eng. Publ. PROC-271, pp. 95–98. Am. Soc.
Agric. Eng., St. Joseph.

Terman, G.L. 1965 Volatilization loss of nitrogen as ammonia from surface-applied fertilizers.
Agric. Chem. 8, 6, 8–9, 13–14.

Terman, G.L. 1979 Volatilization losses of nitrogen as ammonia from surface-applied fertilizers,
organic amendments, and crop residues. Adv. Agron. 31, 189–223.

Terman, G.L. and Hunt, C.M. 1964 Volatilization losses of nitrogen from surface-applied
fertilizers, as measured by crop response. Soil Sci. Soc. Am. Proc. 28, 667–672.

Tomlinson, T.E. 1967 Inhibition of urea hydrolysis in soils. Brit. Pat. 1, 094, 802. December 13, 8 pp. (Chem. Abstr. 68, 48637.).

Touchton, J.T. and Boswell, F.C. 1980 Performance of nitrification inhibitors in the Southeast. In: Stelly, M., Meisinger, J.J., Randal, G.W. and Vitosh, M.L. (eds.), Nitrification Inhibitors – Potentials and Limitations, pp. 63–74. American Society of Agronomy, Madison.

Warnock, R.E. 1966 Ammonia application in irrigation water. In: McVickar, M.H., Martin, W.P., Miles, I.E. and Tucker, H.H. (eds.), Agricultural Anhydrous Ammonia, Technology and Use, pp. 115–124. Agricultural Ammonia Institute, Memphis and American Society of Agronomy, Madison.

Wells, B.R. and Shockley, P.A. 1975 Conventional and controlled-release nitrogen sources for rice. Soil Sci. Soc. Am. Proc. 39, 549–551.

Wetselaar, R. and Farquhar, G.D. 1980 Nitrogen losses from tops of plants. Adv. Agron. 33, 263–302.

Wilson, J.K. 1943 Nitrous acid and the loss of nitrogen. N.Y. Agric. Exp. Stn, Ithaca, Mem No. 253, 25 pp.

Yeck, R.G., Smith, L.W. and Calvert, C.C. 1975 Recovery of nutrients from animal wastes – an overview of existing options and potentials for use in feed. In: Managing Livestock Wastes. Proc. 3rd Intl. Symp. on Livestock Wastes. Amer. Soc. Agric Eng. Publ. PROC-275, pp. 192–194, 196. Am. Soc. Agric. Eng., St. Joseph.

SUBJECT INDEX